ARTIFICIAL INTELLIGENCE IN HEALTHCARE AND MEDICINE

ARTIFICIAL INTELLIGENCE IN HEALTHCARE AND MEDICINE

Edited by
Kayvan Najarian, Delaram Kahrobaei,
Enrique Domínguez, and Reza Soroushmehr

CRC Press
Taylor & Francis Group
Boca Raton London New York

CRC Press is an imprint of the
Taylor & Francis Group, an **informa** business

First edition published 2022
by CRC Press
6000 Broken Sound Parkway NW, Suite 300, Boca Raton, FL 33487-2742

and by CRC Press
4 Park Square, Milton Park, Abingdon, Oxon, OX14 4RN

CRC Press is an imprint of Taylor & Francis Group, LLC

© 2022 selection and editorial matter, Kayvan Najarian, Delaram Kahrobaei,
Enrique Domínguez, Reza Soroushmehr; individual chapters, the contributors

Library of Congress Cataloging-in-Publication Data
Names: Najarian, Kayvan, editor. | Kahrobaei, Delaram, 1975- editor. |
Dominguez, Enrique, editor. | Soroushmehr, Reza, editor.
Title: Artificial intelligence in healthcare and medicine / edited by Kayvan Najarian,
Delaram Kahrobaei, Enrique Dominguez, Reza Soroushmehr.
Description: First edition. | Boca Raton : CRC Press, 2022. | Includes bibliographical
references and index. | Summary: "It is widely believed that Artificial Intelligence (AI) and
its applications will revolutionize healthcare and medicine. This book provides a
comprehensive overview on the recent developments on clinical decision support systems,
precision health and data science in medicine"-- Provided by publisher.
Identifiers: LCCN 2021050834 (print) | LCCN 2021050835 (ebook) | ISBN
9780367619176 (hardback) | ISBN 9780367638405 (paperback) |
ISBN 9781003120902 (ebook)
Subjects: LCSH: Artificial intelligence--Medical applications. | Medical informatics.
Classification: LCC R859.7.A78 A765 2022 (print) | LCC R859.7.A78 (ebook) |
DDC 610.285--dc23/eng/20220120
LC record available at https://lccn.loc.gov/2021050834
LC ebook record available at https://lccn.loc.gov/2021050835

ISBN: 978-0-367-61917-6 (hbk)
ISBN: 978-0-367-63840-5 (pbk)
ISBN: 978-1-003-12090-2 (ebk)

DOI: 10.1201/9781003120902

Typeset in Times
by MPS Limited, Dehradun

Contents

Editor Biographies

Kayvan Najarian is a Professor in the Department of Computational Medicine and Bioinformatics, Department of Electrical Engineering and Computer Science, and Department of Emergency Medicine at the University of Michigan, Ann Arbor.

Delaram Kahrobaei is the University Dean for Research at City University of New York (CUNY), a Professor of Computer Science and Mathematics, Queens College CUNY and the former Chair of Cyber Security, University of York.

Enrique Domínguez is an associate professor at the department of Computer Science at the University of Malaga and a member of Biomedic Research Institute of Malaga.

Reza Soroushmehr is a Research Assistant Professor in the Department of Computational Medicine and Bioinformatics and a member of Michigan Center for Integrative Research in Critical Care, University of Michigan, Ann Arbor.

Contributors

Kiana Aran
Medical Diagnostics and Therapeutics
Keck Graduate Institute (KGI)
Claremont, CA, USA

Jonas Bianchi
Department of Orthodontics
University of the Pacific
San Francisco, CA, USA

Lucia Cevidanes
Department of Orthodontics
School of Dentistry
University of Michigan
Ann Arbor, MI, USA

Jonalyn H. DeCastro
Medical Diagnostics and Therapeutics
Keck Graduate Institute (KGI)
Claremont, CA, USA

Romain Deleat-Besson
CPE Lyon University
Lyon, France

Enrique Domínguez
Department of Computer Science
University of Malaga
Malaga, Spain

Negar Farzaneh
Department of Emergency Medicine
University of Michigan
Ann Arbor, MI, USA

C. Alberto Figueroa
Edward B. Diethrich,
  Biomedical Engineering
University of Michigan
Ann Arbor, MI, USA

Zijun Gao
Department of Computational
  Medicine and Bioinformatics
University of Michigan
Ann Arbor, MI, USA

Jose M. García-Pinilla
Heart Failure and Familial
  Cardiomyopathies Unit
  Cardiology Department Hospital
Universitario Virgen de la Victoria
IBIMA, Malaga, Spain
and
Ciber-Cardiovascular
Instituto de Salud Carlos III
Madrid, Spain

Brett Goldsmith
Cardea Bio Inc
San Diego, CA, USA

Jonathan Gryak
Department of Computational
  Medicine and Bioinformatics
University of Michigan
Ann Arbor, MI, USA

Kritika Iyer
Biomedical Engineering Department
University of Michigan
Ann Arbor, MI, USA

Zahra Keshavarz-Motamed
Department of Mechanical Engineering
McMaster University
Hamilton, ON, Canada

Mark Lawford
Department of Computing and Software
McMaster, University
Hamilton, ON, Canada

Celia Le
CPE Lyon University
Lyon, France

Guillermo López-García
Computer Science Department
University of Malaga
Malaga, Spain

Domingo López-Rodríguez
Department of Applied Mathematics
University of Malaga
Malaga, Spain

Ezequiel López-Rubio
Computer Science Department
University of Malaga
Malaga, Spain

Rafael M. Luque-Baena
Computer Science Department
University of Malaga
Malaga, Spain

Rosa Maza-Quiroga
Computer Science Department
University of Malaga
Malaga, Spain

Guy Meyer
McMaster University
Hamilton, ON, Canada

Cristian Minoccheri
Department of Computational
    Medicine and Bioinformatics
University of Michigan
Ann Arbor, MI, USA

Miguel A. Molina-Cabello
Computer Science Department
University of Malaga
Malaga, Spain

Kayvan Najarian
Department of Computational
    Medicine and Bioinformatics
University of Michigan
Ann Arbor, MI, USA

Brahmajee K. Nallamothu
Department of Internal Medicine
University of Michigan
Ann Arbor, MI, USA

Francisco Ortega-Zamorano
Department of Computing and Systems
University of Las Palmas de Gran Canarias
Malaga, Spain

Jonathan Parkinson
StemoniX
San Diego, USA

Behnam Rahmatikaregar
Department of Electrical Engineering
    and Computer Science
McMaster University
Hamilton, ON, Canada

Kourosh Sabri
Faculty of Health Sciences,
McMaster University
Hamilton, ON, Canada

Shahram Shirani
Department of Electrical and
    Computer Engineering
McMaster University
Hamilton, ON, Canada

Reza Soroushmehr
Department of Computational
    Medicine and Bioinformatics
University of Michigan
Ann Arbor, MI, USA

Karl Thurnhofer-Hemsi
Computer Science Department
University of Malaga
Malaga, Spain

Najla Al Turkestani
Department of Orthodontics
School of Dentistry
University of Michigan
Ann Arbor, MI, USA

Francisco Lopez Valverde
Computer Science Department
University of Malaga
Malaga, Spain

Francisco J. Veredas
Computer Science Department
University of Malaga
Malaga, Spain

Lu Wang
Department of Computational
    Medicine and Bioinformatics
University of Michigan
Ann Arbor, MI, USA

Alan Wassyng
Department of Computing
    and Software
McMaster University
Hamilton, ON, Canada

Craig Williamson
Neurology Department
University of Michigan
Ann Arbor, MI, USA

Heming Yao
Department of Computational
    Medicine and Bioinformatics
University of Michigan
Ann Arbor, MI, USA

Winston Zhang
Department of Computational
    Medicine and Bioinformatics
University of Michigan
Ann Arbor, MI, USA

Introduction

In recent years, artificial intelligence (AI) and, in particular, machine learning (ML) techniques have become a popular tool in many applications, e.g., for the analysis of biomedical datasets, with dramatic growth in volume and complexity. AI also helps clinicians by providing them with a quantitative understanding of structures and functions of biological systems, predictable outcomes of a treatment plan, diagnosis, and prognosis. In this book, we summarize different machine learning methods and their applications in multiple medical scenarios. We also review advanced mathematical techniques that could be used in designing machine learning models.

Chapter 1 shares a perspective on the barriers of complex datasets, the tools available to deal with them, key precautions for the application of machine learning in healthcare and an assessment of the outlook for this field. This review shares a perspective on the unique obstacles that face traditional biomedical approaches to understanding and diagnosing disease, the tools available to deal with them, key precautions for application of machine learning in healthcare, and an assessment of the future outlook for this field.

Chapter 2 to Chapter 5 focus on cardiovascular disease that is the leading cause of death worldwide and is projected to remain the first cause of death in the future. The disease condition can rapidly affect the pumping action of the heart and can ultimately lead to heart failure, which affects more than 26 million people worldwide and is increasing in prevalence with high mortality and morbidity. These three chapters introduce machine learning methods applied to different applications related to heart disease, including automatic segmentation for cardiac, ventricle, coronary vessel, and AI techniques for heart failure.

Chapters 6 and 7 introduce computational techniques for enhancing and analyzing brain images. Chapter 6 provides an overview of recent advances in super-resolution for higher-dimensional magnetic resonance images, from traditional super-resolution methods to deep learning methods. Chapter 7 outlines multiple techniques for computed tomography (CT) image pre-processing and segmentation methods to segment hematoma caused by a traumatic brain injury (TBI) which is a global health priority and could significantly increase the comorbidity of the patients, requiring complex and expensive medical care. Moreover, the commonly used performance evaluation metrics are explained in detail and discussed the current challenges in hematoma segmentation tasks.

Chapter 8 encompasses a collection of attributes of common tools for visually impaired internet users. This chapter reviews published works and products that deal with providing accessibility to visually impaired online users. Due to the variety of tools that are available to computer users, the chapter focuses on search engines as a primary tool for browsing the web and provides readers with a set of references for existing applications, along with practical insight and recommendations for accessible design. Finally, the necessary considerations for future developments and summaries of important focal points are highlighted.

Chapter 9 reviews state-of-the-art artificial intelligence techniques developed for data management, integration and in-depth analysis in the field of dentistry, and the parameters that affect the generalizability of such techniques. In this chapter, AI techniques are categorized into imaging and integrative data and the methods applied to each category are reviewed. Additionally, the evaluation metrics to rigorously test the robustness of the AI approaches are highlighted.

Chapter 10 describes the framework of machine learning and deep learning techniques that aim at automating wound diagnosis. Since the number of wound image samples available in a regular dataset is usually very small, the direct application of deep learning-based segmentation techniques is limited. In this chapter, a method is proposed to deal with this issue by generating a significantly larger dataset, and a multiclass classification model based on convolutional networks is trained.

Since many biomedical datasets in machine learning problems can be naturally organized in tensor forms, Chapter 11 aims at providing a brief introduction to the main mathematical tools used to handle tensors. It provides an overview of biomedical areas where these ideas have found fruitful applications in recent years and leveraging this multilinear structure usually has led to better results. The chapter introduces the mathematics behind tensor techniques, as well as resources for the reader interested in more technical details, and a comprehensive overview of how these methods have been used in clinical informatics.

1 Machine Learning for Disease Classification: A Perspective

*Jonathan Parkinson, Jonalyn H. DeCastro,
Brett Goldsmith, and Kiana Aran*

CONTENTS

1.1 THE GROUNDWORK OF MACHINE LEARNING FOR DISEASE MODELING

Recent years have witnessed dramatic growth in the volume and complexity of data available to biomedical research scientists (Bui & Van Horn, 2017). Deriving actionable insights that can improve diagnosis and treatment from these large heterogeneous datasets presents a formidable challenge for the community. Machine learning (ML) techniques have become a popular tool for the analysis of biomedical datasets, albeit not one yet widely used in medicine (Deo, 2015). Indeed, while the application of machine learning to disease classification holds considerable promise, it also faces unique obstacles arising from the nature of the data and from stakeholder expectations. A comprehensive review of machine learning algorithms

DOI: 10.1201/9781003120902-1

1

and their applications in disease classification would be an ambitious task and one we will not attempt here. Rather, this perspective will provide an accessible high-level introduction to machine learning for disease classification: the mechanics of some popular algorithms, the challenges, and pitfalls this field confronts, and some examples and insights from recent literature.

It is important to realize that machine learning is not some kind of mathematical alchemy that can transform irreproducible data into golden insights. ML is subject to the same "garbage in = garbage out" limitation that applies elsewhere in modeling; hence, good data curation is key for success (Beam & Kohane, 2018; Rajkomar et al., 2019). Nor is machine learning a new field of study; modern deep learning algorithms, for example, are an extension of perceptron models first proposed in the 1950s and 1960s (Schmidhuber, 2015). Instead, it is probably best to view machine learning as an extension of statistical modeling techniques, with the main difference that while statistics seeks to make inferences about populations, machine learning tries to find patterns that provide predictive power and that can be generalized (Bzdok et al., 2018).

Machine learning problems can be broadly classified into three paradigms. Unsupervised learning techniques like clustering and dimension reduction seek structure in unlabeled data and hence are often useful in data exploration or hypothesis generation. Unsupervised learning techniques would for example likely be useful for seeking subsets of a patient population that share many common features. In a study of this type, subgroups identified via clustering could then be further studied to determine whether they respond differently to a treatment. Supervised learning techniques by contrast learn to predict an output variable from an input vector, matrix or a multidimensional array or tensor. In disease classification, a common task for a supervised learning algorithm might be to determine whether an image from an MRI or a histology image indicates the presence or absence of disease; in this example, the output variable to be predicted would be the category, while the input would be the pixel values of the image. Finally, in the reinforcement learning paradigm, the algorithm is provided with a set of choices and is offered "rewards" when its choice leads to a better outcome. Although unsupervised learning and clustering is often a powerful tool for analysis of multi-omics data, in this review, we will focus on supervised learning as the most directly relevant task for disease classification.

1.2 THE "BIG BROTHER" OF PREDICTIONS: SUPERVISED LEARNING

Supervised learning problems are those where an output or label y, sometimes called the "ground truth", must be correctly predicted based on an input. The output y for disease classification is typically a category but may in some instances be a real-valued or complex number (regression), or a ranked category (ordinal regression). In some cases, the output may also be a real-valued vector (e.g., prediction of dihedral angles in a protein, based on an amino acid sequence). Successfully applying supervised learning to disease classification requires selecting and defining the right prediction problem. This may involve careful consideration of the labels to be predicted, the structure of existing workflows and pipelines, and the availability of relevant data.

It is of course crucial that the data is labeled correctly, and this may represent a challenge for biomedical applications in general. International Classification of Disease (ICD) codes, for example, are often used to indicate diagnoses in electronic health records (EHR). Errors in ICD code assignment are however not infrequent – estimates of the accuracy of ICD coding vary widely – and may arise from multiple possible sources of error (O'Malley et al., 2005). Physicians sometimes for example use abbreviations in their notes whose meaning may be ambiguous to the medical coder responsible for selecting and entering an appropriate diagnosis code (Sheppard et al., 2008).

Another more subtle problem can arise when there is a mismatch between the categories used to label the data and the ultimate objective of the study or of stakeholders. This problem is perhaps best illustrated with an example. Cancers that share the same tissue of origin exhibit a striking level of genetic diversity both within a single patient and across patients (Mroz & Rocco, 2017). It is well-established that different cancer subtypes exhibit different prognoses and may respond differently to the same treatment – indeed, many drug discovery efforts have focused on the development of drugs that target cancers with specific genetic features (Haque et al., 2012; Yersal, 2014). Breast cancers, for example, have been divided into five "intrinsic subtypes" (Howlader et al., 2018). More complex classification schemes and a variety of other risk markers have been proposed, since substantial diversity in genetic and transcriptomic profiles and in outcomes are observed within subtypes (Bayani et al., 2017; Dawson et al., 2013; Curtis et al., 2012; Russnes et al., 2017). If a model is trained to classify breast cancers based on genetic, transcriptomic, and/or other information is desired, clearly, the categorization chosen should be appropriate for the ultimate goal of the study: in other words, the labels that are generated should be clinically useful.

When defining metrics for predictive models, is it equally important to take into consideration existing workflows currently in practice. Ideally, a model should be chosen to solve a problem that takes advantage of the strengths and minimizes the weaknesses of existing pipelines. Steiner et al. (2018), for example, found that their deep learning algorithm exhibited a reduced false-negative rate for identification of breast cancer metastases in lymph nodes when compared with human pathologists. The algorithm, however, also exhibited an increased false-positive rate, especially if acquired images were out of focus. To overcome this problem, they designed a machine learning-assisted pipeline whereby the deep learning algorithm color-highlighted regions of interest for review by the pathologist, where different colors indicated different levels of confidence. This pipeline significantly improved both accuracy and speed compared to identification performed by unaided pathologists, thereby improving rather than reinventing the existing pipeline.

In addition to these considerations, the model should require only data that will be readily available at the time when the prediction will be made (Chen et al., 2019). In some cases, accurate diagnosis may require time-consuming lab tests whose results will seldom be available at the time of admission. A model that relies on late-arriving information may be severely limited in scope, while a model that can provide an accurate diagnosis without it may in such instances offer a key advantage. In 2019, for example, Yelin et al. used patient data from over 700,000 urinary tract infections (UTIs) to build a gradient boosted trees model and a logistic regression model to predict antibiotic resistance category solely based on patient history (Yelin et al., 2019). Their models were able to significantly outperform physicians and dramatically reduce the rate of incorrect prescriptions (i.e., situations where a patient has prescribed an antibiotic to which their infection is resistant). Since only patient history data was required, their approach can choose an antibiotic at the time of admission without waiting for antibiotic susceptibility testing results, which may require several days or more (Van Camp et al., 2020).

Availability of relevant data is a key challenge for developing machine learning models for disease classification. Healthcare datasets are in general both highly heterogenous and highly fragmented. A wide variety of EHR systems are marketed; there is little standardization across systems and software packages, so that pooling data acquired on different systems is inherently challenging (DeMartino & Larsen, 2013; Miller, 2011). EHR systems are often designed to prioritize the needs of medical billers and the insurance payors with whom they will communicate, so that the data is seldom formatted in a manner conducive to the needs of researchers or even physicians, many of whom report dissatisfaction with their healthcare system's EHR software (Agrawal & Prabakaran, 2020; Gawande, 2018). Physicians frequently record their observations in the form of notes that cannot easily be translated into encoded input suitable for modeling purposes (DeMartino & Larsen, 2013). Pooling. Furthermore, the pooling and sharing of data between healthcare providers and different sources are substantially hindered by patient privacy and regulatory concerns (Agrawal et al., 2020).

Ultimately, these issues combine to ensure that assembling and pre-processing healthcare datasets are necessary for predictive modeling which may incur substantial effort and expense. Even once such datasets have been assembled, they may appear to be large and yet contain data for a wide array of conditions, so that only a handful of datapoints relevant to a particular disorder or outcome of interest appear in the dataset. Adibuzzaman et al., for example, report their experience with the Medical Information Mart for Intensive Care (MIMIC III) from Beth Israel Deaconess Hospital. This superficially large dataset contains data for some 50,000 patient encounters; yet if a researcher interested in drug-drug interactions were to query it for patients on antidepressants also taking an antihistamine, for example, they would retrieve a mere 44 datapoints (Adibuzzaman et al., 2017). Finally, most healthcare datasets contain missing values such that key information available for some patients is unavailable for others (Allen et al., 2014).

For all these reasons, organizing healthcare data to improve access for biomedical researchers interested in disease classification is a nontrivial task that will require the active participation of the healthcare community. Already many efforts are underway to tackle many aspects of this problem, ranging from repositories like the UK Biobank, which stores genetic and health data for hundreds of thousands of volunteers (Allen et al., 2014), to the development of open-source EHR systems (Purkayastha et al., 2019). However, at present, the major challenges of access to relevant data and organizational issues continue to persist within the field.

1.3   A DIFFERENT LANGUAGE IS A DIFFERENT VISION OF LIFE: BIOMEDICAL DATA FEATURE SELECTION

Data for disease classification may comprise a wide array of different data types, including measurements from lab tests, text from EHRs, images like chest x-rays or CT scans and increasingly genetic, transcriptomic, or proteomic data (i.e., gene sequence data, relative measurements of gene expression and measurements of concentrations of many different proteins). Much of this data cannot be fed into a machine-learning algorithm in raw form but must first be encoded. The input to a supervised learning algorithm is generally a vector or matrix of numeric values or "features" associated with each datapoint. Sometimes the choice of encoding is obvious, for example, when dealing with biomedical image data, the input data is a matrix of pixel values. In other cases, the choice is not so straightforward. Protein or DNA sequences and text from documents are an example of cases in which feature selection is more indirect. For feature selection, protein, or DNA sequences must be converted to numeric vectors or matrices to apply machine learning or statistical modeling techniques. Frequently categorical inputs are converted to numeric vectors by using "one-hot encoding". In this scheme, a categorical input is represented as a vector of length k for k possible categories; all entries of the vector are set to 0 except the entry corresponding to the category for this datapoint. For protein sequences, for example, there are 20 possible amino acids at each position, so each position can be represented as a length 20 one-hot vector, where a 1 at a given element of this vector indicates the amino acid present at that position. This is not the only possible encoding for categorical inputs but is a common default (Rodríguez et al., 2018).

One famous alternative strategy for encoding categorical data is the Word2Vec algorithm developed at Google (Mikolov et al., 2013). The key insight is that the meanings of words are often reflected in their context. In this scheme, a supervised learning algorithm (a shallow neural network) is trained to predict the words likely to occur close to a one-hot encoded-word of interest, using a large training set derived from webscraping online documents. The weights or parameters from a layer of the trained model are used as learned representations of words that can be fed into other models; each is a real-valued vector containing far fewer entries than the one-hot encoding that would have been required to represent the same word. The learned representations generated by Word2Vec show many of the same relationships as the words themselves. For example, if the vector corresponding to the word "man" is subtracted from the vector corresponding to the word "king" and the vector corresponding to the word "woman" is added, the resulting vector is approximately the same as the learned representation for the word "queen". Many variations on this strategy have been developed (Pennington et al., 2014), including Google's BERT or Bi-directional Encoder Representations from Transformers (BERT) model, which achieved state-of-the-art performance on many natural language processing tasks (Devlin et al., 2019) and is currently used in Google search (Nayak, 2019). Efforts have been made to develop analogous representations for protein and DNA sequences (Asgari & Mofrad, 2015; Yang et al., 2018).

These approaches fall into a broader category of so-called "representation learning", where vector representations of non-numerical input (text, sequence data, etc.) useful across different datasets or prediction tasks are generated by a supervised or unsupervised learning approach. By capturing and encoding relationships between discrete variable inputs – between different words, for example – the learned representation may simplify the task confronting a model trained to perform a prediction task and enable it to learn useful relationships from smaller datasets (Bengio et al., 2014). Representation learning for discrete data has sometimes found applications in data pre-processing for disease classification; for example, Banerjee et al. report the use of a Word2Vec related algorithm for processing radiology reports as inputs to a model for diagnosis of pulmonary embolism (Banerjee et al., 2018). Although the field is in its early days, promising early results suggest representation learning may be especially important for the use and analysis of sequence data (DNA, protein, etc.) (Iuchi et al., 2021).

In the absence of a learned representation appropriate for a given discrete data type, one-hot encoding is frequently a good default for unranked (nominal) categorical inputs. For numeric inputs, there are other considerations. Patient weight in kilograms and blood cholesterol in mg/dL, for example, are not on the same scale. Using raw measurements such as these as the inputs to a machine learning model may result in poor performance depending on the algorithm (Hastie et al., 2009). Unless it is already predetermined that a scale-invariant model such as random forest is the best for the task of interest, it is usually preferable to rescale the numeric data. Two common approaches are standardization (subtract the mean of each feature and divide by the standard deviation) or minmax normalization (rescale the data to the 0–1 range). Regardless of the approach used, the scaling factors should be calculated using the training data only and applied to any validation or tests sets. In some cases, the choice of what information to include as an input to the machine learning model, aka "feature selection", requires careful consideration. If features necessary to predict the outcome of interest are excluded, the model may not be able to achieve optimal performance. Yet including too many features may be undesirable depending on the amount of data available. In a data-rich scenario, a deep learning model may be able to "learn" to disregard those features which provide minimal predictive power (Kuhn and Johnson, 2013; Rajkomar et al., 2019). In this type of situation, it may be best to err on the side of including features that may or may not be relevant. For small datasets, by contrast, the presence of numerous features irrelevant to the outcome can increase the risk of overfitting and degrade performance (Kuhn & Johnson, 2013; Teschendorff, 2019).

Some care should be taken to exclude or understand the impact of spurious confounders – variables whose apparent correlation to the outcome of interest is an artifact rather than a cause-and-effect relationship. Machine learning models can be inadvertently trained to generate predictions based on spurious confounders, resulting in superficially good initial performance followed by real-world predictions that are disastrously wrong. In one (in)famous example, neural network models trained to predict the risk of death in pneumonia patients predicted lower risk for asthmatic patients. This completely erroneous conclusion learned by the models stemmed from hospital policies, which dictated that patients with both asthma and pneumonia should be routed to the ICU for immediate treatment because they were at higher risk. The models "learned" the spurious correlation between asthma and improved likelihood of survival caused by an underlying variable (hospital policy) and incorrectly predicted lower risk for asthmatic patients (Caruana et al., 2015; Yonas et al., 2020).

COVID diagnosis furnishes another illustration of the challenges posed by feature selection. Studies have shown that elevated concentrations of certain proteins involved in the inflammation process (e.g., interleukins such as IL-1, IL-2, IL-6, TNF- α IL-1β, IL-6, IL-12) are associated with pulmonary distress (difficulty in breathing) caused by SARS-CoV-2 and other closely related coronaviruses (Yonas et al., 2020). Measuring concentrations of these proteins might assist in predicting disease progression and severity, but these measurements will not generally be available at the time the patient is admitted when predictions from the model would be most useful. Moreover, elevated concentrations of these proteins are observed in many other conditions, such as patients undergoing immunotherapies, that may present within the ICU (where most COVID-19 patients are assessed and admitted) (Berraondo et al., 2019; Tanaka et al., 2014; Yonas et al., 2020). Consequently, elevated levels of these proteins might indicate either the severity of the COVID case or the presence of a comorbidity. Building a model to predict COVID prognosis might therefore require working closely with the clinical users to assess both which features are most likely to prove important, which features introduce the risk that predictions will rely on spurious confounders, what type of predictions might prove most useful and the information that will be available.

1.4   OH ME, OH MY, OMICS: REDUCTION TECHNIQUES FOR TACKLING LARGE OMICS DATASETS

The increasing use of genetic, transcriptomic, and metabolomic data for disease classification poses some other special challenges. "Omics" datasets inherently contain many features, for example, when presented with gene expression data, the results may include thousands of genes from RNA-Sequencing or metabolite profiles for numerous metabolites. There may be little or no prior knowledge on which features are important, and in many cases, the number of features may be much greater than the number of datapoints – a situation where overfitting is hard to avoid. In such cases, features are sometimes chosen out of those available based on which features exhibit the strongest correlations with the outcome of interest, but this approach presents pitfalls. If selecting a model via cross-validation, for example, features should be selected based only on the subset used for training in each iteration, not the whole training set, since otherwise the cross-validation procedure no longer provides a reliable indication of the model's ability to generalize (Hastie et al., 2009). More crucially, univariate correlations between a single variable and the outcome may fail to capture situations where an interaction between two variables is a strong predictor even though neither variable alone exhibits a strong relationship (e.g., the datapoint belongs to class c if both a and b are present but class d if only one of a and b or neither) (Hastie et al., 2009; Murphy, 2012).

In cases where there are significantly more features than datapoints, dimension reduction may assist in reducing the risk of overfitting. Although dimension reduction techniques are commonly considered a form of unsupervised learning that can aid in revealing hidden structures in a dataset, they can also prove useful for reducing the number of features preparatory to fitting a supervised learning model. Principal component analysis (PCA) is perhaps the most classic dimension reduction technique. It finds a new orthogonal set of variables, or "principal components", ranked in order of how much of the total variance is explained by each. Frequently with high-dimensional datasets, most of the variance can be explained by a small subset of the principal components and only these need to be retained, thereby dramatically reducing the number of features (Murphy, 2012).

PCA is a linear dimension reduction technique; it builds new features which are linear combinations of the original ones. Clearly, there are situations where a nonlinear technique may be more applicable. A wide variety of nonlinear dimension reduction techniques have been described in the literature. We will not cover them all here, although it is perhaps worth mentioning two that have proven popular in recent years. Uniform Manifold Approximation and Projection or UMAP is typically used for visualization but can be used for dimension reduction for inputs to a classification model. This algorithm basically builds a weighted graph in the original high-dimensional space (a weighted graph is a description of the original data as a list of nodes and "edges" connecting nodes, where the weight on each edge indicates the likelihood the nodes at either end are connected). Next, an analogous low-dimensional graph is constructed, using cross-entropy – an information-theoretic measure of how different two probability distributions are – to optimize the construction of the low-dimensional graph so that it is as similar to the high-dimensional one as possible (Becht et al., 2019; McInnes et al., 2020). The low-dimensional representation that is achieved is highly sensitive to the analyst's parameter selections (Xiang et al., 2021). Autoencoders are neural network models which consist of two modules. The encoder module converts the input to a low-dimensional learned representation, while the decoder module tries to accurately reconstruct the input from the learned representation. Essentially, the autoencoder tries to find a "compressed" version of the input that contains all the information needed to reconstruct the input (Tschannen et al., 2018). In some of the more popular versions of this approach, the cost function (the function used to assess divergence between output and objective) for training is penalized in such a way that the learned representation tends to exhibit specific desired properties. Regularized autoencoders, for example, are trained to learn sparse or smooth representations of their input (Ghosh et al., 2020). In variational autoencoders, the learned representation takes the form of the parameters of a specified probability distribution, and the decoder uses samples from this distribution to reconstruct the original input (Kingma & Welling, 2014).

There are many examples of feature selection and dimension reduction for disease classification. One instance appears in Gárate-Escamila et al. (2020) in determining classification models for heart disease prediction using data from the UCI Machine Learning Repository (Gárate-Escamila et al., 2020). From the original raw data set, 74 features were selected using univariate statistical significance tests.

The selected features were in turn were used to generate a dimension-reduced set of features through PCA, which was then used as input for a random forest model. Feature selection and dimension reduction in this case improved the performance of the model.

A major drawback for dimension reduction techniques is that each feature in the new post-reduction feature set is a combination of the original features. Consequently, if dimension reduction is used to process data for training a supervised learning model, the model may become much harder to interpret, i.e., it may become more challenging to explain why the model made the predictions that it did or what variables were most important in its decision. For disease classification lack of interpretability can be a major drawback; owing to the high costs of making a mistake, clinicians and patients are unlikely to trust a model whose predictions cannot be meaningfully explained (Ahmad et al., 2018). We will return to the problem of model interpretability in disease classification shortly.

1.5 LET'S GET READY TO RUMBLE! TRAINING AND TESTING OF DISEASE MODEL PREDICTIONS

Once features have been selected and data partitioned, a model can be trained. A key consideration here is the so-called bias-variance trade-off. High-variance models have been overfitted; they have fit to the noise in the dataset rather than capturing real trends. The model may appear to perform well on its training set but will generalize poorly to new data. A high-bias model, by contrast, makes incorrect assumptions about the distribution of the data that cause it to miss real relationships and underfit the data. As a broad generalization, choosing a less flexible model with fewer parameters tends to reduce the risk of overfitting while increasing the risk of underfitting and vice versa (Theodoridis, 2015).

Hyperparameters are parameters selected by the analyst rather than learned during fitting that controls the flexibility of the model. Adjusting the hyperparameters of the model enables the analyst to seek settings that neither under- nor overfit the data and achieve the optimal performance possible. In many cases there is an upper bound to the accuracy that can be achieved given the information available to the modeler; in the classification setting, this theoretical limit is called the Bayes error rate. Since for real-world datasets this best-possible accuracy is not known, the accuracy of a model is generally judged by comparison with other state-of-the-art models, human expert performance, or the level of accuracy needed for a given task (Hastie et al., 2009; Murphy, 2012).

To ensure that performance on the training set can be reproduced, it is crucial to test the performance of any trained machine learning model on data not made available to it during the fitting process. Consequently, data is typically separated into a training set and a test set. Ideally, the test set should consist of data acquired independently of the training set. The test set cannot be used to assess algorithm or hyperparameter selection. If by any chance the testing data set is used for the "learning" algorithm or the hyperparameter selection, it becomes part of the training set and no longer provides an independent assessment of model performance. Rather, algorithm and hyperparameter selection is performed using the training set only, typically using a resampling procedure. Cross-validation is the most common resampling procedure used for model selection, although bootstrapping is another possible choice. Once an appropriate algorithm and hyperparameters have been chosen, the final model is trained on the full training set and scored on the test set. If the test set is representative, the score on the test set(s) provides an indication of how the model will likely perform on new datasets. It is crucial that the training set and test set should be representative samples of the population for which predictions are desired (Hastie et al., 2009; Murphy, 2012).

The importance of rigorous model evaluation is hard to overstate. Cutting corners here leads to models which are clinically useless and may hinder adoption in healthcare by creating a perception that the promise of machine learning techniques for healthcare has been overhyped (Ross, 2021). Remarkably, many published studies do not evaluate the performance of their model with sufficient rigor to demonstrate clinical utility. A 2021 study in Nature Machine Intelligence examined some 320 papers describing applications of machine learning for detection and diagnosis of COVID-19 from chest x-ray and CT scan images. Of these, none were clinically useful; all exhibited critical methodological errors, ranging from failing to report key details of their data pre-processing and model evaluation procedure to the absence of a held-out test set. A true external validation, where the model was tested on data acquired from a different source than the training data, was uncommon (Roberts et al., 2021).

The importance of external evaluation is underscored in another 2021 Nature Machine Intelligence paper from DeGrave et al., which recreated Convolutional Neural Network (CNN) architectures described in the literature for COVID-19 diagnosis from chest radiographs. The models were trained on a dataset frequently used in the literature that contains images provided by a group of hospitals. All models exhibited good performance on a held-out test set (mean AUC-ROC > 0.99, where AUC-ROC is a measure of classification performance that measures the sensitivity and specificity of the model using different thresholds – the closer the AUC-ROC to 1, the better). When evaluated on images from another hospital, however, performance dropped dramatically (mean AUC-ROC 0.76). Reversing the two datasets and training on the second dataset then evaluating on the first exhibited the same outcome. Further examination suggested the models had "learned" to generate predictions at least in part based on spurious confounders unique to each dataset (DeGrave et al., 2021). Consistent with their findings, Zech et al. reported in 2018 that unless appropriate precautions are taken, deep learning models used to predict the presence or absence of pneumonia from chest x-ray images frequently underperform significantly on data from hospitals that do not present in their training set, again due to the presence of spurious cofounders present in datasets from some hospitals (Zech et al., 2018). As these examples illustrate, rigorous evaluation on representative and preferably external test sets is crucial to ensuring the model can truly generalize to new datasets and should be an essential part of any study involving machine learning for disease classification. This may seem basic, but the literature suggests this advice has unfortunately sometimes been ignored and is worth restating.

1.6   THE MODEL RHYTHM FOR YOUR ALGORITHM

Once appropriate representations and features have been chosen, the next task is clearly selecting an appropriate model. There are a wide variety of algorithms that have been developed for supervised learning; we will not attempt to describe all of these in detail here but will rather focus on several of the most popular options and some applications of interest.

It is worth noting that there is probably no "best" algorithm that universally outperforms all others across all problems (Farrelly, 2017; Wolpert & Macready, 1997). Generally, each algorithm has both advantages and disadvantages and maybe optimally suited for specific types of tasks (Hastie et al., 2009). The choice of the algorithm and the model structure is thus generally driven by the type of data, the amount of data available, and desired performance characteristics. In disease classification, the interpretability of the model may frequently be a key consideration. A deep learning model that generates highly accurate predictions may ultimately prove less useful than a slightly less accurate model that can indicate what information is used to generate a given prediction. Given the high cost of making an incorrect call, clinicians and patients are unlikely to trust a model whose predictions cannot be justified or explained in a language that they can understand – and are therefore unlikely to use or rely on non-interpretable models (Ahmad et al., 2018; Katuwal & Chen, 2016). Additionally, interpretable models that can indicate which variables were used to make a decision can assist researchers in generating new hypotheses that could open fruitful avenues for research. It is also true however that various stakeholders in the healthcare decision-making process may define interpretability differently, so that understanding stakeholder expectations and what "interpretable model" means to end users may play a role in model selection as well (Rudin, 2019).

Carauna et al. of Microsoft Research present two examples of instances in machine learning for disease classification where a simpler model was preferred to a more complex and slightly more accurate one, including a collaboration with a large hospital involving data for 195,901 patients. In both cases, the simpler model provided easy-to-interpret information on which inputs were most important for making a given prediction, and this information provided valuable insights to physicians, therefore making the simpler model preferable in both cases (Caruana et al., 2015). It is worth noting that there is not always a trade-off between model interpretability and model accuracy, and that in some cases it may be possible to achieve both aims simultaneously (Rudin, 2019).

1.7   SEEING THROUGH THE BRUSH: DECISIONS TREES

Among the most interpretable of models in machine learning is an algorithm called a decision tree. A decision tree for classification is built by splitting the dataset into two subpopulations, then splitting each subpopulation into two further subpopulations and so on until a subpopulation can be split no further (it contains < a minimum number of datapoints or all datapoints in that subpopulation belong to the same class) or a maximum depth has been reached. At each split point or node, the tree seeks a feature and a split point for that feature that will maximize the "purity" of the child subpopulations where "purity" is quantified using either the information gain or, more commonly, change in Gini impurity associated with a candidate split. The Gini impurity tells us what the probability is of misclassifying an observation and helps decide how to "split" the data within the decision tree. The lower the Gini impurity value, the better the split and the lower the likelihood of misclassification at that node (Hastie et al., 2009). Decision trees are easy to fit and interpret, thus for tasks where the practitioner must be able to explain why the model made the prediction it did, they possess some obvious advantages. They are however highly prone to overfitting and are seldom robust (Kuhn & Johnson, 2013).

Given their limitations, decision trees in machine learning are most often used as components of ensemble models called random forests and gradient boosted trees, which overcome the limitations of the decision tree model by using many decision trees. In the random forest algorithm, each tree is trained on a bootstrapped re-sampling of the training set, and at each split point, the tree is permitted to access only a randomly selected subset of the available features. Consequently, each tree sees a different view of the data. At prediction time, the trees "vote" to classify any new datapoint presented to them. While any individual tree is a high-variance classifier, averaging over the trees reduces the variance of the overall model without causing the model bias to increase significantly. This "bootstrap aggregation" or bagging technique greatly reduces the risk of overfitting resulting in a remarkably robust model that is relatively easy to implement and train and frequently achieves good performance on classification tasks (Hastie et al., 2009; Kuhn & Johnson, 2013; Theodoridis, 2015).

If compared with other popular algorithms, random forests are generally inferior to deep learning for image and sequence processing but frequently win out on more generic classification tasks with highly structured input. In a 2014 evaluation of 179 classifiers (including neural networks) on 121 datasets, Fernandez-Delgado et al. found that random forest models were the top performer among the algorithms evaluated and concluded random forest may be a good "default option" for classification (Fernandez-Delgado et al., 2014). Additionally, while random forests are much less interpretable than traditional statistical models like logistic regression or discriminant analysis, they do provide good interpretability compared to deep learning models – it is easy to determine the importance of a given feature by determining how frequently it was used. Important model hyperparameters include the number of trees, the maximum depth of each tree, and the number of features each tree are permitted to access at a given split.

Gradient boosted trees algorithms operate through a boosting strategy. In this approach, a single decision tree is fitted to make predictions for each datapoint, then a cost function or loss function evaluates how far off the model is from its objective. The cost is calculated and the negative gradient of the cost function in function space is evaluated. A new tree is then trained to predict the negative gradient of the cost function, the predictions are updated based on the new tree, subsequently, the gradient is re-evaluated based on the new predictions, and so forth. At prediction time, each tree adds to and corrects the output of the previous tree and thereby "boosts" the model. Since each tree compensates to some extent for the deficiencies of its predecessors, the overall model is more reliable than the relatively weak models from which it is built (Natekin & Knoll, 2013; Wyner et al., 2017).

Consider a simple example in the regression setting to illustrate. If the loss for a given datapoint is as follows:

$$L(y, \ F(x)) = \frac{1}{2}(y - F(x))^2$$

The gradient in function space becomes:

$$\frac{\partial L(y, \ F(x))}{\partial F(x)} = F(x) - y$$

Each new tree is trained to predict the negative gradient of the previous tree – which in this instance is clearly the residuals of the previous tree. In other words, each tree is "improving" on the predictions of previous trees. Gradient boosted trees generally require more hyperparameter tuning than random forests but frequently offer excellent performance on classification tasks (Kuhn & Johnson, 2013; Natekin & Knoll, 2013). Hyperparameters that require tuning include the number of trees, the learning rate (a constant by which the negative gradient is multiplied) and the subsampling fraction, i.e., the percentage of the dataset that is randomly selected to train any given tree.

Gradient boosted trees and random forests have both seen widespread use in disease classification. As mentioned previously, Yelin et al. used a gradient boosted trees model to significantly reduce the rate of incorrect antibiotic prescription compared to physician selections (Yelin et al., 2019). To cite one of many other examples, Cobb et al. evaluated the use of both random forest and gradient boosting algorithms to predict survival for burn patients using a 31,350-patient dataset and identify the variables most important for predicting the outcome (Cobb et al., 2018).

Another popular class of algorithms converts linear models into powerful tools capable of approximating nearly any relationship. Consider two groups of data that lie in the x-y plane; the datapoints from one category encircle the datapoints from the other. Each datapoint in this coordinate system is represented by an x-coordinate and a y-coordinate – its coordinates with respect to two basis vectors using the Cartesian coordinate system. Clearly the data if used in the form given above cannot be separated by a linear boundary. Imagine however that we now specify for each datapoint an additional coordinate given by, $x^2 + y^2$ where x and y are the original x and y coordinates. In this enlarged three-dimensional space, the two groups can be separated by a plane, and a linear model may now be employed.

1.8 BUTTERED UP APPROACH: KERNEL METHOD

The insight in the previous example is that relationships that are nonlinear in low-dimensional spaces may be approximated by a linear model in a higher-dimensional space. Finding the appropriate basis expansion, however, may be nontrivial and may incur an increased risk of overfitting. Computing similarity between datapoints post-basis expansion may also be computationally expensive. Kernel methods use a kernel function to implicitly map the data into a higher dimensional (or even infinite-dimensional) space, achieving the same end effect as an explicit mapping without the computational cost – an approach sometimes called the "kernel trick". The kernel function assesses the similarity between any two datapoints. A variety of kernel functions including the squared exponential kernel, the Matern kernel, and the spectral mixture kernel have been described in the literature (Genton, 2001; Wilson & Adams, 2013).

Several algorithms use the kernel method, including kernel ridge regression (the kernel trick applied to linear regression), kernel support vector machines (the kernel trick applied to linear support vector machines models) and Gaussian process regression (GPR), which can be thought of as the kernel trick applied to Bayesian linear regression (Rasmussen & Williams, 2006). Each algorithm has some unique features. GPR, for example, provides a well-calibrated estimate of the model's uncertainty for each prediction, which for some applications may be invaluable. All of them do however share certain advantages and disadvantages.

When equipped with an appropriate kernel function, kernel methods are flexible enough to "learn" and approximate nearly any relationship present in the training set. But their flexibility leaves them prone to overfitting, which must be avoided through careful hyperparameter tuning. For Gaussian process regression, good hyperparameters can be "learned" from the data during training (Rasmussen & Williams, 2006); for the other algorithms, hyperparameters must generally be tuned through cross-validation. The performance of kernel methods is heavily reliant on selecting the right kernel, and the choice of kernel function can also be challenging for many types of data. It is not immediately clear, for example, what kernel function is most appropriate to assess the similarity of two chest x-ray images or of two patient histories. Finally, kernel methods exhibit poor scaling with dataset size. A straightforward implementation leads to training time complexity of $O(N^3)$ – in other words, the number of calculations required to fit the model increases in proportion to the cube of the number of datapoints. This limitation arises from the need to generate an N-by-N covariance matrix for N datapoints and compute its Cholesky decomposition. This drawback has traditionally limited the application of kernel methods to datasets containing at most a few tens of thousands of datapoints (Titsias, 2009). Since a kernel method must assess the similarity between a new datapoint and all datapoints in its training set before making a prediction, prediction times increase with the size of the training set as well.

Thus far, a variety of approximation methods that improve scaling for Gaussian processes have been described. Perhaps currently most important are variational approximations. Rather than measuring the similarity of each datapoint with all other datapoints, variational techniques use a set of so-called "inducing points" and compute the similarity between these and the other datapoints in the training set (Titsias, 2009). The number of inducing points is chosen by the analyst for a particular problem; if it is much less than the number of datapoints, the computational cost is dramatically reduced. The locations of the inducing points are optimized by minimization of a cost function through an iterative process during training. In stochastic variational inference, "noisy" estimates of the gradient of the cost function are calculated using a small subset of the training set on each iteration, significantly reducing computational cost (Hoffman et al., 2013). Stochastic variational inference enables the fitting of Gaussian process models to datasets containing hundreds of thousands or millions of datapoints, thereby removing one of the biggest obstacles to the application of this approach (Hensman et al., 2014; Izmailov et al., 2018).

While Gaussian processes are clearly less "interpretable" than a traditional statistical model like logistic regression, their predictions are in one sense easily explained. Under logistic regression models, the predicted value for new datapoints is generated using a weighted average of the observed values in the training set, with weights determined by the similarity of the new datapoint to the training set datapoints (or inducing points in the case of variational models), where "similarity" is defined and quantified by the kernel function. The confidence interval generated by the model also provides valuable information. In a regression case, for instance, a wide confidence interval may indicate that there are few datapoints in the training set similar to the new datapoint, or that similar datapoints in the training set to exhibit a wide range of observed values; the wide interval therefore automatically flags a model prediction that should be treated as suspect. One drawback for Gaussian processes is that they assume normally distributed residuals. A closely related model, the Student's t-process, is more robust because it assumes an increased likelihood of outliers and may therefore be more suitable for some datasets with distant outliers (Tracey & Wolpert, 2018). Another drawback previously mentioned is the challenging task of choosing a kernel, since performance depends on the appropriateness of the kernel for the problem.

Kernel methods are less common in recent disease classification and healthcare applications than deep learning models or random forests but have seen some use, especially for modeling time series. Clifton et al. (2013), for example, demonstrated the use of Gaussian processes to model data acquired from wearable sensors (Clifton et al., 2013). By fitting time series data from sensors to a Gaussian process model, the authors were able to estimate patient status for intervals where data was missing and to determine when values had strayed outside an expected range, providing early warning of deterioration in patient health. Futoma et al. reported the use of Gaussian processes to model clinical time series data and in turn use the outputs of their Gaussian process models as inputs to a deep learning model to provide early detection for sepsis. The Gaussian process model converted noisy, intermittently sampled clinical data into evenly gridded data, filling in the missing values from regions of the time series where no patient test data was available. In cases where few clinical test results are available, the wide confidence interval on the value predicted by the Gaussian process indicated its high level of uncertainty in its prediction (Futoma et al., 2017).

1.9   DEEP BLUE SEA OF PREDICTIONS: DEEP LEARNING AND NEURAL NETWORKS

The last group of algorithms whose applications we will discuss is so-called deep learning and neural network algorithms, which trace their origins to the perceptron algorithm first proposed in the 1950s and have come to dominate the image classification, protein structure prediction, and natural language processing fields (Sengupta et al., 2020). Their many advantages include good scaling to large datasets, fast predictions for new datapoints and the ability to handle inputs of variable dimensionality – images or protein sequences, for example – in a straightforward way. They are particularly useful when the input data exhibits hierarchical structure as in the case of images, where the value of a single-pixel is by itself not generally meaningful but when combined with information about neighboring pixels may be informative (Murphy, 2012).

Recent years have witnessed dramatic growth in the diversity of neural network model structures described in the literature. The simplest structure is the multi-layer perceptron (MLP), sometimes also called a fully connected neural network. It consists of a sequential arrangement of layers; the first layer processes the input, while subsequent layers process the output of the previous layer. Each layer follows the general form:

$$output = G\left(Wx + b\right)$$

Where G is a nonlinear function selected from a variety of "activation functions", W is a weight matrix, x is an input column vector and b is a vector of bias terms (analogous to a y-intercept). Common activation functions in modern neural networks include the sigmoid function, the rectified linear unit (ReLU) and the softmax function (Goodfellow et al., 2016). SoftMax function. The term deep learning usually refers to neural networks with multiple layers. In many neural network architectures, the output of a layer is normalized before being passed to the next layer, which has been shown to speed up convergence during training and can reduce the risk of overfitting (Ioffe & Szegedy, 2015).

During training, the output of the final layer is assessed with respect to a cost or loss function – mean squared error in the case of regression, for example. Like all neural networks, MLP are designed to be end-to-end differentiable, meaning that the gradient of the cost function with respect to all elements of the weight matrices and bias terms can be evaluated analytically and the parameters of the model can be updated using gradient descent. Since training sets may contain hundreds of thousands or millions of datapoints, calculating the gradient using the whole dataset is generally impractical and therefore usually the training set is broken up into "minibatches" of hundreds of datapoints or less per minibatch. During training, predictions are generated for a minibatch, the gradient is calculated, and the parameters are updated using the gradient. This process is then repeated with respect to another minibatch until all minibatches in the training set have been evaluated where a full cycle of this kind is called an epoch. While standard gradient descent as described here exhibits slow convergence, more modern versions of the algorithm make use of information about past gradients to modify the current gradient and generally achieve much better results. Two of the most popular modified gradient descent algorithms for training neural networks include Adam (Kingma & Ba, 2017) and RMSProp (Ruder, 2017).

One of the many advantages of neural network & deep learning models is their flexibility. A wide variety of alternative model structures have been developed to cope with different types of input data, and the development of new architectures is an area of highly active research. We will not attempt a comprehensive review but will provide a high-level introduction to some popular architectures. In CNNs, rather than multiplying the input by a weight matrix as in MLPs, CNNs perform convolution (or a closely related operation called cross-correlation) on the input it received with a set of weight matrices, aka filters or kernels. Frequently the resulting output is downsampled using a pooling operation. These operations provide good translational equivariance, meaning that a pattern can be detected regardless of where it occurs. Consequently, CNNs have become the dominant model architecture for image recognition and are also sometimes used for sequence or time-series data (Rawat & Wang, 2017).

CNNs are widely used in image processing, and examples of applications for disease classification abound. One especially impressive example is CheXNet, described in Rajpurkar et al., where a 121-layer dense convolutional network was trained to detect pneumonia in chest x-ray images. The authors used the ChestX-ray14 dataset for training; this dataset includes over 100,000 images with 14 different disease states (Rajpurkar et al., 2017). Normally accurate interpretation of chest x-ray data is a challenging assignment requiring substantial experience and expertise, since the images are often blurred, and a variety of other abnormalities can mimic the appearance of pneumonia in an x-ray image. CheXNet, however, outperformed four experienced academic radiologists on a held-out test set, achieving both better accuracy and significantly improved throughput.

In another well-known example, Esteva et al. trained a deep convolutional network on over 129,000 images of over 2,000 separate skin disorders. Their model was designed to classify skin lesions as either malignant or benign. Their model achieved performance competitive with that of 21 dermatologists on a held-out test set, with substantially improved throughput compared to expert classification (Esteva et al., 2017).

Recurrent neural networks (RNNs) are a second common family of network architectures. RNNs traverse a sequence or time series while maintaining an "internal state" which they update as they run using the new information provided by each input element in the sequence. The internal state enables the network to "remember" what elements it has seen previously and take that "memory" into account when processing the most recent element in the sequence. The most popular recurrent network architecture, the long-short term memory network or LSTM, solves problems with gradient-based parameter updates encountered in older RNN architectures (Lipton et al., 2015). The LSTM maintains two states: an external or "cell" state and an internal or "hidden" state. It contains four "gates", each of which is a fully connected layer similar to the fully connected layers in an MLP with a learned weight matrix, bias term, and activation function. As the LSTM moves along the sequence, at each step it takes the most recent encoded input element, concatenates it with the LSTM's hidden state, and processes the concatenated input and hidden state through all four gates to generate an updated hidden state and cell state. The hidden state and cell state preserve information about the portion of the sequence the model has seen so far. The cell state at the end of the sequence or at any desired point can be used as the output of the model or passed through additional fully connected or recurrent layers, depending on the prediction task. As in other architectures, the cost function assesses the divergence between model predictions and labels, and the gradient of the cost function is used to update the LSTM's gate parameters. This structure possesses obvious advantages for time-series or signal data, natural language processing and DNA or protein sequence data (Lipton et al., 2015).

LSTMs have recently been used in many physiological time-series analyses for clinical applications such as those for classifying EEG patterns for seizure predictions and emotion predictions, sleep states, and mental workload (Craik et al., 2019). The LSTM architecture has also been utilized for optimizing wearable sensors in assessing biomechanical movements and therapeutic plans for those undergoing stroke recovery (Tortora et al., 2020).

In the last three years, an alternative architecture developed at Google called the Transformer has begun to outperform LSTM- and RNN-based models for natural language and sequence data processing. The Transformer architecture makes use of a so-called "attention" mechanism which is also incorporated into some LSTM-based sequence-to-sequence models. Using the attention mechanism, the model can access any element of the sequence and uses learned parameters to weigh the importance of each element of the input – i.e., which elements it should pay "attention" to – rather than trying to incorporate information about all previous elements into a hidden state like an LSTM (Vaswani et al., 2017). Many current state-of-the-art natural language processing models employ variations on the Transformer architecture. Prominent applications include Google's BERT model for representation learning (mentioned above) and the AlphaFold model for protein structure prediction developed by DeepMind (Jumper et al., 2021). Given the great success enjoyed by variations on this model architecture for sequence processing in natural language processing and bioinformatics, applications in healthcare and disease classification are almost certainly in its future.

In the context of deep learning models, clearly some of the most important choices are the design of the model architecture, the number of layers and size and type of the layers, which may often require some experimentation to optimize. It is important to bear in mind though that if the test set is used to compare different model architectures, it is no longer a test set; experimentation should only be performed using the training data (or a subset of the training data). For smaller datasets, overfitting can quickly become a challenging issue, and typically deep learning is probably not a suitable technique for small datasets in general. If overfitting is encountered, it may be helpful to use a smaller network or add appropriate batch normalization between layers (if not already applied). The dropout technique has sometimes been employed as well. In this approach, on each mini-batch a randomly selected fraction of the output values from a given layer are zeroed out and are randomly "dropped out" before being passed to the next layer. However, if the fraction of output values from the layer that is dropped is too high, the model will become unstable, and performance will rapidly deteriorate. Introducing a low level of dropout can reduce the risk of overfitting on some tasks.

Dropout is more seldom used in modern deep learning architectures due to the impact on performance (Garbin et al., 2020).

1.10   DISEASE MODEL TO DISEASE SPECIALIST: MODEL INTERPRETABILITY FOR HEALTHCARE STAKEHOLDERS

As we have argued, model interpretability is especially important for disease classification and healthcare applications of machine learning. This tends to be a weak point for deep learning models (Ahmad et al., 2018). In a deep model with dozens or hundreds of layers, the input undergoes many transformations in the process of being converted into a prediction; it is very hard to convert this process into a justification for the prediction that might make sense to a patient or a physician. The development of "explainable AI" deep learning models that can provide some explanation for their predictions is a large and growing field of research; we will not review this in-depth but will note a few highlights.

Many model explanation techniques generate so-called "saliency maps", which estimate and provide a nice visualization of the relative importance of specific features in the input; these are especially popular for image classification models (Adebayo et al., 2020). Some techniques use "gradient explanations"; there are many variations on this theme but in general these evaluate the extent to which a change in the inputs affects the model's outputs to determine relative importance (Adebayo et al., 2020; Sayres et al., 2019). The LIME technique fits a local linear model to provide a simple local approximation of the full much more complex model. Perturbing the inputs reveals the importance of specific features (Ribeiro et al., 2016). This approach obviously has limitations in cases where the region of interest exhibits highly nonlinear behavior. The SHAP approach unites LIME and several related techniques with quantitative measures of feature importance that approximate the effect of removing the feature from the model (Lundberg & Lee, 2017). As an additional benefit, SHAP and LIME are not unique to deep learning and can be used with other model types.

The approaches described above merely indicate the importance of individual features. Stakeholders may however expect a more coherent explanation than feature importance; "the model rated these ten genes as very important" may or may not be satisfactory to some patients and physicians. Counterfactual techniques seek to improve on feature importance metrics by determining how the inputs would need to have changed to have generated a different prediction; different algorithms have been developed for generating these explanations (Verma et al., 2020). As with other techniques for generating post-hoc explanations for deep learning models, this is an area of very active research and is by no means a solved problem. Some authors of course have argued that in cases where model interpretability is critical (as will often be the case for disease classification), "black box" models like deep learning models are best avoided (Rudin, 2019), although a decision of this kind may sometimes involve sacrificing accuracy instead and will clearly require a solid understanding of stakeholder expectations. For a recent review on "explainable AI" techniques, an interested reader can refer Barredo et al. (2020).

Aside from the issue of interpretability, one other drawback for deep learning lies in the accurate estimation of uncertainty on model predictions. Neural network models are trained using a maximum likelihood approach such that the parameters are adjusted to maximize the likelihood of the observations, and the model does not offer a well-calibrated measure of the uncertainty associated with its predictions; rather, deep learning models are prone to "overconfidence" (Blundell et al., 2015). For disease classification and healthcare, where the risks associated with a bad call are high, it may prove beneficial for models to provide a more accurate measure of the reliability of a given prediction. Bayesian methodology offers one possible solution. As with many complex models in Bayesian statistics, the posterior distribution over the parameters and predictive posterior over future datapoints cannot be calculated analytically and must instead be approximated (Papamarkou et al., 2021).

The numerical Markov chain Monte Carlo (MCMC) techniques widely used for parameter estimation in Bayesian models in statistics can be and have been applied to approximate predictive intervals in deep learning models (Heek & Kalchbrenner, 2019; Jospin et al., 2020; Papamarkou et al., 2021). This approach provides an improved estimation of model uncertainty and improved accuracy but faces various algorithmic challenges and incurs tremendous computational cost due to the large number of parameters in most deep learning models (Jospin et al., 2020; Papamarkou et al., 2021). Variational techniques are an alternative from Bayesian statistics that approximates a complicated posterior distribution using a simpler distribution selected for its tractability and seeks to minimize the divergence between the two; variational techniques are much faster than MCMC but tend to be "overconfident" and provide only a rough estimate of uncertainty (Jospin et al., 2020; Murphy, 2012). Nonetheless, Bayesian neural networks using a variational approximation were introduced by Blundell et al. of Google DeepMind and while not common at this time have found a variety of applications (Bhattacharya et al., 2020; Blundell et al., 2015; Jospin et al., 2020). As with much else in deep learning, improved and efficient quantitation of uncertainty remains an area of active research.

1.11 THE FUTURE IS NOT SOMETHING TO PREDICT

Setting aside the limitations and advantages of specific algorithms, it is perhaps hard to overstate the immense success machine learning algorithms have achieved over the last decade. Problems like speech recognition, image classification, and protein structure prediction that were once regarded as intractable now lie well within the realm of the practical (LeCun et al., 2015). Given that many of these problems are either directly or indirectly relevant to disease classification, and given the substantial effort now directed towards machine learning applications in biomedical research, it seems likely that machine learning techniques are poised to play an increasing role in the diagnosis of disease. Just as in any field of machine learning research, rigorous model evaluation is crucial to earning end-user trust. Realizing the promise of machine learning for disease classification additionally requires an understanding of the special problems posed by this field: highly heterogeneous and noisy data, patient privacy concerns, the acute cost of incorrect predictions, the importance of interpretable models, and high stakeholder expectations. Overcoming these challenges will require ingenuity and collaboration between biologists, data scientists, and clinicians. While much work remains to be done, we believe this is a field that holds considerable promise for improving patient care and building the healthcare of the future.

REFERENCES

Adebayo, J., Gilmer, J., Muelly, M., Goodfellow, I., Hardt, M., & Kim, B. (November 6, 2020). Sanity Checks for Saliency Maps. ArXiv:1810.03292 [Cs, Stat]. http://arxiv.org/abs/1810.03292.

Adibuzzaman, M., DeLaurentis, P., Hill, J., & Benneyworth, B. D. (2017). Big Data in Healthcare - The Promises, Challenges and Opportunities from a Research Perspective: A Case Study with a Model Database. AMIA Annual Symposium 2017, 384–392.

Agrawal, R., & Prabakaran, S. (April 2020). Big Data in Digital Healthcare: Lessons Learnt and Recommendations for General Practice. *Heredity*, *124*(4), 525–534. https://doi.org/10.1038/s41437-020-0303-2.

Ahmad, M. A., Teredesai, A., & Eckert, C. (2018). Interpretable Machine Learning in Healthcare. 2018 IEEE International Conference on Healthcare Informatics (ICHI). New York, NY: IEEE. https://doi.org/10.1109/ICHI.2018.00095.

Allen, N. E., Sudlow, C., Peakman, T., Collins, & on behalf of UK Biobank (February 19, 2014). UK Biobank Data: Come and Get It. *Science Translational Medicine*, *6*(224), 224ed4–224ed4. https://doi.org/10.1126/scitranslmed.3008601.

Asgari, E., & Mofrad, M. R. K. (2015). Continuous Distributed Representation of Biological Sequences for Deep Proteomics and Genomics. *PLOS ONE*, 10(11), e0141287. https://doi.org/10.1371/journal.pone.0141287.

Banerjee, I., Chen, M. C., Lungren, M. P., & Rubin, D. L. (January 2018). Radiology Report Annotation Using Intelligent Word Embeddings: Applied to Multi-Institutional Chest CT Cohort. *Journal of Biomedical Informatics*, *77*, 11–20. https://doi.org/10.1016/j.jbi.2017.11.012.

Barredo Arrieta, A., Díaz-Rodríguez, N., Ser, J. D., Bennetot, A., Tabik, S., Barbado, A., Garcia, S., et al. (June 2020). Explainable Artificial Intelligence (XAI): Concepts, Taxonomies, Opportunities and Challenges toward Responsible AI. *Information Fusion*, *58*, 82–115. https://doi.org/10.1016/j.inffus.2019.12.012.

Bayani, J., Yao, C. Q., Quintayo, M. A., Yan, F., Haider, S., D'Costa, A., Brookes, C. L., et al. (December 2017). Molecular Stratification of Early Breast Cancer Identifies Drug Targets to Drive Stratified Medicine. *Npj Breast Cancer*, *3*(1), 3. https://doi.org/10.1038/s41523-016-0003-5.

Beam, A. L., & Kohane, I. S. (April 3, 2018). Big Data and Machine Learning in Health Care. *JAMA*, *319*(13), 1317. https://doi.org/10.1001/jama.2017.18391.

Becht, E., McInnes, L., Healy, J., Dutertre, C.-A., Kwok, I. W. H., Ng, L. G., Ginhoux, F., & Newell, E. W. (January 2019). Dimensionality Reduction for Visualizing Single-Cell Data Using UMAP. *Nature Biotechnology*, *37*(1), 38–44. https://doi.org/10.1038/nbt.4314.

Bengio, Y., Courville, A., & Vincent, P. (April 23, 2014). Representation Learning: A Review and New Perspectives. ArXiv:1206.5538 [Cs]. http://arxiv.org/abs/1206.5538.

Berraondo, P., Sanmamed, M. F., Ochoa, M. C., Etxeberria, I., Aznar, M. A., Pérez-Gracia, J. L., Rodríguez-Ruiz, M. E., Ponz-Sarvise, M., Castañón, E., & Melero, I. (January 2019). Cytokines in Clinical Cancer Immunotherapy. *British Journal of Cancer*, *120*(1), 6–15. https://doi.org/10.1038/s41416-018-0328-y.

Bhattacharya, S., Liu, Z., & Maiti, T. (November 18, 2020). Variational Bayes Neural Network: Posterior Consistency, Classification Accuracy and Computational Challenges. ArXiv:2011.09592 [Cs, Math, Stat]. http://arxiv.org/abs/2011.09592.

Blundell, C., Cornebise, J., Kavukcuoglu, K., & Wierstra, D. (May 21, 2015). Weight Uncertainty in Neural Networks. ArXiv:1505.05424 [Cs, Stat]. http://arxiv.org/abs/1505.05424.

Bui, A. A.T., & Van Horn, J. D. (May 2017). Envisioning the Future of 'Big Data' Biomedicine. *Journal of Biomedical Informatics, 69*, 115–117. https://doi.org/10.1016/j.jbi.2017.03.017.

Bzdok, D., Altman, N., & Krzywinski, M. (April 2018). Statistics versus Machine Learning. *Nature Methods, 15*(4), 233–234. https://doi.org/10.1038/nmeth.4642.

Caruana, R., Lou, Y., Gehrke, J., Koch, P., Sturm, M., & Elhadad, N. (2015). Intelligible Models for HealthCare: Predicting Pneumonia Risk and Hospital 30-Day Readmission. Proceedings of the 21th ACM SIGKDD International Conference on Knowledge Discovery and Data Mining, 1721–1730. Sydney NSW Australia: ACM. https://doi.org/10.1145/2783258.2788613.

Chen, P.-H. C., Liu, Y., & Peng, L. (May 2019). How to Develop Machine Learning Models for Healthcare. *Nature Materials, 18*(5), 410–414. https://doi.org/10.1038/s41563-019-0345-0.

Clifton, L., Clifton, D. A., Pimentel, M. A. F., Watkinson, P. J., & Tarassenko, L. (January 2013). Gaussian Processes for Personalized E-Health Monitoring With Wearable Sensors. *IEEE Transactions on Biomedical Engineering, 60*(1), 193–197. https://doi.org/10.1109/TBME.2012.2208459.

Cobb, A. N., Daungjaiboon, W., Brownlee, S. A., Baldea, A. J., Sanford, A. P., Mosier, M. M., & Kuo, P. C. (March 2018). Seeing the Forest beyond the Trees: Predicting Survival in Burn Patients with Machine Learning. *The American Journal of Surgery, 215*(3), 411–416. https://doi.org/10.1016/j.amjsurg.2017.10.027.

Craik, A., He, Y., & Contreras-Vidal, J. L. (June 1, 2019). Deep Learning for Electroencephalogram (EEG) Classification Tasks: A Review. *Journal of Neural Engineering, 16*(3), 031001. https://doi.org/10.1088/1741-2552/ab0ab5.

Curtis, C., Shah, S. P., Chin, S.-F., Turashvili, G., Rueda, O. M., Dunning, M. J., et al. (June 2012). The Genomic and Transcriptomic Architecture of 2,000 Breast Tumours Reveals Novel Subgroups. *Nature, 486*(7403), 346–352. https://doi.org/10.1038/nature10983.

Dawson, S.-J., Rueda, O. M., Aparicio, S., & Caldas, C. (February 8, 2013). A New Genome-Driven Integrated Classification of Breast Cancer and Its Implications. *The EMBO Journal, 32*(5), 617–628. https://doi.org/10.1038/emboj.2013.19.

DeGrave, A. J., Janizek, J. D., & Lee, S.-I. (July 2021). AI for Radiographic COVID-19 Detection Selects Shortcuts over Signal. *Nature Machine Intelligence, 3*(7), 610–619. https://doi.org/10.1038/s42256-021-00338-7.

DeMartino, J. K., & Larsen, J. K. (April 2013). Data Needs in Oncology: 'Making Sense of The Big Data Soup.' *Journal of the National Comprehensive Cancer Network, 11*(2), S-1–S-12. https://doi.org/10.6004/jnccn.2013.0214.

Deo, R. C. (November 17, 2015). Machine Learning in Medicine. *Circulation, 132*(20), 1920–1930. https://doi.org/10.1161/CIRCULATIONAHA.115.001593.

Devlin, J., Chang, M.-W., Lee, K., & Toutanova, K. (May 24, 2019). BERT: Pre-Training of Deep Bidirectional Transformers for Language Understanding. ArXiv:1810.04805 [Cs]. http://arxiv.org/abs/1810.04805.

Esteva, A., Kuprel, B., Novoa, R. A., Ko, J., Swetter, S. M., Blau, H. M., & Thrun, S. (February 2, 2017). Dermatologist-Level Classification of Skin Cancer with Deep Neural Networks. *Nature, 542*(7639), 115–118. https://doi.org/10.1038/nature21056.

Farrelly, C. M. (August 21, 2017). Deep vs. Diverse Architectures for Classification Problems. ArXiv:1708.06347 [Cs, Stat]. http://arxiv.org/abs/1708.06347.

Fernandez-Delgado, M., Cernadas, E., Barro, S., & Amorim, D.(2014). Do We Need Hundreds of Classifiers to Solve Real World Classification Problems?. *Journal of Machine Learning Research, 15*(3), 3133–3182. https://doi.org/10.1117/1.JRS.11.015020

Futoma, J., Hariharan, S., Sendak, M., Brajer, N., Clement, M., Bedoya, A., O'Brien, C., & Heller, K. (August 19, 2017). An Improved Multi-Output Gaussian Process RNN with Real-Time Validation for Early Sepsis Detection. ArXiv:1708.05894 [Stat]. http://arxiv.org/abs/1708.05894.

Gárate-Escamila, A. K., Hassani, A. H. E., & Andrès, E. (2020). Classification Models for Heart Disease Prediction Using Feature Selection and PCA. *Informatics in Medicine Unlocked, 19*, 100330. https://doi.org/10.1016/j.imu.2020.100330.

Garbin, C., Zhu, X., & Marques, O. (May 2020). Dropout vs. Batch Normalization: An Empirical Study of Their Impact to Deep Learning. *Multimedia Tools and Applications, 79*(19–20), 12777–12815. https://doi.org/10.1007/s11042-019-08453-9.

Gawande, A.. (November 12, 2018). Why Doctors Hate Their Computers. *The New Yorker*. https://www.newyorker.com/magazine/2018/11/12/why-doctors-hate-their-computers.

Genton, M. G. (December 1, 2001). Classes of Kernels for Machine Learning: A Statistics Perspective. *Journal of Machine Learning Research, 2*, 299–312.

Ghosh, P., Sajjadi, M. S. M., Vergari, A., Black, M., & Schölkopf, B. (May 29, 2020). From Variational to Deterministic Autoencoders. ArXiv:1903.12436 [Cs, Stat]. http://arxiv.org/abs/1903.12436.

Goodfellow, I., Bengio, Y., & Courville, A.(2016). *Deep Learning. Adaptive Computation and Machine Learning*. Cambridge, MA: The MIT Press.

Haque, R., Ahmed, S. A., Inzhakova, G., Shi, J., Avila, C., Polikoff, J., Bernstein, L., Enger, S. M., & Press, M. F. (October 2012). Impact of Breast Cancer Subtypes and Treatment on Survival: An Analysis Spanning Two Decades. *Cancer Epidemiology Biomarkers & Prevention, 21*(10), 1848–1855. https://doi.org/10.1158/1055-9965.EPI-12-0474.

Hastie, T., Tibshirani, R., & Friedman, J. (2009). *The Elements of Statistical Learning: Data Mining, Inference and Prediction*. 2nd Edition. New York: Springer Science+Business Media.

Heek, J., & Kalchbrenner, N. (August 9, 2019). Bayesian Inference for Large Scale Image Classification. ArXiv:1908.03491 [Cs, Stat]. http://arxiv.org/abs/1908.03491.

Hensman, J., Matthews, A., & Ghahramani, Z. (November 7, 2014). Scalable Variational Gaussian Process Classification. ArXiv:1411.2005 [Stat]. http://arxiv.org/abs/1411.2005.

Hoffman, M. D., Blei, D. M., Wang, C., & Paisley, J. (2013). Stochastic variational inference. *Journal of Machine Learning Research, 14*(5).

Howlader, N., Cronin, K. A., Kurian, A. W., & Andridge, R. (June 2018). Differences in Breast Cancer Survival by Molecular Subtypes in the United States. *Cancer Epidemiology Biomarkers & Prevention, 27*(6), 619–626. https://doi.org/10.1158/1055-9965.EPI-17-0627.

Ioffe, S., & Szegedy, C. (March 2, 2015). Batch Normalization: Accelerating Deep Network Training by Reducing Internal Covariate Shift. ArXiv:1502.03167 [Cs]. http://arxiv.org/abs/1502.03167.

Iuchi, H., Matsutani, T., Yamada, K., Iwano, N., Sumi, S., Hosoda, S., Zhao, S., Fukunaga, T., & Hamada, M. (February 27, 2021). Representation Learning Applications in Biological Sequence Analysis. *Preprint. Bioinformatics, 19*, 3198–3208. https://doi.org/10.1101/2021.02.26.433129.

Izmailov, P., Novikov, A., & Kropotov, D. (January 17, 2018). Scalable Gaussian Processes with Billions of Inducing Inputs via Tensor Train Decomposition. ArXiv:1710.07324 [Cs, Stat]. http://arxiv.org/abs/1710.07324.

Jospin, L. V., Buntine, W., Boussaid, F., Laga, H., & Bennamoun, M. (July 14, 2020). Hands-on Bayesian Neural Networks – a Tutorial for Deep Learning Users. ArXiv:2007.06823 [Cs, Stat]. http://arxiv.org/abs/2007.06823.

Jumper, J., Evans, R., Pritzel, A., Green, T., Figurnov, M., Ronneberger, O., Tunyasuvunakool, K., et al. (July 15, 2021). Highly Accurate Protein Structure Prediction with AlphaFold. *Nature.* 596(7873), 583–589. https://doi.org/10.1038/s41586-021-03819-2.

Katuwal, G. J., & Chen, R. (October 27, 2016). Machine Learning Model Interpretability for Precision Medicine. ArXiv:1610.09045 [q-Bio]. http://arxiv.org/abs/1610.09045.

Kingma, D. P., & Ba, J. (January 29, 2017). Adam: A Method for Stochastic Optimization. ArXiv:1412.6980 [Cs]. http://arxiv.org/abs/1412.6980.

Kingma, D. P., & Welling, M. (May 1, 2014). Auto-Encoding Variational Bayes. ArXiv:1312.6114 [Cs, Stat]. http://arxiv.org/abs/1312.6114.

Kuhn, M., & Johnson, K. (2013). *Applied Predictive Modeling.* New York: Springer Science +Business Media.

LeCun, Y., Bengio, Y., & Hinton, G. (May 28, 2015). Deep Learning. *Nature, 521*(7553), 436–444. https://doi.org/10.1038/nature14539.

Lipton, Z. C., Berkowitz, J., & Elkan, C. (October 17, 2015). A Critical Review of Recurrent Neural Networks for Sequence Learning. ArXiv:1506.00019 [Cs]. http://arxiv.org/abs/1506.00019.

Lundberg, S., & Lee, S.-I. (November 24, 2017). A Unified Approach to Interpreting Model Predictions. ArXiv:1705.07874 [Cs, Stat]. http://arxiv.org/abs/1705.07874.

McInnes, L., Healy, J., & Melville, J. (September 17, 2020). UMAP: Uniform Manifold Approximation and Projection for Dimension Reduction. ArXiv:1802.03426 [Cs, Stat]. http://arxiv.org/abs/1802.03426.

Mikolov, T., Chen, K., Corrado, G., & Dean, J. (September 6, 2013). Efficient Estimation of Word Representations in Vector Space. ArXiv:1301.3781 [Cs]. http://arxiv.org/abs/1301.3781.

Miller, R. S. (July 2011). Electronic Health Record Certification in Oncology: Role of the Certification Commission for Health Information Technology. *Journal of Oncology Practice, 7*(4), 209–213. https://doi.org/10.1200/JOP.2011.000330.

Mroz, E. A., & Rocco, J. W. (March 15, 2017). The Challenges of Tumor Genetic Diversity. *Cancer, 123*(6), 917–927. https://doi.org/10.1002/cncr.30430.

Murphy, K. (2012). *Machine Learning: A Probabilistic Approach.* Cambridge, England: The MIT Press.

Natekin, A., & Knoll, A. (2013). Gradient Boosting Machines, a Tutorial. *Frontiers in Neurorobotics, 7.* https://doi.org/10.3389/fnbot.2013.00021.

Nayak, P. (October 25, 2019). Understanding Searches Better than Ever Before. Google: The Keyword (blog), https://blog.google/products/search/search-language-understanding-bert.

O'Malley, K. J., Cook, K. F., Price, M. D., Wildes, K. R., Hurdle, J. F., & Ashton, C. M. (October 2005). Measuring Diagnoses: ICD Code Accuracy. *Health Services Research, 40*(5p2), 1620–1639. https://doi.org/10.1111/j.1475-6773.2005.00444.x.

Papamarkou, T., Hinkle, J., Young, M. T., & Womble, D. (February 23, 2021). Challenges in Markov Chain Monte Carlo for Bayesian Neural Networks. ArXiv:1910.06539 [Cs, Stat]. http://arxiv.org/abs/1910.06539.

Pennington, J., Socher, R., & Manning, C. (2014). Glove: Global Vectors for Word Representation. Proceedings of the 2014 Conference on Empirical Methods in Natural Language Processing (EMNLP), 1532–1543. Doha, Qatar: Association for Computational Linguistics. https://doi.org/10.3115/v1/D14-1162.

Purkayastha, S., Allam, R., Maity, P., & Gichoya, J. W. (2019). Comparison of Open-Source Electronic Health Record Systems Based on Functional and User Performance Criteria. *Healthcare Informatics Research, 25*(2), 89. https://doi.org/10.4258/hir.2019.25.2.89.

Rajkomar, A., Dean, J., & Kohane, I. (April 4, 2019). Machine Learning in Medicine. *New England Journal of Medicine 380*(14), 1347–1358. https://doi.org/10.1056/NEJMra1 814259.

Rajpurkar, P., Irvin, J., Zhu, K., Yang, B., Mehta, H., Duan, T., Ding, D., et al. (December 25, 2017). CheXNet: Radiologist-Level Pneumonia Detection on Chest X-Rays with Deep Learning." ArXiv:1711.05225 [Cs, Stat]. http://arxiv.org/abs/1711.05225.

Rasmussen, C. E., & Williams, C. K. I. (2006). *Gaussian Processes for Machine Learning. Adaptive Computation and Machine Learning.* Cambridge, MA: MIT Press.

Rawat, W., & Wang, Z. (September 2017). Deep Convolutional Neural Networks for Image Classification: A Comprehensive Review. *Neural Computation, 29*(9), 2352–2449. https://doi.org/10.1162/neco_a_00990.

Ribeiro, M. T., Singh, S., & Guestrin, C. (August 9, 2016). 'Why Should I Trust You?': Explaining the Predictions of Any Classifier. ArXiv:1602.04938 [Cs, Stat]. http://arxiv.org/abs/1602.04938.

Rodríguez, P., Bautista, M. A., Gonzàlez, J., & Escalera, S. (July 2018). Beyond One-Hot Encoding: Lower Dimensional Target Embedding. *Image and Vision Computing, 75,* 21–31. https://doi.org/10.1016/j.imavis.2018.04.004.

Roberts, Michael,Driggs, Derek, Thorpe, Matthew, Gilbey, Julian, Yeung, Michael, Ursprung, Stephan, Aviles-Rivero, Angelica I., Etmann, Christian, McCague, Cathal, Beer, Lucian, Weir-McCall, Jonathan R., Teng, Zhongzhao, Gkrania-Klotsas, Effrossyni, Rudd, James H. F., Sala, Evis, & Schönlieb, Carola-Bibiane (2021). Common pitfalls and recommendations for using machine learning to detect and prognosticate for COVID-19 using chest radiographs and CT scans. Nature Machine Intelligence, 3, 199–21710.1038/s42256-021-00307-0.

Ross, C. (June 2, 2021). Machine Learning Is Booming in Medicine. It's Also Facing a Credibility Crisis. *StatNews.* https://www.statnews.com/2021/06/02/machine-learning-ai-methodology-research-flaws/.

Ruder, S. (June 15, 2017). An Overview of Gradient Descent Optimization Algorithms. ArXiv:1609.04747 [Cs]. http://arxiv.org/abs/1609.04747.

Rudin, C. (September 21, 2019). Stop Explaining Black Box Machine Learning Models for High Stakes Decisions and Use Interpretable Models Instead. ArXiv:1811.10154 [Cs, Stat]. http://arxiv.org/abs/1811.10154.

Russnes, H. G., Lingjærde, O. C., Børresen-Dale, A.-L., & Caldas, C. (October 2017). Breast Cancer Molecular Stratification. *The American Journal of Pathology, 187*(10), 2152–2162. https://doi.org/10.1016/j.ajpath.2017.04.022.

Sayres, R., Taly, A., Rahimy, E., Blumer, K., Coz, D., Hammel, N., Krause, J., et al. (April 2019). Using a Deep Learning Algorithm and Integrated Gradients Explanation to Assist Grading for Diabetic Retinopathy. *Ophthalmology, 126*(4), 552–564. https://doi.org/10.1016/j.ophtha.2018.11.016.

Schmidhuber, J. (January 2015). Deep Learning in Neural Networks: An Overview. *Neural Networks, 61,* 85–117. https://doi.org/10.1016/j.neunet.2014.09.003.

Sengupta, S., Basak, S., Saikia, P., Paul, S., Tsalavoutis, V., Atiah, F., Ravi, V., & Peters, A. (April 2020). A Review of Deep Learning with Special Emphasis on Architectures, Applications and Recent Trends. *Knowledge-Based Systems, 194,* 105596. https://doi.org/10.1016/j.knosys.2020.105596.

Sheppard, J. E., Weidner, L. C. E., Zakai, S., Fountain-Polley, S., & Williams, J. (March 1, 2008). Ambiguous Abbreviations: An Audit of Abbreviations in Paediatric Note Keeping. *Archives of Disease in Childhood, 93*(3), 204–206. https://doi.org/10.1136/adc.2007.128132.

Steiner, D. F., MacDonald, R., Liu, Y., Truszkowski, P., Hipp, J. D., Gammage, C., Thng, F., Peng, L., & Stumpe, M. C. (December 2018). Impact of Deep Learning Assistance on the Histopathologic Review of Lymph Nodes for Metastatic Breast Cancer. *The American Journal of Surgical Pathology*, *42*(12), 1636–1646. https://doi.org/10.1097/PAS.0000000000001151.

Tanaka, T., Narazaki, M., & Kishimoto, T. (October 1, 2014). IL-6 in Inflammation, Immunity, and Disease. *Cold Spring Harbor Perspectives in Biology*, *6*(10), a016295–a016295. https://doi.org/10.1101/cshperspect.a016295.

Teschendorff, A. E. (May 2019). Avoiding Common Pitfalls in Machine Learning Omic Data Science. *Nature Materials*, *18*(5), 422–427. https://doi.org/10.1038/s41563-018-0241-z.

Theodoridis, S. (2015). *Machine Learning: A Bayesian and Optimization Perspective*. London: Elsevier Academic Press.

Titsias, M. (2009). Variational learning of inducing variables in sparse Gaussian processes. *Journal of Machine Learning Research*, 5, 567–574.

Tortora, S., Ghidoni, S., Chisari, C., Micera, S., & Artoni, F. (July 13, 2020). Deep Learning-Based BCI for Gait Decoding from EEG with LSTM Recurrent Neural Network. *Journal of Neural Engineering*, *17*(4), 046011. https://doi.org/10.1088/1741-2552/ab9842.

Tracey, B. D., & Wolpert, D. H. (January 8, 2018). Upgrading from Gaussian Processes to Student's-T Processes. 2018 AIAA Non-Deterministic Approaches Conference. https://doi.org/10.2514/6.2018-1659.

Tschannen, M., Bachem, O., & Lucic, M. (December 12, 2018). Recent Advances in Autoencoder-Based Representation Learning. ArXiv:1812.05069 [Cs, Stat]. http://arxiv.org/abs/1812.05069.

Van Camp, P.-J., Haslam, D. B., & Porollo, A. (May 25, 2020). Prediction of Antimicrobial Resistance in Gram-Negative Bacteria From Whole-Genome Sequencing Data. *Frontiers in Microbiology*, *11*, 1013. https://doi.org/10.3389/fmicb.2020.01013.

Vaswani, A., Shazeer, N., Parmar, N., Uszkoreit, J., Jones, L., Gomez, A. N., Kaiser, L., & Polosukhin, I. (December 5, 2017). Attention Is All You Need. ArXiv:1706.03762 [Cs]. http://arxiv.org/abs/1706.03762.

Verma, S., Dickerson, J., & Hines, K. (October 20, 2020). Counterfactual Explanations for Machine Learning: A Review. ArXiv:2010.10596 [Cs, Stat]. http://arxiv.org/abs/2010.10596.

Wilson, A. G., & Adams, R. P. (December 31, 2013). Gaussian Process Kernels for Pattern Discovery and Extrapolation. ArXiv:1302.4245 [Cs, Stat]. http://arxiv.org/abs/1302.4245.

Wolpert, D. H., & Macready, W. G. (April 1997). No Free Lunch Theorems for Optimization. *IEEE Transactions on Evolutionary Computation*, *1*(1), 67–82. https://doi.org/10.1109/4235.585893.

Wyner, A. J., Olson, M., Bleich, J., & Mease, D. (2017). Explaining the Success of AdaBoost and Random Forests as Interpolating Classifiers. *Journal of Machine Learning Research*, *18*, 1–33.

Xiang, R., Wang, W., Yang, L., Wang, S., Xu, C., & Chen, X. (March 23, 2021). A Comparison for Dimensionality Reduction Methods of Single-Cell RNA-Seq Data. *Frontiers in Genetics*, *12*, 646936. https://doi.org/10.3389/fgene.2021.646936.

Yang, K. K., Wu, Z., Bedbrook, C. N., & Arnold, F. H. (August 1, 2018). Learned Protein Embeddings for Machine Learning. *Bioinformatics*, *34*(15), 2642–2648. https://doi.org/10.1093/bioinformatics/bty178.

Yelin, I., Snitser, O., Novich, G., Katz, R., Tal, O., Parizade, M., Chodick, G., Koren, G., Shalev, V., & Kishony, R. (July 2019). Personal Clinical History Predicts Antibiotic Resistance of Urinary Tract Infections. *Nature Medicine*, *25*(7), 1143–1152. https://doi.org/10.1038/s41591-019-0503-6.

Yersal, O. (2014). Biological Subtypes of Breast Cancer: Prognostic and Therapeutic Implications. *World Journal of Clinical Oncology*, *5*(3), 412. https://doi.org/10.5306/wjco.v5.i3.412.

Yonas, E., Alwi, I., Pranata, R., Huang, I., Lim, M. A., Yamin, M., Nasution, S. A., Setiati, S., & Virani, S. S. (November 2020). Elevated Interleukin Levels Are Associated with Higher Severity and Mortality in COVID 19 – A Systematic Review, Meta-Analysis, and Meta-Regression. *Diabetes & Metabolic Syndrome: Clinical Research & Reviews*, *14*(6), 2219–2230. https://doi.org/10.1016/j.dsx.2020.11.011.

Zech, J. R., Badgeley, M. A., Liu, M., Costa, A. B., Titano, J. J., & Oermann, E. K. (November 6, 2018). Variable Generalization Performance of a Deep Learning Model to Detect Pneumonia in Chest Radiographs: A Cross-Sectional Study. Edited by Sheikh, A. *PLOS Medicine*, *15*(11), e1002683. https://doi.org/10.1371/journal.pmed.1002683.

2 A Review of Automatic Cardiac Segmentation using Deep Learning and Deformable Models

Behnam Rahmatikaregar, Shahram Shirani, and Zahra Keshavarz-Motamed

CONTENTS

DOI: 10.1201/9781003120902-2

2.1   INTRODUCTION

Cardiovascular disease is the leading cause of death worldwide and is projected to remain the first cause of death by 2030 (Mozaffarian et al., 2015) Complex and mixed valvular, vascular, and ventricular disease (Complex VVVD) is one of the most acute and chronic cardiovascular disease conditions in which multiple valvular, vascular, and ventricular pathologies have mechanical interactions with one another. Physical phenomena associated with each pathology amplify the effects of others on the cardiovascular system (Ben-Assa et al., 2019; Keshavarz-Motamed, 2020; Keshavarz-Motamed et al., 2020, 2014). Examples of components of C3VI include: valvular disease (e.g., aortic valve stenosis, mitral valve stenosis, aortic valve regurgitation, and mitral valve insufficiency), ventricular disease (e.g., left ventricle dysfunction and heart failure), vascular disease (e.g., hypertension), paravalvular leaks, LV outflow tract obstruction in patients with implanted cardiovascular devices (e.g., transcatheter valve replacement), changes due to surgical procedures for C3VI (e.g., valve replacement and left ventricular reconstructive surgery) and etc. (Ben-Assa et al., 2019; Blanke et al., 2017; Elmariah et al., 2013; Généreux et al., 2013; Keshavarz-Motamed et al., 2020; Nombela-Franco et al., 2014). In Complex VVVD, the disease condition can rapidly affect the pumping action of the heart and can ultimately lead to heart failure. Heart failure affects more than 26 million people worldwide and is increasing in prevalence with high mortality and morbidity (Savarese et al., 2017).

Segmentation of the anatomical structures inside the heart including the left ventricle (LV), right ventricle (RV), left atrium (LA), aorta (AO), and great vessels such as coronary arteries, etc. are very important in the diagnosis of Complex VVVDs. For example, delineation of the different layers of the ventricle's wall is essential in the assessment of the left and right ventricle's function, volume, and mass, however, manual segmentation is time-consuming, requires the presence of clinicians, and is prone to human error. To overcome the limitations of manual segmentation, many automatic and semi-automatic approaches have been proposed for the delineation of the ventricles by using medical image segmentation methods.

Among the automatic segmentation methods, deep learning is currently the most frequently used approach, however, learning-based approaches have become popular in the last decade. Previously traditional segmentation methods were more common.

Traditional medical image segmentation methods can be categorized into five major sets as described below.

1. Image-driven methods such as thresholding, edge-based segmentation, region-based segmentation, and partial differential equation (PDE) based segmentation. These methods require user interaction and are unable to accept strong prior knowledge.
2. Deformable models including parametric deformable models (also known as active contours) and geometric deformable models (also known as level sets) have very good performance for automatic segmentation of the left ventricle due to the energy function flexibility which can accept strong or weak prior knowledge. Limitations of this method include the initialization of the contours and designing an energy function that works robustly for all different images, both of which require user interaction.
3. Pixel classification methods assign a label to each pixel of the image based on extracted features. Classification methods including K-means and fuzzy cluster means can be unsupervised, whereas some deep learning and machine learning-based classification approaches require supervision.
4. Atlas-guided methods register a pre-segmented atlas to each image and apply the same deformation to the atlas segments in order to delineate each segment of the input images. Due to the lack of general atlases for the cardiovascular system, these methods are less common for LV segmentation.
5. Active shape/appearance modeling (ASM/AAM) are statistical models based on the shape of objects which iteratively deform to fit an example of

the object in a new image. Segmentation is performed by placing the model on the image, and iteratively estimating rotation, translation and scaling parameters using least square estimation, while constraining the weights of the instance shape to stay within suitable limits for similar shapes. ASM have been extended to gray level modeling, yielding active appearance models (AAM) (Cootes et al., 2001), that represent both the shape and texture variability seen in a training set (Petitjean & Dacher, 2011). Model-based methods have many benefits, however, designing a robust model that can function for all variations of different tissues is challenging.

Application of image segmentation methods for medical images has been reviewed by many papers (Angenent et al., 2006; Jain et al., 1998; McInerney & Terzopoulos, 1996; Montagnat et al., 2001; D. L. Pham et al., 2000; Suri et al., 2002; Withey & Koles, 2008). Among all medical segmentation methods, only some review papers have focused on the segmentation of cardiac images (C. Chen et al., 2020; Frangi et al., 2001; Heimann & Meinzer, 2009; Noble & Boukerroui, 2006; Peng et al., 2016; Petitjean et al., 2015; Petitjean & Dacher, 2011; Tavakoli & Amini, 2013; Zhuang, 2013). For example, Petitjean & Dacher (2011) has done a comprehensive review of 70 papers using traditional segmentation methods on the short axis MRI scans.

In the past decade, deep learning has developed a new category for image segmentation due to its significant performance. Many deep learning-based medical image segmentation approaches have been offered, a review of them is available in C. Chen et al. (2020), Hesamian et al. (2019), Litjens et al. (2017), and Shen et al. (2017). Due to the lack of labeled medical images, deep learning approaches are not robust and are prone to overfitting. Some approaches aim to design medical segmentation architectures that work with a small set of labeled data such as U-Net (Ronneberger et al., 2015) and its variants (Çiçek et al., 2016; Milletari et al., 2016); however, this remains an open problem.

In the presence of sufficient labeled data, an end-to-end deep learning design can have the best performance, otherwise adding some feature-engineering is required. To solve this issue, hybrid methods are offered that combine deep learning along with additional approaches (Avendi et al., 2016; Duan et al., 2018; Ngo et al., 2017; Rohé et al., 2018; Rupprecht et al., 2016; Veni et al., 2018; D. Yang et al., 2018). Several segmentation approaches are offered by combining these methods. The performance of deformable models among traditional segmentation methods makes them a good candidate to solve the problem of limited data associated with deep learning.

2.2 BACKGROUND

2.2.1 MEDICAL BACKGROUND

The heart is made up of four chambers and four valves. The chambers include the left and right atrium (upper chambers), as well as the left and right ventricle (lower chambers). The valves include the mitral, aortic, tricuspid, and pulmonary valves.

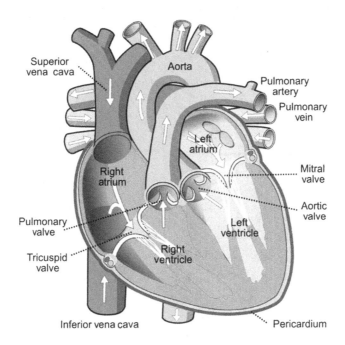

FIGURE 2.1   The anatomy of the heart. https://en.wikipedia.org/wiki/Heart.

The right ventricle receives the deoxygenated blood from the right atrium and pumps it to the pulmonary artery which carries blood to the lungs for oxygenation. The left ventricle collects oxygen-rich blood from the left atrium and pumps it through the aorta and coronary arteries to the rest of the body. Figure 2.1 shows the heart's structures.

Early and accurate diagnosis of Complex VVVD requires the assessment of both the left and right ventricles which carry out vital functions. Function, volume, and mass of the left and right ventricles during end-diastole and end-systole phases are important factors required for the accurate assessment of the cardiovascular system.

The right ventricle has a thinner wall and more complex geometry compared to the circular shape of the left ventricle. These complexities make delineation of the right ventricle challenging, even for physicians. Most of the automatic segmentation approaches are therefore limited to left ventricle annotation; Figure 2.2 shows the geometry of the LV and RV.

Segmentation of the walls of both ventricles is an initial step for the assessment of the Complex VVVD. The heart wall is divided into three different layers, the epicardium, myocardium, and endocardium, from outermost to innermost, respectively. Different segmentation approaches will focus on the annotation of different layers.

The cardiac cycle is defined as a sequence of alternating contractions and relaxations of the atria and ventricles in order to pump blood throughout the body. This cycle starts at the beginning of one heartbeat and ends at the start of the next. Each cardiac cycle has a diastolic phase (also called diastole) and a systolic phase

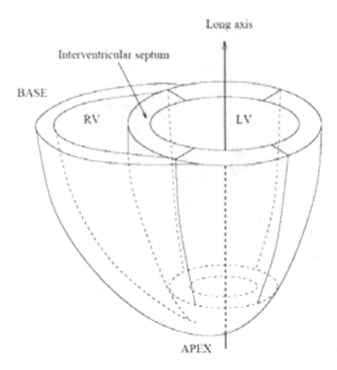

FIGURE 2.2 LV and RV geometry (Jain et al., 1998). LV has a circular shape while RV has a thinner wall and more complex geometry.

(also called systole). Diastole occurs when the heart muscles relax, and the chambers are able to fill with blood. Systole occurs when the ventricles contract, pushing blood out of the right and left ventricles into the lungs and the rest of the body, respectively. Since manual delineation of ventricle contours in all cardiac phases is not possible, physicians focus only on end-diastole and end-systole phases for assessment of the cardiovascular system.

Volume, function, and mass of the ventricles at the end-diastole (ED) and end-systole (ES) should be evaluated. Volume is calculated by integrating surfaces obtained on the endocardium, whereas mass is computed by integration of both endocardial and epicardial surfaces.

The obtained volumes at the ED and ES will be used for calculation of the ejection fraction (EF). EF is a measurement of the amount of blood the left ventricle pumps out with each contraction expressed as a percentage. It is defined as the ratio of the difference in heart volume in ED and ES to the heart volume during ED.

2.2.2 IMAGE MODALITIES

Over the past decade, the use of different medical imaging modalities has exponentially increased. The main image modalities used for cardiovascular diagnosis are:

1. **Magnetic resonance imaging (MRI):** MRI allows for 3D visualization and measurement of complex anatomy, e.g., Complex VVVD. On the downside, MRI has a lower temporal resolution (20 ms highest) compared to Doppler echocardiography, and is the most costly of the compared imaging modalities (Orwat et al., 2014; Picano, 2005; Watson et al., 2018). Most importantly, its use is limited in patients with implanted medical devices as they remain a major risk during the examination (Watson et al., 2018). There are different MRI protocols, each giving specific information (for example having different contrasts, scanning different tissues, or having different image planes). The common MRI protocols for cardiac imaging include cine CMR, late gadolinium enhancement (LGE) CMR, flow CMR, perfusion CMR, and tagged CMR (Peng et al., 2016).

   Cine CMR provides functional (dynamic) scans from the heart with good spatiotemporal resolution. It generally provides between 20 and 30 scans from a cardiac cycle. Each scan contains between 10 and 15 2D slices and the distance between those slices is in order of 10–20 mm. Total 2D images in a Cine MRI scan is the multiplication of the number of 2D slices (10–15) by the number of time points (20–30).

   The images can be obtained in any orientation but in the case of cardiac imaging they are generally captured along two perpendicular axes: short axis and long axis. The long axis goes across the LV through apex to base and the short axis is perpendicular to the long axis (Figure 2.2).

   In **LGE CMR** gadolinium-based contrast agents are injected and delivered to the myocardium and the imaging is performed at least 10 minutes after the injection. This protocol is important for the estimation of scar tissue in the myocardium.

   Flow CMR allows measuring the velocity as well as structure. For example, in a 4D flow CMR the velocity of flowing blood at each voxel in the volume is encoded which enables fluid dynamics to be visualized using specialist software.

2. **Computed tomography (CT):** CT scans allow for 3D visualization and measurement of complex anatomy as well as flexible structures at high spatial resolution (Fleischmann et al., 2008; Villarraga-Gómez et al., 2018). Dual-source CT has poor temporal resolution with the highest resolution output of 83 ms, which is the lowest of the compared modalities, thus requiring slow and steady heart rates to yield a clear image (Lin & Alessio, 2009; Watson et al., 2018).

3. **Doppler echocardiography (DE):** DE provides functional, real-time information regarding cardiac geometry, instantaneous flow, and pressure gradients (Anavekar & Oh, 2009). DE can detect structural abnormalities as well as assess contractility and ejection fraction at an excellent temporal resolution of <4 ms, and has an infinitesimal risk-to-benefit ratio (Papolos et al., 2016). Despite DE's versatility and potential, DE images cannot describe the 3D visualization of cardiovascular anatomy including Complex VVVD.

2.2.3 DATASETS AND CHALLENGES

For training and validation of the learning-based heart segmentation approaches (such as Convolutional Neural Networks (CNN) or deep learning), thousands of labeled images are required. Labeled images are usually obtained by manual delineation of the LV, RV, or any other region of interest.

Different segmentation methods rely on different image modalities and datasets. For the case of LV and RV, there are some common datasets used by many researchers. When evaluated, approaches that use the same dataset are easier to compare.

Sunnybrook (Radau, Perry et al., 2009) LVSC (Suinesiaputra et al., 2014), RVSC (Petitjean et al., 2015), ACDC (Bernard et al., 2018), HVSMR (Pace et al., 2015), MM-WHS (Zhuang et al., 2019), York University dataset (Andreopoulos & Tsotsos, 2008), CETUS (Bernard et al., 2016), and CAMUS (Leclerc et al., 2019) are the most common datasets for segmentation of the heart. CETUS and CAMUS are based on Ultrasound scans, and MM-WHS provides both MRI and CT scans, while the other datasets mentioned above are based on cine CMR. MM-WHS and HVSMR provide dataset for whole heart segmentation while other mentioned datasets focus solely on segmentation of the ventricles. Many competitions and challenges have been held for the automatic segmentation of the cardiovascular images, most of them from the STACOM (statistical atlases and computational modeling of the heart). The challenges focus on automatic segmentation of different heart structures and provide datasets with ground truth contours and standard evaluation metrics for the competitors.

MICCAI 2009 left ventricle segmentation challenge focused on the segmentation of the left ventricle's myocardium. The dataset that was provided for this challenge was Sunnybrook cardiac data (SCD) which included 45 short-axis cine-MRI images from a mix of patients with different pathologies: healthy, hypertrophic, heart failure with infarction, and heart failure without infarction. Some of the deep learning-based approaches competed in this challenge are compared in Table 2.1.

Tran (2017) which has the best performance among the approaches validated on the MICCAI 2009 challenge uses an end-to-end FCN trained, fine-tuned, and tested on different datasets. Zreik et al. (2016) designed a two-stage segmentation approach: ROI selection followed by a voxel classification CNN. Avendi et al. (2016) and Ngo et al. (2017) combined deep learning and deformable models for the segmentation.

Left ventricle segmentation challenge (LVSC) that was held in 2011, focused on the segmentation of the left ventricle's myocardium from the cine MRI scans. The provided dataset consists of short-axis cine MRI scans from 200 patients with coronary artery disease and myocardial infarction.

Right ventricle segmentation challenge (RVSC) was held in 2012 with the aim of automatic segmentation of the right ventricle's epicardium and endocardium. The dataset includes short-axis cine-MRI from 48 patients with various cardiac pathologies.

TABLE 2.1

Comparison of the Approaches that Used Deep Learning-Based Segmentation Methods and Validated on Sunnybrook Dataset

Method	APD (mm)		Good contour (%)		Dice index (mm)		Method
(Tran, 2017)	Endo 1.73(0.35)	Epi 1.65(0.31)	Endo 98.48(4.06)	Epi 99.17(2.20)	Endo 0.92(0.03)	Epi 0.96(0.01)	FCN
(Zreik et al., 2016)	Endo 2.3(1.1)	Epi -	Endo 97.9	Epi -	Endo 0.88(0.1)	Epi -	Other CNN
(Avendi et al., 2016)	1.81		96.69		0.94		Hybrid
(Ngo et al., 2017)	2.26(0.46)		93.23(9.84)		0.89(0.03)		Hybrid

HVSMR (2016) challenge focused on the segmentation of the heart's myocardium and great vessels from 3D MRI datasets of patients with congenital heart disease. The dataset included 3D MRI scans from 20 patients and manual segmentation of the blood pool and ventricular myocardium were provided.

Automated cardiac diagnosis challenge (ACDC) which was held in 2017 focused on the automatic segmentation of the left ventricular endocardium and myocardium as well as the right ventricular endocardium from the cine MRI scans. There is another task in the challenge which includes the classification of the examinations in five classes (normal case, heart failure with infarction, dilated cardiomyopathy, hypertrophic cardiomyopathy, abnormal right ventricle).

The provided dataset includes short-axis Cine MRI from 150 patients diagnosed with previous diseases including myocardial infarction, dilated cardiomyopathy, hypertrophic cardiomyopathy, abnormal right ventricles, as well as normal controls. Bernard et al. (2018) and Isensee et al. (2018) reviewed the approaches presented in this challenge, the results of different approaches are compared in Table 2.2. The segmentation accuracy of ten methods evaluated on the test dataset of the ACDC challenge has been compared in terms of dice index. Segmentation results of the left ventricle's endocardium, right ventricle's endocardium, and the left ventricle's myocardium are computed at the end-systolic and end-diastolic phases.

Among all the approaches validated on the ACDC dataset, Isensee et al. (2018) had the highest dice index for all the structures (LV endocardium, RV endocardium, LV myocardium) and all the cardiac phases (ED, ES). Some approaches used the U-Net for the segmentation task (Baumgartner et al., 2017; Isensee et al., 2018; Jang et al., 2018) while some others designed their segmentation networks inspired by the U-Net architecture (Baumgartner et al., 2017; Isensee et al., 2018; Jang et al., 2018).

TABLE 2.2

Segmentation Accuracy of Methods Validated on ACDC Dataset (Bernard et al., 2018)

Method	LV(ED)	LV(ES)	RV(ED)	RV(ES)	Myo(ED)	Myo(ES)
(Isensee et al., 2018)	0.968	0.931	0.946	0.899	0.902	0.919
(Baumgartner et al., 2017)	0.963	0.911	0.932	0.883	0.892	0.901
(Jang et al., 2018)	0.959	0.921	0.929	0.885	0.875	0.895
(Zotti et al., 2017)	0.957	0.905	0.941	0.882	0.884	0.896
(Khened et al., 2018)	0.964	0.917	0.935	0.879	0.889	0.898
(Wolterink et al., 2017)	0.961	0.918	0.928	0.872	0.875	0.894
(Patravali et al., 2017)	0.955	0.885	0.911	0.819	0.882	0.897
(Rohé et al., 2018)	0.957	0.900	0.916	0.845	0.867	0.869
(Tan, 2017)	0.948	0.865	0.863	0.743	0.794	0.801
(X. Yang et al., 2018)	0.864	0.775	0.789	0.770	–	–

Isensee et al. (2018) implemented an ensemble of 2D and 3D U-Nets. Baumgartner et al. (2017) Compared segmentation results of three different networks including FCN-8, 2D U-Net and 3D U-Net. Patravali et al. (2017) also compared the performance of 2D and 3D U-Nets. Tan (2017) implemented a 3D U-Net for segmentation. Jang et al. (2018) designed "M-Net" based on the architecture of U-Net, Khened et al. (2018) performed segmentation by combining inception modules with 2D U-Nets, and Zotti et al. (2017) designed "grid-Net" which corresponds to a U-Net with convolutional layers along the skip connections.

Multi-modality whole heart segmentation challenge (MM-WHS) which was held in 2017 and aimed to segment seven structures of the heart including: LV blood cavity, RV blood cavity, LA blood cavity, RA blood cavity, LV myocardium, AO trunk from the aortic valve, and the PA trunk from the pulmonary valve. 120 3D cardiac scans including 60 MRI and 60 CT scans were provided by the challenge. Zhuang et al. (2019) reviewed the approaches competed in this challenge. The results of different approaches are compared in Table 2.3. That includes the segmentation accuracy of seven different approaches evaluated on the MM-WHS test dataset in terms of dice index for different structures.

In general, the evaluated approaches had better results on the CT images compared to MRI. Most of the approaches used a two-stage algorithm, an ROI detection stage for localization of each structure and another stage for final segmentation of each structure. Most of the approaches used U-Nets or modified them as a part of the segmentation task (Payer et al., 2018; Tong et al., 2018; Wang & Smedby, 2018; Xu et al., 2018; Ye et al., 2019).

Payer et al. (2018) used a U-Net like FCN for localization of each structure, and another FCN for segmentation of each structure. X. Yang et al., (2018) performed the segmentation based on the 3D FCNs. Mortazi et al. (2017) used an encoder-decoder CNN for the segmentation Tong et al. (2018) used a 3D U-Net for localization of each structure and a deeply supervised 3D U-Net for segmentation of each structure. Ye et al. (2019) used 3D deeply supervised U-Net with multi-depth fusion. Xu et al. (2018) used a fast R-CNN based approach for localization of each structure and a 3D U-Net for segmentation of each structure. Wang et al. (2018) used a modified 3D U-Net performing ROI localization, followed by another modified 3D U-Net for segmentation of each ROI.

MICCAI 2009, LVSC, RVSC, ACDC, HVSMR, and MM-WHS challenges are summarized in Table 2.4. Structures of interest in each challenge, number of training and validation datasets provided by the challenge, modality of the images, and cardiac conditions of the subjects provided by the challenge are reviewed in Table 2.4.

2.2.4 EVALUATION METRICS

Results of the automatic segmentation should be compared to the ground truth contours delineated manually. There are different metrics for evaluation of the segmentation results. In this paper, some of the most frequently used metrics are discussed.

TABLE 2.3

Segmentation Accuracy of Methods Validated on MM-WHS Dataset. Reported Numbers are Dice Scores (CT/MRI) (C. Chen et al., 2020). The Bold Number in each Column Represents the Highest Score for the Corresponding Structure on CT Images

Method	LV	RV	LA	RA	Myo	AO	PA	WHS
(Payer et al., 2018)	91.8/91.6	**90.9**/86.8	92.9/85.5	**88.8**/88.1	88.1/77.8	93.3/88.8	84.0/73.1	**90.8**/86.3
(X. Yang et al., 2018)	92.3/75.0	85.7/75.0	**93.0**/82.6	87.1/85.9	85.6/65.8	89.4/80.9	83.5/72.6	89.0/78.3
(Mortazi et al., 2017)	90.48/87.1	88.3/83.0	91.6/81.1	83.6/75.9	85.1/74.7	90.7/83.9	78.4/71.5	87.9/81.8
(Tong et al., 2018)	89.3/70.2	81.06/68.0	88.9/67.6	81.2/65.4	83.7/62.3	86.8/59.9	69.8/47.0	84.9/67.4
(Wang et al., 2018)	80.0/86.3	78.6/84.9	90.4/85.2	79.4/74.4	72.9/74.4	87.4/82.4	64.8/78.8	80.6/83.2
(Ye et al., 2019)	**94.4**/-	89.5/-	91.6/-	87.8/-	**88.9**/-	**96.7**/-	**86.2**/-	90.7/-
(Xu et al., 2018)	87.9/-	90.2/-	83.2/-	84.4/-	82.2/-	91.3/-	82.1/-	85.9/-

TABLE 2.4

Summary of Challenges Relevant to the Automatic Segmentation of the LV or Whole Heart

Challenge	Structures	# Training data	# Validation data	Modality	Subjects description
MICCAI 2009	LV myocardium	15 training	15 validation 15 online	SAX	mix of cardiac conditions: healthy, hypertrophy, heart failure with infarction, and heart failure without infarction
LVSC	LV myocardium	100	100	SAX	coronary artery disease and myocardial infarction
RVSC	RV	16 training	16 test1 16 test2	SAX	various cardiac pathologies
ACDC	LV, RV, LV myocardium	100	50	SAX	myocardial infarction, dilated cardiomyopathy, hypertrophic cardiomyopathy, abnormal right ventricles, and normal controls
HVSMR 2016	WHS: myocardium and blood cells	10	10	3D CMR	complex congenital heart disease (CHD)
MM-WHS	WHS: LV, RV, LARA, AA, PALV myo	20 CT 20 MRI	40 CT 40 MRI	3D CMR 3D CT	Congenital heart disease, Atrial fibrillation, others (various pathologies)

2.2.4.1 Dice Index

Having the area of the manually delineated contour and automatically segmented contour, the dice index is defined as $2 \times \frac{the\ Area\ of\ Overlap}{the\ total\ number\ of\ pixels\ in\ both\ images}$. Figure 2.3 demonstrates the concept.

2.2.4.2 Average Perpendicular Distance (APD)

The distance between two manual delineated and automatically generated contours, measured along a line that is perpendicular to one or both.

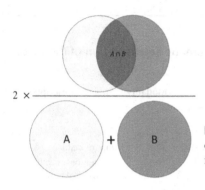

FIGURE 2.3 Dice index is calculated as the area of the overlap divided by the total number of pixels in both contours.

2.2.4.3 Percentage of Good Contours

The percentage of automatically generated contours that have an APD measured to be less than 5 mm.

2.2.4.4 Jaccard Index

An additional metric that computes the overlap between the two contours (same as APD) with a different equation. Consider A and B the manually delineated and automatically segmented contours, the Jaccard index is formulated below.

$$J(A, B) = \frac{(A \cap B)}{(A \cup B)} = \frac{(A \cap B)}{A + B - (A \cap B)} \tag{2.1}$$

2.2.4.5 Sensitivity, Specificity, Positive Predictive Value (PPV), Negative Predictive Value (NPV)

Sensitivity or the "true positive rate," is equal to $\frac{TP}{(TP + FN)}$. Specificity or the "true negative rate," is calculated as $\frac{TN}{(FP + TN)}$. PPV "positive predictive value" is equal to $\frac{TP}{(TP + FP)}$. NPV "negative predictive value" is calculated as $\frac{TN}{(FN + TN)}$. Where TP (true positive) is the number of pixels that are correctly predicted as belonging to the object (which is the LV here), TN (true negative) is the number of pixels that are correctly predicted as belonging to the background. FP is the number of pixels that belong to background but are misclassified as belonging to the left ventricle. FN is the number of pixels belonging to the LV that are misclassified as belonging to the background.

2.2.4.6 Hausdorff Distance

A symmetric measure of the distance between the two manually delineated and automatically generated contours from each other. Equation 2.2 demonstrates how to calculate the Hausdorff distance between two contours A, and B.

$$H(A, B) = \max \left(\max_{i \in A}(\min_{j \in B} d(i, j)), \max_{i \in B}(\min_{j \in A} d(i, j)) \right) \tag{2.2}$$

d(i,j) is the Euclidean distance.

2.3 DEEP LEARNING-BASED APPROACHES

In the past decade, deep learning has developed a new category for image segmentation due to its significant performance. Pixel classification using deep learning is the most common approach for image segmentation. Long et al. (2015) proposed fully convolutional neural networks for semantic segmentation. Due to the lack of labeled medical images, deep learning-based segmentation of medical images is still challenging and prone to overfitting, however, some approaches aim to design medical segmentation architectures that function with a small set of labeled data such as U-net (Ronneberger et al., 2015) and its variants (Çiçek et al., 2016; Milletari et al., 2016; Zhou et al., 2018).

Several review papers have focussed on deep learning and its applications in image processing (Guo et al., 2016). Some papers focus on the applications of deep learning in the processing of the medical images (C. Chen et al., 2020; Hesamian et al., 2019; Litjens et al., 2017; Shen et al., 2017).

In this section, deep learning and convolutional neural networks are briefly described as well as the review of approaches that use deep learning-based methods for automatic segmentation of the LV.

2.3.1 DEEP LEARNING PRE-REQUISITES

2.3.1.1 Artificial Neural Network (ANN)

Neural networks are a type of supervised learning algorithm and are the basis of deep learning frameworks. A neural network includes a set of nodes (neurons) that are connected to each other via weighted links and are distributed in multiple layers. Each neural network has an input layer, several hidden layers, and an output layer. Figure 2.4 shows an example of an artificial neural network.

Considering a_i^j as the j_{th} node of the i_{th} layer, the value of this node is calculated as the linear combination of the nodes in the previous layer ($i - 1_{th}$ layer) with respect to the weights of the links, followed by a non-linear activation function. Equation 2.3 shows how to compute the value of a single neuron a_i^j (Figure 2.5).

$$a_i^j = \sigma \left(\sum_k W_{i-1}^{kj} a_{i-1}^j + b_i \right) \quad (2.3)$$

Considering A_i a vector containing all the neurons in the i_{th} layer, W_i matrix of the weights between the $i - 1_{th}$ and i_{th} layer, and bi vector of biases in the i_{th} layer, the values of the neurons in the i_{th} layer are formulated in equation 2.4.

$$Ai = \begin{bmatrix} a_i^1 \\ a_i^2 \\ \vdots \\ a_i^{ni} \end{bmatrix} \quad Wi = \begin{bmatrix} w_i^{11} & w_i^{12} & \cdots \\ w_i^{21} & w_i^{22} & \cdots \\ \vdots & \vdots & \ddots \end{bmatrix}$$

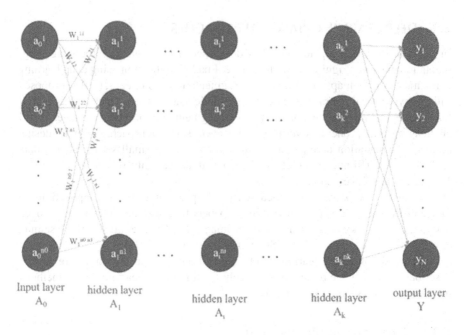

FIGURE 2.4 An example of a neural network including an input layer, hidden layers, and output layer. Nodes are connected to each other via weighted links.

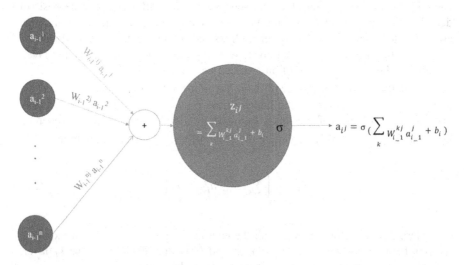

FIGURE 2.5 The value of a single node in the network is calculated as a linear weighted combination of the nodes in the previous layer followed by a non-linear activation function.

$$A_i = \sigma(W_i^T A_{i-1} + b_i) \qquad (2.4)$$

The activation function (σ) is a non-linear function that detects the neuron's output at the end (In other words, it detects whether the neuron should fire or not).

The most common activation functions are sigmoid, Tanh, and Relu. Figure 2.6 shows these activation functions.

2.3.1.2 Training the Network

Deep learning is a supervised learning approach that obtains information through the labeled datasets. Considering that the output labels ($y^\wedge$) for a training dataset are known, a trained network should generate the output layer (y) as close as possible to the labels ($y^\wedge$). A cost function is defined as the summation of differences between the generated outputs (y) and the labels ($y^\wedge$) over all training datasets. In the training phase, the parameters of the network (such as weights and biases) are trained by minimizing the cost function.

$$W^* = \mathrm{argmin} \sum_{n=1}^{N} loss \ (y^{(n)}, y^{\wedge(n)}) \tag{2.5}$$

Loss function ($loss\,(y^{(n)}, y^{\wedge(n)})$) is a criterion that computes the difference between the output layer y and the labels $y^\wedge$ for each input data. The summation of the loss function on all training datasets will result in the cost function. Cross-entropy is the most common loss function for image classification and segmentation tasks.

Rather than solely having good performance on the training data, a neural network should have good performance and accuracy on the test data, otherwise, overfitting occurs. In order to reduce the possibility of overfitting, weight regularization, augmenting the training dataset, reducing the complexity of the model, using dropouts (Baldi & Sadowski, 2013; Hinton et al., 2012) or its improved variants (McAllester, 2013; Park et al., 2018; Srivastava et al., 2014; Wager et al., 2013; Wan et al., 2013.; Wang et al., 2018; Warde-Farley et al., 2014), and ensemble learning can all be used.

2.3.1.3 Convolutional Neural Networks

For image processing tasks, due to a large number of the input layer's units (pixels), a deep neural network will have many weights, which ultimately leads to a complicated and non-efficient network. Convolutional neural networks, as are discussed in (Albawi et al., 2017; O'Shea & Nash, 2015) are used to solve this problem. In a CNN, instead of a weighted combination of nodes, convolutional filters are convolved with the images. The same convolutional filters are further used for all of the pixels in the image, therefore, this approach reduces the number of parameters drastically by sharing the parameters.

In addition to input and output layers, CNNs also include convolutional layers, pooling layers, and fully connected layers.

In the input layer, the input images are fed to the network. The output layer contains the results that we are seeking from the network. For example, in an image classification task, the output of the network assigns a label to the whole image whereas in a pixel classification task, the network assigns a label to each pixel of the input image.

CNNs perform a variety of tasks, including classification, segmentation, registration, object detection, localization, etc. The most common approach for image

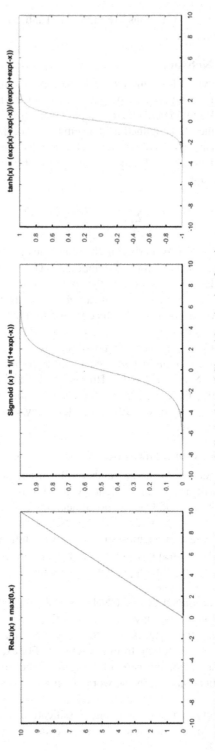

FIGURE 2.6 Rectified linear unit (ReLu), tangent hyperbolic (tanh), and Sigmoid activation functions.

segmentation is pixel classification. In a pixel classification task, the output is a label map that assigns a label to each pixel of the input image.

2.3.1.4   Convolutional Layer

At each convolutional layer, multiple convolutional filters are convolved with the image. The results are then added to bias values and passed to a non-linear activation function. The convolution results will be similar to equation 2.6 since the convolutional kernels in the i_{th} layer, the biases and the result of the k_{th} convolution are considered to be $\{W_1^i, W_2^i, \dots W_k^i\}$, $\{b_1^i, b_2^i, \dots b_k^i\}$ and A_K^i, respectively.

$$A_k^i = \sigma(W_k^{i-1} * A_{i-1} + b_k^{i-1})$$ (2.6)

Convolutional kernels are typically small in dimensions. Each kernel is used for the entire image and will reduce the complexity of the network. For example, considering a $253 \times 253 \times 3$ input image and considering the kernel to be 6×6, for each convolutional filter in the first layer, only $6 \times 6 \times 3 = 108$ weights should be learned. Considering the same image in an ANN, the input includes $253 \times 266 \times 3 = 196608$ nodes that will lead to millions of weights in the first layer (Figure 2.7).

2.3.1.5   Convolution Operation

Having an image I, and a 2D $f{\times}f$ convolutional filter F, the 2D convolution is formulated in equation 2.7 (Figure 2.8).

$$z[i, j] = \sum_{m=1}^{f} \sum_{n=1}^{f} F[m, n]I[i + m - 1, j + n - 1]$$ (2.7)

If the input and convolutional filters are 3D, convolution over volume is used. For each input channel and the corresponding convolutional kernel, the convolution operation is first performed followed by the summation of the results. Figure 2.9 shows this concept.

2.3.1.6   Pooling Layer

Pooling layers usually follow a convolutional layer. They are supposed to reduce the spatial size of the representations to reduce the number of parameters and computations in the network. The most common pooling layer is Max-pooling. As shown in Figure 2.10, in Max-pooling layer with a stride of s and a filter size of f, in each $f{\times}f$ region the maximum pixel is chosen, and the filter will then move s pixels. Another common pooling layer is referred to as average pooling. In average pooling, instead of choosing the maximum, the average of each window is calculated.

2.3.1.7   Fully Connected Layer

Each CNN architecture includes a number of fully connected layers at the end which converts the 2D feature maps into a 1D feature vector. Similar to ANNs, a network can have multiple fully connected layers in order to produce the final result.

Figure 2.7 shows a simple convolutional neural network architecture.

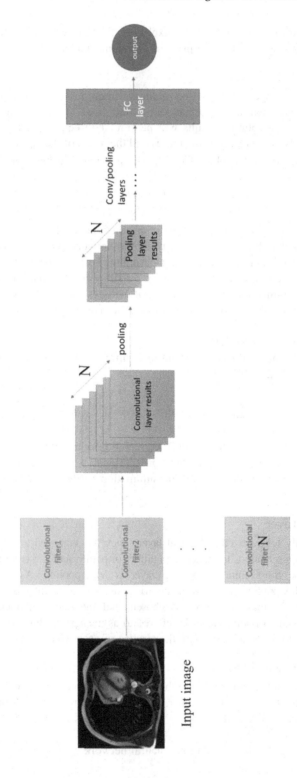

FIGURE 2.7 CNN architecture. N convolutional kernels are used in the first convolutional layer, after multiple convolution/ReLU units fully connected layer is used.

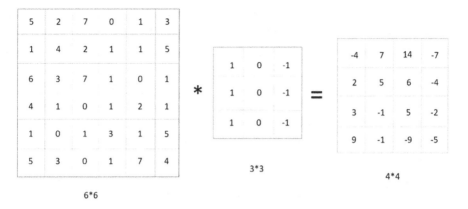

FIGURE 2.8 2D convolutional operation. The left table is the image, the middle one is the convolutional filter, and the right table shows the convolution result.

2.3.1.8 Fully Convolutional Neural Networks (FCN)

Long et al. (2015) suggested the idea of using fully convolutional networks for semantic segmentation for the first time. FCNs are a type of CNN that do not have any fully connected layers and are able to generate segmentation maps in the output with the same size of the input image through convolutional layers. These networks are among the commonly used methods for the automatic segmentation of cardiovascular structures.

Segmentation is performed by combining semantic information (from the deep layers) and appearance information (from the shallow layers). In other words, the deep layers include information about "what" structure exists in an image (i.e., classification) whereas the shallow layers include "where" each structure is in the image.

As it is shown in Figure 2.11, the feature maps from different layers are combined together, Figure 2.12 shows the results of each combination.

2.3.1.9 U-Net

Inspired by FCN architectures, (Ronneberger et al., 2015) designed U-net for the segmentation of medical images in the absence of enough labeled data. In order to use the available data more efficiently, this method relies on data augmentation techniques. To the best of our knowledge, U-Net and its variants are the most commonly used deep learning approaches for automatic segmentation of cardiovascular images.

As it is shown in Figure 2.13, the network includes a contracting path that extracts the context and an expanding path for extracting the geometry. The feature maps from these paths are combined with each other to perform segmentation.

In the left path (contracting path), 3 × 3 convolution filters are performed, each followed by a ReLU. At each step, the output maps are passed to the right path while in the falling path, max-pooling layers are performed for down sampling.

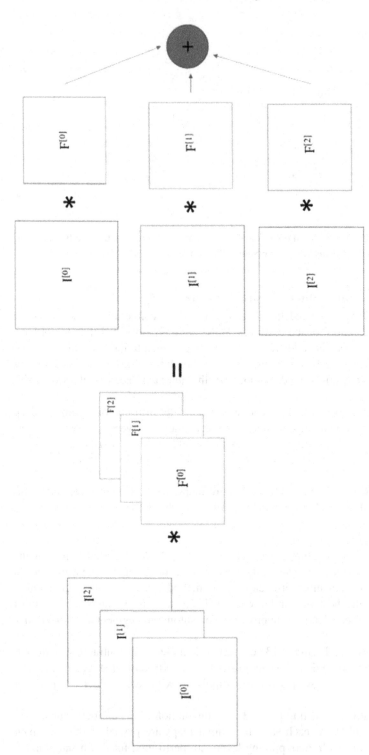

FIGURE 2.9 Convolution over volume. At first, 2D convolutions are performed followed by the summation of the results.

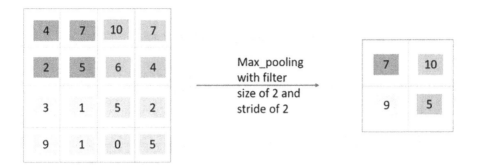

FIGURE 2.10 Result of applying a Max-pooling with a filter size of 2 and stride of 2. For each 2 × 2 window, the maximum is extracted and the window is moved 2 pixels.

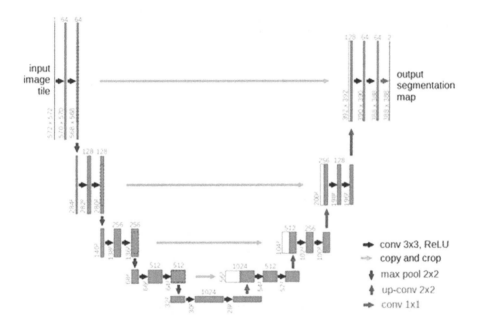

FIGURE 2.11 U-Net architecture (Ronneberger et al., 2015).

In the upward path at the right (expansive path), the feature maps are upsampled, followed by a 2 × 2 convolution (up-convolution) that halves the number of feature channels. The results are then concatenated with output maps from the left path, and 3 × 3 convolutional filters are applied to the output maps followed by ReLU units. Feature maps are cropped before concatenation because border pixels are decreased in every convolution.

The final layer is a 1 × 1 convolutional layer that maps each pixel to the desired number of classes. Input size should be chosen in a way that all 2 × 2 max-pooling layers have even x-size and even y-size. Random elastic deformation of the training samples is used for the purpose of data augmentation.

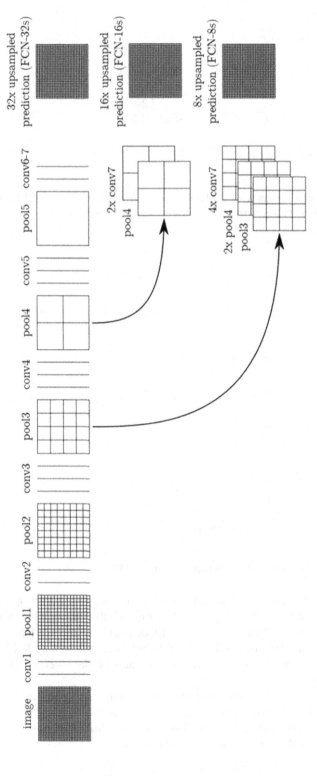

FIGURE 2.12 First row (FCN-32s) does not use any skip. Second row (FCN-16s): combines the predictions from the final layer and the pool4 layer. Third row (FCN-8s): combines the predictions from the last layer, pool4 layer, and pool3 layer (Long et al., 2015).

FCN-32s   FCN-16s   FCN-8s   Ground truth

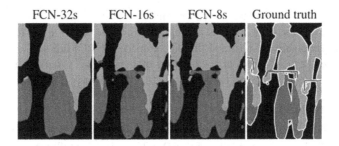

FIGURE 2.13 Results of the segmentation maps for different networks. FCN-32s, FCN-16s, FCN-8s, and ground truth (Long et al., 2015).

2.3.2 DEEP LEARNING FOR SEGMENTATION OF THE CARDIOVASCULAR IMAGES

2.3.2.1 The Main Deep Learning-Based Cardiovascular Image Segmentation Methods

FCN: Fully convolutional neural networks are very common and successful in segmentation of the cardiovascular images (Bai et al., 2018; Baumgartner et al., 2017; Jang et al., 2018; Lieman-Sifry et al., 2017; Patravali et al., 2017; Payer et al., 2018; Tran, 2017; Yu et al., 2017). Long et al. (2015) used FCN for semantic segmentation for the first time, different approaches design their own FCN architecture for getting better results.

U-Net: It is a type of FCN which is very common for medical image segmentation due to the data augmentation techniques and having a pre-designed network. Many approaches use U-net (or its variants) for the segmentation solely or as a part of their segmentation algorithm (Baumgartner et al., 2017; Isensee et al., 2018; Khened et al., 2018, p. 33; Moreno et al., 2019; Oktay et al., 2018; Patravali et al., 2017; Tao et al., 2019; Wang & Smedby, 2018; Zheng et al., 2018; Zyuzin & Chumarnaya, 2019). Changing the U-Net architecture, modifying the loss function, using 3D U-Net or combining different U-Net variants with each other are common approaches for getting better results.

Other CNN architectures: Although FCNs including U-Net are more common for segmentation, some approaches use other CNN architectures (Moreno et al., 2019; Tan et al., 2016; Wolterink et al., 2017; Zreik et al., 2016). For example, some approaches classify each of the pixels (or voxels) to achieve a segmentation label map.

Hybrid: Methods that combine more than one method for the segmentation are called hybrid methods (Avendi et al., 2016; Carneiro et al., 2012; Ghesu et al., 2016; Wolterink et al., 2016). A traditional segmentation method such as deformable models, atlas-based method, or AAM/ASM method is usually combined with a deep learning framework. The combination of deformable models with CNNs is one of the most common hybrid approaches which is discussed in Chapter 4.

2.3.2.2 Single-Stage and Multi-Stage Segmentation

Some approaches segment the cardiac images in a single stage. In a single-stage approach, the input images are fed into an end-to-end network, and the segmentation label map is generated in the output.

Some approaches segment the cardiac images in multi-stage frameworks. For example, some prior work extracts the region of interest (ROI) in the first stage to reduce the computational cost and the pixels inside the ROI are segmented in the second stage. ROI selection or localization of the cardiac structures using deep learning is offered by papers (Emad et al., 2015; Krizhevsky et al., 2017).

2.3.2.3 Structures of Interest

Left ventricle (LV), right ventricle (RV), left atrium (LA), right atrium (RA), aorta (AO), and great vessels including coronary arteries (CA), pulmonary artery (PA) are the most common structures of interest for segmentation.

Different approaches have focused on segmentation of different structures, however, in this chapter left ventricle segmentation and whole heart segmentation approaches have been considered.

2.3.2.4 Modality

Deep learning-based approaches can also be categorized based on their modality. Some approaches segment the MRI images (Avendi et al., 2016; Bai et al., 2018; Baumgartner et al., 2017; Carneiro et al., 2012; Emad et al., 2015; Fahmy et al., 2019; Isensee et al., 2018; Jang et al., 2018; Khened et al., 2018; Lieman-Sifry et al., 2017; Molaei et al., 2017; Moreno et al., 2019; Patravali et al., 2017; Tan, 2017; Tan et al., 2016; Tao et al., 2019; Tran, 2017; Wolterink et al., 2016, 2017; Yu et al., 2017; Zheng et al., 2018), some segment CT scans (Lessmann et al., 2016; Moradi et al., 2016; Wolterink et al., 2016; Zreik et al., 2016), others focus on segmenting ultrasound scans (Carneiro & Nascimento, 2013; H. Chen et al., 2016; Ghesu et al., 2016; Smistad et al., 2017), and some will focus on multi-modality segmentation (Mortazi et al., 2017; Payer et al., 2018; Tong et al., 2018; Wang & Smedby, 2018; Xu et al., 2018; X. Yang et al., 2018a; Ye et al., 2019).

Some of the prior work focused on the automatic segmentation of the LV or whole heart are reviewed in Table 2.5. The terms that are reviewed for each approach include: the structures of interest, the modality of the training and test images provided, a brief description of the segmentation framework used in each approach, end-to-end/ multi-stage approach, the test dataset used by the approach, and the segmentation methods used. The approaches discussed are the competitors of MICCAI 2009, LVSC, RVSC, ACDC, HVSMR 2016, and MM-HWS challenges and focus on segmentation different structures including LV, RV, LA, aorta, and WHS.

In the rest of this section, some of the successful deep learning-based segmentation approaches are described in detail.

2.3.3 Some Approaches With More Detailed

2.3.3.1 Approach L1

2.3.3.1.1 Overview

Tran (2017) proposed a 15 layer fully convolutional neural network for the segmentation of both left and right ventricles. The CNNs are trained end-to-end in a single learning stage.

TABLE 2.5

Summary of the Methods Used for the Automatic Segmentation of the LV or Whole Heart

Method	Structures	Modality	Description	End-to-end or Multi-stage	Test dataset	Method
(Bai et al., 2018)	LA, RA, LV, RV	SAX MRI	FCN trained on a large dataset (>5000 subjects)	end-to-end	none	FCN
(Moreno et al., 2019)	LV	SAX and LAX MRI	1. ROI selection using a modification of YOLO net (Redmon et al., 2016) 2. LV segmentation inside the ROI using 2D U-Net.	Multi-stage	none	Other CNN, U-Net
(Isensee et al., 2018)	LV, RV	SAX MRI	Ensemble of 2D and 3D U-Nets	End-to-end	ACDC	U-Net
(Tao et al., 2019)	LV	Cine MRI	2D U-Net trained on a large multi-vendor multi-center dataset (596 datasets)	end-to-end	none	U-Net
(Baumgartner et al., 2017)	LV, RV	SAX MRI	Compared FCN-8 (Long et al., 2015) 2D U-Net and 3D U-Net. Networks were trained and tested each with three different loss functions.	End-to-end	ACDC	FCN, U-Net
(Khened et al., 2018)	LV, RV	SAX MRI	1. ROI is selected using Fourier and Hough transforms. 2. Segmentation of ROI by combining Inception module with 2D Dense U-Net.	Multi-stage	ACDCLVSC	U-Net
(Smistad et al., 2017)	LV	US	2D U-Net is used for segmentation. Training dataset is labeled automatically using a Kalman Filter based automatic segmentation algorithm (no manual labeling)	End-to-end	none	U-Net

(Continued)

TABLE 2.5 (Continued)
Summary of the Methods Used for the Automatic Segmentation of the LV or Whole Heart

Method	Structures	Modality	Description	End-to-end or Multi-stage	Test dataset	Method
(Tran, 2017)	LV, RV	SAX MRI	FCN trained, fine-tuned, and tested on different datasets.FCN is trained on LVSC dataset and fine-tuned on the Sunnybrook and RVSC datasets	End-to-end	MICCAI	2009LVS-CRVSC
FCN						
(Jang et al., 2018)	LV, RV	SAX MRI	2D M-Net based on FCN	End-to-end	ACDC	FCN
(Lieman-Sifry et al., 2017)	LV, RV	SAX MRI	2D E-Net based on FCN	End-to-end	none	FCN
(Zheng et al., 2018)	LV, RV	SAX MRI	1. ROI is selected using a modified 2D U-Net. 2. ROI is segmented using another modified U-Net that propagates information from the adjacent slices.	Multi-stage	ACDCMICCAI 2009RVSC	U-Net
(Zreik et al., 2016)	LV	CT	1. a 3D ROI is extracted around LV using three CNNs (each detecting the presence of LV in all slices of an image plane). 2. Another CNN performing voxel classification segments the LV from the ROI.	Multi-stage	none	Other CNN
(Tan et al., 2016)	LV	SAX MRI	1. A CNN detects the LV centroid (in a polar system). 2. Another CNN detects the endocardial radius of the LV centered around the LV centroid.	Multi-stage	MICCAI 2009	Other CNN
(Patravali et al., 2017)	LV, RV	SAX MRI	proposed 2D and 3D U-Nets with varying multi-slice inputs and compared them	End-to-end	ACDC	U-Net

Reference	Structures	Imaging	Description	Pipeline	Dataset	Network
(Yu et al., 2017)	WHS: Myocardium, Blood pool	3D CMR	Densely connected 3D FCN	Multi-stage	HVSMR 2016	FCN
(Zyuzin & Chumarnaya, 2019)	LV	US	Compared 2D U-Net and U-Net++ for LV segmentation	End-to-end	none	U-Net
(Wolterink et al., 2017)	WHS: Myocardium, Blood pool	3D CMR	Patch-based voxel classification CNN: A CNN labels each voxel in an image based on the classification of three orthogonal patches centered at the voxel	End-to-end	HVSMR 2016	Other CNN
(Wang & Smedby, 2018)	WHS: LV, RV, LA, RA, AA[1], PA[2]LV myo	3D CT, 3D CMR	2D U-Nets combined with shape context.	Multi-stage	MM-WHS	U-Net
(Wang et al., 2018)	WHS: LV, RV, LA, RA, AA, PALV myo	3D CT, 3D CMR	Dynamic ROI detection using modified 3D U-Net followed by segmentation of each ROI using another modified 3D U-Net.	Multi-stage	MM-WHS	U-Net
(Payer et al., 2018)	WHS: LV, RV, LA, RA, AA, PALV myo	3D CT, 3D CMR	1. A U-Net like FCN localizes each substructure. 2. A second FCN segments the voxels inside each region.	Multi-stage	MM-WHS	FCN
(Tong et al., 2018)	WHS: LV, RV, LA, RA, AA, PALV myo	3D CT, 3D CMR	1. 3D U-Net for localization of each structure. 2. deeply supervised 3D U-Net for segmentation of each structure.	Multi-stage	MM-WHS	U-Net
(Ye et al., 2019)	WHS: LV, RV, LA, RA, AA, PALV myo	3D CT, 3D CMR	3D deeply supervised U-Net with multi-depth fusion	End-to-end	MM-WHS	U-Net

(Continued)

TABLE 2.5 (Continued)
Summary of the Methods Used for the Automatic Segmentation of the LV or Whole Heart

Method	Structures	Modality	Description	End-to-end or Multi-stage	Test dataset	Method
(Xu et al., 2018)	WHS: LV, RV, LA, RA, AA, PA LV myo	3D CT 3D CMR	1. Fast R-CNN-based approach for localization of each structure. 2. 3D U-Net for segmentation of each structure.	Multi-stage	MM-WHS	U-Net
(Avendi et al., 2016)	LV	SAX MRI	1. ROI detection CNN. 2. Stacked auto-encoder for LV initialization. 3. Chan-Vese deformable model segmentation using the initial contour.	Multi-stage	MICCAI 2009	Hybrid
(Wolterink et al., 2016)	LV	SAX MRI	1. ROI detection using DBN 2. segmentation using level-sets	Multi-stage	MICCAI 2009	Hybrid
(Carneiro et al., 2012)	LV	MRI	1. Initial segmentation using a combination of 2D and 3D U-Nets. 2. Final segmentation of the initial segment using a multi-component deformable model.	Multi-stage	none	Hybrid
(Ghesu et al., 2016)	LV	US	1. 2D U-Net initial segmentation 2. Final segmentation using level sets. U-Net segment is used as initialization and shape prior.	Multi-stage	none	Hybrid

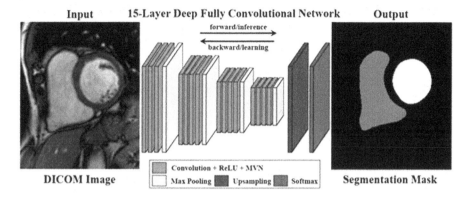

FIGURE 2.14   Left: input image. Middle: network architecture. Right: segmentation mask (LV,RV, background) (Tran, 2017).

As shown in Figure 2.14 2D cine MRI images are fed to the network as input. For each input image, the network generates a dense heat map in the output showing the prediction for pixels belonging to each class (LV, background, RV).

2.3.3.1.2 Training/Test

To train the network, the weights are first randomly assigned using the "Xavier" method. The pre-trained network with the weights obtained from this training step is fed into a transfer learning algorithm with the same network but with a different training dataset. This fine-tuning procedure improves the results significantly.

2.3.3.2   Approach L2

2.3.3.2.1 Overview

Zreik et al. (2016) proposed a CNN architecture for automatic segmentation of the left ventricle in cardiac CT angiography. The segmentation process is performed in two steps, in the first step a 3D bounding box detects the ROI. A voxel classification CNN will then detect the voxels that belong to the left ventricle from the selected ROI.

2.3.3.2.2 Architecture and Algorithm Details

de Vos et al. (2016) was used for the localization of the left ventricle (ROI detection) in the first CNN. All 2D input images are fed to this CNN, and as a result, a 3D bounding box enclosing the LV is extracted from the dataset.

In the second step, voxels inside the 3D ROI are classified one by one using a CNN to annotate the left ventricle. As shown in the input layer in Figure 2.15, three patches, including axial, coronal, and sagittal slices centered at each voxel are extracted and fed into the network. The network's output is a probability map displaying the likelihood of each voxel belonging to the left ventricle.

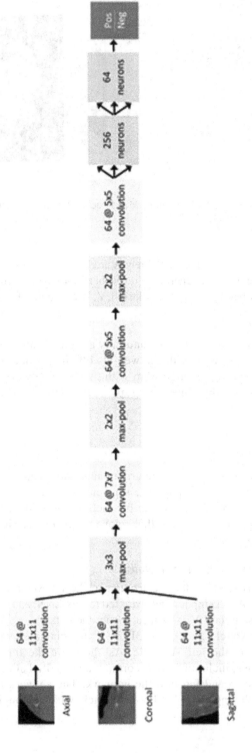

FIGURE 2.15 Three 48 × 48 patches, including axial, coronal, and sagittal slices centered at each voxel of the ROI, are extracted, and fed to the network. The network's output is a label showing the voxel belongs to LV or not (Zreik et al., 2016).

2.3.3.3   Approach L3

2.3.3.3.1 Overview

Tan et al. (2016) proposed a two-stage algorithm for determining the LV en-docardial wall. One convolutional neural network is designed for the localization of the LV centroid and another is used for annotation of the LV contour in terms of the endocardial radius in a polar radial system.

2.3.3.3.2 Architecture and Algorithm Details

Since the heart is a moving organ, in order to reduce unnecessary computations, for each input image, a 1D FFT is performed across the temporal domain of each pixel, and the first harmonic is saved. Keeping the first harmonic and discarding the others will result in an image "I_F" that highlights the organs in motion and removes stationary organs. The original image "I_M" is then added to the "I_F" and is fed to both CNNs as described above.

For each input slice, the ($I_M + I_F$) image is fed to the first CNN which detects the coordinates of the LV centroid (x,y). The ($I_M + I_F$) image is then remapped to a polar space and is fed to the second CNN as the input. The output of this CNN is the endocardial border in the polar system which is in terms of 96 evenly spaced points centered around the LV centroid.

2.3.3.4   Approach L4

2.3.3.4.1 Overview

Tan (2017) proposed an algorithm for the detection and segmentation of the left ventricle using a multi-vendor, multi-center dataset. Their network performs two tasks, LV detection that determines the presence of the LV in an input image, and LV segmentation that identifies the border of the LV myocardium. They considered the idea that using a large set of training data covering sufficient variability, accuracy, and generalizability of the CNN will be improved.

2D U-Nets were trained using three different datasets with an increasing variability, ultimately resulting in three trained networks. All trained networks were tested on the same test dataset, and the results were compared.

2.3.3.5   Approach L5

2.3.3.5.1 Overview

Moreno et al. (2019) developed a two-stage deep learning approach for the segmentation of the left ventricle. Segmentation is performed in two stages; at first, the region of interest is selected using a YOLO net (Redmon et al., 2016), a U-net architecture will then annotate the LV from the ROI.

2.3.3.5.2 Architecture and Algorithm Details

At first, the region of interest defined as a bounding box containing both the left and right ventricles is detected using a modified version of the YOLO algorithm. The result is then fed to a U-net architecture for segmentation of the LV.

Two networks (YOLO-T and UNET) are trained and validated separately and then connected to each other as shown in Figures 2.16 and 2.17.

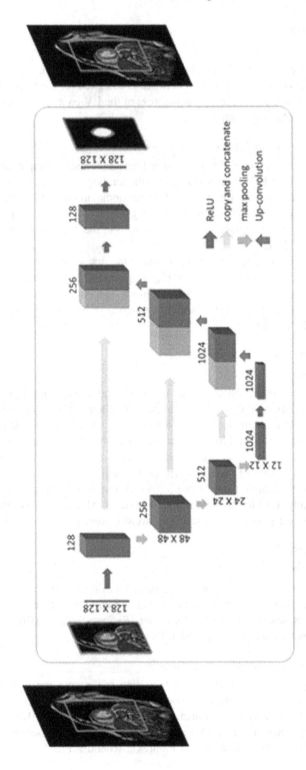

FIGURE 2.16 Left: preprocessing. Middle: U-Net. Right: post-processing (Tao et al., 2019).

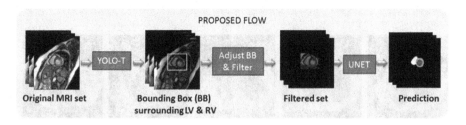

FIGURE 2.17 YOLO-T network is used for ROI detection. U-Net is used for the segmentation of the ROI (Moreno et al., 2019).

2.3.3.6   Approach L6

2.3.3.6.1 Overview

de Vos et al. (2016) developed a method for 3D localization of the heart using 2D convolutional neural networks. Three different regions of interest can be extracted using this method including the heart, aortic arch, and descending aorta. Each ROI is detected as a 3D rectangular volume around the anatomical region of interest.

For extracting each anatomical ROI, three CNNs were used, each one identifying the region of interest in an image plane (axial, sagittal, coronal).

2.3.3.6.2 Architecture and Algorithm Details

For each 2D slice in an image plane (axial, sagittal, coronal), an Alex Net (Krizhevsky et al., 2017) was used to detect the probability of the slice being a part of the 3D anatomical ROI and classify that slice as ROI or background. (AlexNet is a CNN structure for image classification, assigning a probability to each input image to be a member of a class). Three AlexNets were used in total (one used for each image plane) for each anatomical ROI. Figure 2.18 shows different slices and their probability to include a specific ROI in each image plane.

2.3.3.6.3 Training/Test

For the training phase, each 2D slice in each image plane (axial, sagittal, coronal) was labeled as negative if it was not a part of the 3D anatomical ROI and labeled as positive if it was a part of the 3D anatomical ROI. In the test phase Output probabilities bigger than 0.5 were extracted. The largest 1D connected component was then retained in each image plane and a 3D bounding box containing the anatomical ROI was created by combining those slices.

2.4   DEFORMABLE MODELS COMBINED WITH DEEP LEARNING

In this section, deformable model-based segmentation methods are discussed. Deformable models, due to their high flexibility, are among the most successful approaches for cardiac image segmentation. Deformable models for medical image segmentation have been reviewed in some previous papers (McInerney & Terzopoulos, 1996; Montagnat et al., 2001;).

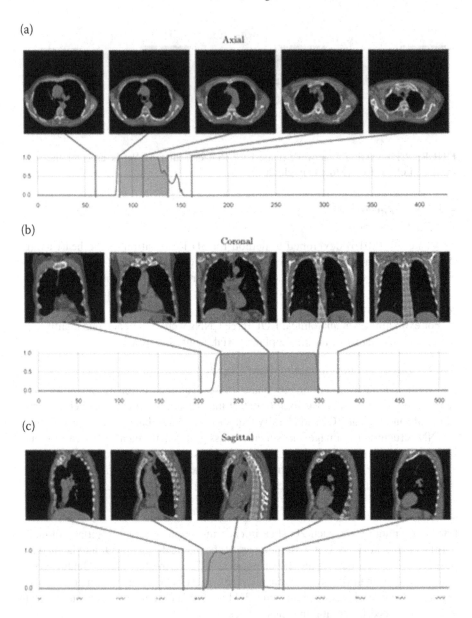

FIGURE 2.18 For each slice in axial, coronal, and sagittal planes the graphs show the probabilities of that slide belonging to the ROI. The blue area is the detected ROI for each image plane (de Vos et al., 2016).

Limitations of this method include the requirement of user interaction for the initialization of the contours and designing an energy function that works robustly for all different images. In the case of cardiovascular images, due to the presence of papillary muscles and poor contrast between the different layers of the heart and blood cells with each other, the deformable contours might fail to yield accurate results. In addition to the mentioned problems, the opening and closing of the structures inside the heart can lead the deformable contour to shrink inward or leak outward by mistake.

To address those challenges, some prior work combines deep learning with deformable models for the segmentation of cardiac images (Avendi et al., 2016; Duan et al., 2018; Ngo et al., 2017; Rupprecht et al., 2016; Veni et al., 2018; D. Yang et al., 2018). Deformable models will add some feature-engineering to the deep learning approaches, resulting in the functioning of segmentation in the presence of a low amount of labeled data and without any user interaction. To conclude, in research combining deep learning and deformable models, initial contour and shape prior for deformable models are defined using deep neural networks. Deformable models will then generate the final refined segmentation contours based on the initial contour and shape priors.

In this section, deformable model-based segmentation theory is first reviewed briefly, however, due to deformable models being a complicated subject, a complete description of the deformable models is out of our scope. A comprehensive review of the deformable model segmentation is available in (Xu, Chenyang et al., 2000). Finally, the prior works which combine deep learning with deformable models are reviewed.

2.4.1 DEFORMABLE MODELS THEORY

Deformable models are curves or surfaces defined within an image domain that can move under the influence of internal forces, which are defined within the curve or surface itself, and external forces, which are computed from the image data. The internal forces are designed to keep the model smooth during deformation. The external forces are defined to move the model toward an object boundary or other desired features within an image. By constraining extracted boundaries to be smooth and incorporating other prior information about the object shape, deformable models offer robustness to both image noise and boundary gaps and allow integrating boundary elements into a coherent and consistent mathematical description.

(Xu, Chenyang et al., 2000)

To run a deformable model, an initial contour is defined (usually) manually. At each point on the contour, an internal and external force exist. These two forces will deform the contour at each point such that the summation of the energy terms is minimized (resulting in a smooth curve that is close to the image boundary). After several iterations, the smooth curve will stick to the image's boundary. Figure 2.19 shows an example of this concept.

Deformable models can be divided into two groups: parametric deformable models (also known as active contours) and geometric deformable models (also known as level sets). Both of them have the same functionality from the user's point of view, but their implementation is completely different. Geometric deformable models are solved on the image space rather than the contoured space.

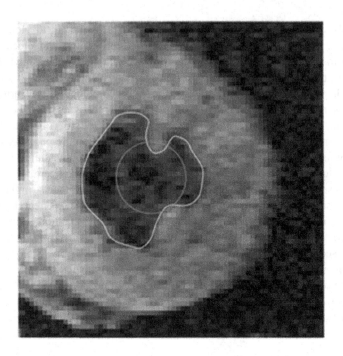

FIGURE 2.19 The initial contour (circle) and the result of deformable models' segmentation (Xu, Chenyang et al., 2000).

2.4.1.1 Parametric Deformable Models (Active Contours)

A smooth moving curve (such as a deformable contour) can be shown as $X(s, t) = [x(s, t), y(s, t)]$ where, t is the time, and s is the curve's length parameter which is between $[0, 1]$.

The parametric deformable model is formulated as the summation of internal energy (force) and external energy (force) which should be minimized on the surface of the curve.

$$\mathcal{E}(X) = \mathbb{S}(X) + \mathbb{P}(X) \qquad (2.8)$$

$\mathbb{S}(X)$ and $\mathbb{P}(X)$ are the internal and external energy functions, respectfully. The internal energy function is designed to keep the curve smooth, which is formulated below.

$$\mathbb{S}(X) = \frac{1}{2} \int_0^1 \alpha(s) \left| \frac{\partial X}{\partial s} \right|^2 + \beta(s) \left| \frac{\partial^2 X}{\partial s^2} \right|^2 ds \qquad (2.9)$$

The first term in the internal energy function will minimize the curve's length, and the second term will discourage bending, resulting in a smooth curve. The integral over the surface of the curve shows the summation of both terms on all the curve's points. The external energy function is designed to move the contour toward the edges of the image and is formulated below.

$$\mathbb{P}(X) = \int_0^1 P(X(s))\, ds \qquad (2.10)$$

This external energy function will minimize a function $P(X)$ over the whole curve's points. The function $P(X)$ is defined below.

$$P(x, y) = -w_e |\nabla [G_\sigma(x, y) * I(x, y)]|^2 \qquad (2.11)$$

Where w_e, G_σ, ∇, and $*$ represent a positive weighting parameter, a two-dimensional Gaussian function with standard deviation, the gradient operator, and the 2D image convolution operator, respectively. This function has its minimum on the points close to the image's boundary and maximum on points located on the flat surfaces. The external energy term is minimized when the contour is close to the image's boundary.

The combination of these two energy functions will find a curve that is smooth and close to the image boundaries. Parametric deformable models aim to find the result $X(s)$ by solving the above equations, which are performed using the Euler-Lagrange solution.

2.4.1.2   Geometric Deformable Models (Level Sets)

Geometric deformable models define the concept of deformable models using a different approach. The problem is formulated using the curve evolution theory (Kimia et al., 1995; Sapiro & Tannenbaum, 1993; Kimmel et al., 1995; Alvarez et al., 1993) and solved using the level set method (Osher & Sethian, 1988; Čupić, 2003).

2.4.1.3   Curve Evolution Theory

The purpose of curve evolution theory is to study the deformation of curves using only geometric measures such as the unit normal and curvature rather than the quantities that depend on parameters such as the derivatives of an arbitrary parameterized curve.

In the deformable model segmentation, the initial curve is moved through the image space until the energy of the curve is minimized. The energy function consists of an internal and external energy term, as described previously. The movement of the curve can be formulated by considering the evolution of the curve along its normal direction, which is known as curve evolution theory.

The deformation of the curve in time and along its normal direction is formulated below.

$$\frac{\partial X}{\partial t} = V(k)\vec{N} \qquad (2.12)$$

$\vec{N}$, and k represent the unit normal vector, the curve's curvature, respectively, and $V(k)$ represents the speed function which determines the speed of the curve evolution through its normal direction. In other words, it represents the amount a point on the curve should shrink or expand along its normal direction. Two principal examples for the speed function are offered below

1. Constant deformation

$$V(k) = V_0 = constant \tag{2.13}$$

The constant deformation will shrink or expand all points on the curve without changing the curve's shape showing that the speed function is uniform at all points, and the curve will change uniformly. The constant deformation continuously inflates or deflates the curve until it is stopped by an opposing force.

2. Curvature deformation

$$V(k) = \alpha \times k \tag{2.14}$$

k represents the curvature of the curve, and α is a coefficient. By choosing an appropriate value for α, curvature deformation can expand the curve on the low curvature points and shrink the curve on the high curvature points. This kind of speed function will try to smoothen the curve and will have the same function as the internal energy term in the active contours.

The basic idea of the geometric deformable model is to couple the speed functions (using curvature and/or constant deformation) with the image data so that the evolution of the curve stops at object boundaries. This combination can segment the image.

$$\frac{\partial X}{\partial t} = C\,X\,(v_0 + \alpha * K)\vec{N} \tag{2.15}$$

$$C = \frac{c_0}{|1 + \nabla(I)|} \tag{2.16}$$

Equation 2.16 combines the above examples with the image data. It has two terms for speed function; constant deformation that shifts the initial curve, and curvature deformation that keeps the curve smooth. Near the edges of the image, the gradient is largely resulting in a small C value. In this condition, the speed function is small and the curve will not move. In contrast, in the absence of an edge, the C function has a large value and the combination of constant deformation and curvature deformation moves the curve while keeping it smooth until it reaches an image boundary.

It is possible to segment an image using equation 2.16, but that should be solved on the curve space. When the curve is shrinking, the curve resolution will decrease, and it is not efficient to solve equation 2.16 in curve space. The level-set method will solve equation 2.16 on the level-set of image space whose resolution will not change during segmentation.

2.4.2 SEGMENTATION APPROACHES USING DEFORMABLE MODELS

2.4.2.1 Level-Set and Active Contour

Both level sets and active contour approaches are common for cardiac segmentation (Chakraborty et al., 1996; Gotardo et al., 2006; Hae-Yeoun Lee et al., 2010; Hajiaghayi et al., 2017; Pluempitiwiriyawej et al., 2005; Ranganath, 1995; Santarelli et al., 2003) and (Angelini et al., 2005; Ben Ayed et al., 2009; Feng et al., 2013; Lynch et al., 2008; Sarti et al., 2005). Some approaches attempt to change the energy terms in a way that can better track the geometry of regions of interest, however, most of the approaches change the data driven term and the regularization term is usually not changed.

2.4.2.2 Shape Prior

While normal regular deformable models are common for cardiac segmentation, some approaches aim to feed strong prior knowledge to the model by adding an extra term (Ecabert et al., 2008, 2011, 2005; Lynch et al., 2006; Paragios et al., 2002; Peters et al., 2007, 2010). This extra term adds an anatomical constraint to the deformable model, for example, distance from a prior contour.

2.4.2.3 2D Deformable Models vs 3D Deformable Models

While the majority of approaches use the regular 2D deformable models, in some approaches, the 3D extension of the deformable models is used (Hajiaghayi et al., 2017; Heiberg et al., 2005; Q. C. Pham et al., 2001; Zhukov et al., 2002).

2.4.3 COMBINING DEFORMABLE MODELS AND DEEP LEARNING

The idea for combining deep learning and deformable models is to generate the deformable models' shape prior, initial contour, or both by deep learning. The deep learning output is used as a reference for deformable models' segmentation that will have the following benefits:

1. Deformable model approaches require an initial contour; deep learning removes the user interaction by generating the deformable model's initial contour.
2. In many deformable model algorithms, the contour will shrink inward or leak outward due to the presence of the LV's papillary muscles or the opening and closing of the anatomical structures. Deep learning will increase the accuracy and robustness of the contour by providing a shape prior for the LV as the reference.
3. In the majority of the medical segmentation problems, there is limited labeled data available, and deep learning frameworks are prone to overfitting. Deformable models will refine and finalize the segmentation which will lead to better results even with few training datasets.

Some of the papers combining deep learning and deformable models (Carneiro et al., 2012; H. Chen et al., 2016; Ghesu et al., 2016; Lessmann et al., 2016; Moradi et al., 2016; Tan et al., 2016; Wolterink et al., 2016) are discussed below.

2.4.3.1 Approach H1

(Avendi et al., 2016) proposed a framework for automatic segmentation of the left ventricle by combining deep learning and the level set deformable models.

Their algorithm consists of three steps, as is shown in Figure 2.20. At first, the stack of 2D MRI data is fed into a deep learning architecture and a region of interest is extracted from each image. In the second step, the extracted ROI images are fed into a second neural network and the left ventricle is segmented as a binary mask. This initial contour will be used for both the initialization and shape infer. The initial contour is then fed into a level set algorithm that will segment the left ventricle, which is depicted in the third box. Finally, the 2D segmented slices are aligned to each other.

2.4.3.2 Approach H2

D. Yang et al. (2018) proposed an automatic segmentation approach using deep neural networks coupled with deformable models for the segmentation of the LV myocardium and blood cells.

The algorithm is depicted in Figure 2.21. Raw MRI images are fed to a 2D-3D U-Net architecture for initial segmentation of the LV. The output of LV muscle segmentation is fed to a multi-component deformable model as a prior shape to generate the final segmentation of the LV endocardium and epicardium walls as well as myocardium muscle.

2.4.3.3 Approach H3

Veni et al. (2018) proposed a framework for segmentation of the LV in echocardiography scans combining level-set and deep learning. The proposed framework is shown in Figure 2.22. Based on the training set images, a U-Net is trained for the segmentation of different anatomical structures of the heart, including the left ventricle. The LV contour is used as a reference for the level set segmentation. The level set energy function is a variant of the Chan-Vese deformable models.

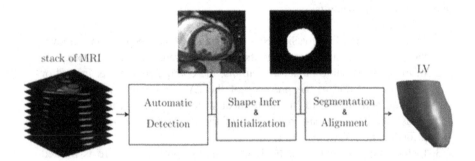

FIGURE 2.20 Stack of MRI images are fed to the network. ROI is detected by a CNN. From the selected ROI, an auto-encoder detects the shape infer and initializes the deformable model for segmentation of the LV (Avendi et al., 2016).

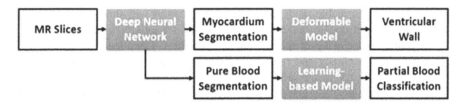

FIGURE 2.21 2D-3D U-Net coupled with deformable models for segmentation of the LV (D. Yang et al., 2018).

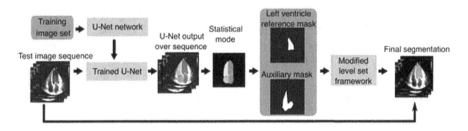

FIGURE 2.22 U-Net combined with level-sets for segmentation of the LV (Veni et al., 2018).

2.5 CONCLUSION

Automatic segmentation approaches of cardiac images can be divided into image-driven methods, deformable models, pixel classification, atlas-guided techniques, active appearance/active shape models, and deep learning-based methods. Deformable model-based, deep learning-based, and hybrid methods combining deep learning along with deformable models for the automatic segmentation of the cardiac images are discussed in this review paper. Deformable models require some user interaction and are typically considered for semi-automatic segmentation, however, in some cases, they are combined with other methods for fully-automatic segmentation. Designing an energy function (for deformable models) that works robustly for all different images is also challenging, especially in the case of cardiovascular images due to the presence of papillary muscles and poor contrast of different layers of the heart and blood cells with each other. Due to these challenges, the deformable contours might fail to have accurate results. Deep learning-based methods, on the other hand, require enough training datasets that may not always be available. Hybrid methods, including the combination of deep learning and deformable models are offered to overcome the shortcomings of separate approaches. In the case of combining deep learning and deformable models, the combination can have the following benefits:

1. Deformable model approaches require an initial contour; deep learning removes the user interaction by generating the deformable model's initial contour.

2. In many deformable model algorithms, the contour will shrink inward or leak outward due to the presence of the LV's papillary muscles or the opening and closing of the anatomical structures. Deep learning will increase the accuracy and robustness of the contour by providing a shape prior for the LV as the reference.
3. In most medical segmentation problems, there is limited labeled data available, and deep learning frameworks are prone to overfitting. Deformable models will refine and finalize the segmentation and lead to better results even with limited training datasets.

It is also worth noting that using a suitable deep learning design with enough labeled data can lead to better results compared to hybrid methods. U-Nets and generally FCN architectures are the most common methods for the segmentation of medical images using deep learning.

NOTES

1 Ascending aorta.
2 Pulmonary artery.

REFERENCES

Albawi, S., Mohammed, T. A., & Al-Zawi, S. (2017). Understanding of a convolutional neural network. *2017 International Conference on Engineering and Technology (ICET)*, 1–6. https://doi.org/10.1109/ICEngTechnol.2017.8308186

Alvarez, Luis, Guichard, Frédéric, Lions, Pierre -Louis, & Morel, Jean -Michel (1993). Axioms and fundamental equations of image processing. *Archive for Rational Mechanics and Analysis*, *123*, 199–257. 10.1007/bf00375127.

Andreopoulos, A., & Tsotsos, J. K. (2008). Efficient and generalizable statistical models of shape and appearance for analysis of cardiac MRI. *Medical Image Analysis*, *12*(3), 335–357. https://doi.org/10.1016/j.media.2007.12.003

Anavekar, Nandan S., & Oh, Jae K. (2009). Doppler echocardiography: A contemporary review. *Journal of Cardiology*, *54*, 347–358. 10.1016/j.jjcc.2009.10.001.

Angelini, E. D., Homma, S., Pearson, G., Holmes, J. W., & Laine, A. F. (2005). Segmentation of real-time three-dimensional ultrasound for quantification of ventricular function: A clinical study on right and left ventricles. *Ultrasound in Medicine & Biology*, *31*(9), 1143–1158. https://doi.org/10.1016/j.ultrasmedbio.2005.03.016

Angenent, S., Pichon, E., & Tannenbaum, A. (2006). Mathematical methods in medical image processing. *Bulletin of the American Mathematical Society*, 43, 365–396. https://doi.org/10.1090/S0273-0979-06-01104-9

Avendi, M. R., Kheradvar, A., & Jafarkhani, H. (2016). A combined deep-learning and deformable-model approach to fully automatic segmentation of the left ventricle in cardiac MRI. *Medical Image Analysis*, *30*, 108–119. https://doi.org/10.1016/j.media.2016.01.005

Bai, W., Sinclair, M., Tarroni, G., Oktay, O., Rajchl, M., Vaillant, G., Lee, A. M., Aung, N., Lukaschuk, E., Sanghvi, M. M., Zemrak, F., Fung, K., Paiva, J. M., Carapella, V., Kim, Y. J., Suzuki, H., Kainz, B., Matthews, P. M., Petersen, S. E., ... Rueckert, D. (2018). Automated cardiovascular magnetic resonance image analysis with fully convolutional networks. *Journal of Cardiovascular Magnetic Resonance*, *20*(1), 65. https://doi.org/10.1186/s12968-018-0471-x

Baldi, P., & Sadowski, P. J. (2013). Understanding dropout. *Advances in Neural Information Processing Systems*, 26(1).

Baumgartner, C. F., Koch, L. M., Pollefeys, M., & Konukoglu, E. (2017). An exploration of 2D and 3D deep learning techniques for cardiac mr image segmentation. *ArXiv:1709.04496 [Cs]*. http://arxiv.org/abs/1709.04496

Ben-Assa, E., Brown, J., Keshavarz-Motamed, Z., de la Torre Hernandez, J. M., Leiden, B., Olender, M., Kallel, F., Palacios, I. F., Inglessis, I., Passeri, J. J., Shah, P. B., Elmariah, S., Leon, M. B., & Edelman, E. R. (2019). Ventricular stroke work and vascular impedance refine the characterization of patients with aortic stenosis. *Science Translational Medicine*, *11*(509), eaaw0181. https://doi.org/10.1126/scitranslmed.aaw0181

Ben Ayed, I., Li, Shuo, & Ross, I. (2009). Embedding overlap priors in variational left ventricle tracking. *IEEE Transactions on Medical Imaging*, *28*(12), 1902–1913. https://doi.org/10.1109/TMI.2009.2022087

Bernard, O., Bosch, J. G., Heyde, B., Alessandrini, M., Barbosa, D., Camarasu-Pop, S., Cervenansky, F., Valette, S., Mirea, O., Bernier, M., Jodoin, P.-M., Domingos, J. S., Stebbing, R. V., Keraudren, K., Oktay, O., Caballero, J., Shi, W., Rueckert, D., Milletari, F., … D'hooge, J. (2016). Standardized evaluation system for left ventricular segmentation algorithms in 3D echocardiography. *IEEE Transactions on Medical Imaging*, *35*(4), 967–977. https://doi.org/10.1109/TMI.2015.2503890

Bernard, O., Lalande, A., Zotti, C., Cervenansky, F., Yang, X., Heng, P.-A., Cetin, I., Lekadir, K., Camara, O., Gonzalez Ballester, M. A., Sanroma, G., Napel, S., Petersen, S., Tziritas, G., Grinias, E., Khened, M., Kollerathu, V. A., Krishnamurthi, G., Rohe, M.-M., … Jodoin, P.-M. (2018). Deep learning techniques for automatic MRI cardiac multi-structures segmentation and diagnosis: Is the problem solved? *IEEE Transactions on Medical Imaging*, *37*(11), 2514–2525. https://doi.org/10.1109/TMI.2018.2837502

Blanke, P., Naoum, C., Dvir, D., Bapat, V., Ong, K., Muller, D., Cheung, A., Ye, J., Min, J. K., Piazza, N., Theriault-Lauzier, P., Webb, J., & Leipsic, J. (2017). Predicting LVOT obstruction in transcatheter mitral valve implantation. *JACC: Cardiovascular Imaging*, *10*(4), 482–485. https://doi.org/10.1016/j.jcmg.2016.01.005

Carneiro, G., & Nascimento, J. C. (2013). Combining multiple dynamic models and deep learning architectures for tracking the left ventricle endocardium in ultrasound data. *IEEE Transactions on Pattern Analysis and Machine Intelligence*, *35*(11), 2592–2607. https://doi.org/10.1109/TPAMI.2013.96

Carneiro, G., Nascimento, J. C., & Freitas, A. (2012). The segmentation of the left ventricle of the heart from ultrasound data using deep learning architectures and derivative-based search methods. *IEEE Transactions on Image Processing*, *21*(3), 968–982. https://doi.org/10.1109/TIP.2011.2169273

Chakraborty, A., Staib, L. H., & Duncan, J. S. (1996). Deformable boundary finding in medical images by integrating gradient and region information. *IEEE Transactions on Medical Imaging*, *15*(6), 859–870. https://doi.org/10.1109/42.544503

Chen, C., Qin, C., Qiu, H., Tarroni, G., Duan, J., Bai, W., & Rueckert, D. (2020). Deep learning for cardiac image segmentation: A review. *Frontiers in Cardiovascular Medicine*, *7*, 25. https://doi.org/10.3389/fcvm.2020.00025

Chen, H., Zheng, Y., Park, J.-H., Heng, P.-A., & Zhou, S. K. (2016). Iterative multi-domain regularized deep learning for anatomical structure detection and segmentation from ultrasound images. *ArXiv:1607.01855 [Cs]*. http://arxiv.org/abs/1607.01855

Chenyang Xu, & Prince, J.L. (1998). Snakes, shapes, and gradient vector flow. IEEE Transactions on Image Processing, 7, 359–36910.1109/83.661186.

Çiçek, Ö., Abdulkadir, A., Lienkamp, S. S., Brox, T., & Ronneberger, O. (2016). 3D U-Net: Learning dense volumetric segmentation from sparse annotation. *ArXiv:1606.06650 [Cs]*. http://arxiv.org/abs/1606.06650

Cootes, T. F., Edwards, G. J., & Taylor, C. J. (2001). Active appearance models 15. Computer vision: ECCV '98 : 5th. European conference on computer vision. Freiburg, Germany.

Čupić, M. (2003). Online communities – Designing usability, supporting sociability. *Journal of Computing and Information Technology*, *11*(1), 77. https://doi.org/10.2498/cit. 2003.01.06

de Vos, B. D., Wolterink, J. M., de Jong, P. A., Viergever, M. A., & Išgum, I. (2016). 2D image classification for 3D anatomy localization: Employing deep convolutional neural networks. SPIE Medical Imaging 2016. San Diego, United States.

Duan, J., Schlemper, J., Bai, W., Dawes, T. J. W., Bello, G., Doumou, G., De Marvao, A., O'Regan, D. P., & Rueckert, D. (2018). Deep nested level sets: Fully automated segmentation of cardiac MR images in patients with pulmonary hypertension. *ArXiv:1807.10760 [Cs]*. http://arxiv.org/abs/1807.10760

Ecabert, O., Peters, J., Lorenz, C., von Berg, J., Vembar, M., Subramanyan, K., Lavi, G., & Weese, J. (2005). Towards automatic full heart segmentation in computed-tomography images. *Computers in Cardiology*, *2005*, 223–226. https://doi.org/10.1109/CIC.2005. 1588077

Ecabert, O., Peters, J., Schramm, H., Lorenz, C., von Berg, J., Walker, M. J., Vembar, M., Olszewski, M. E., Subramanyan, K., Lavi, G., & Weese, J. (2008). Automatic model-based segmentation of the heart in CT images. *IEEE Transactions on Medical Imaging*, *27*(9), 1189–1201. https://doi.org/10.1109/TMI.2008.918330

Ecabert, O., Peters, J., Walker, M. J., Ivanc, T., Lorenz, C., von Berg, J., Lessick, J., Vembar, M., & Weese, J. (2011). Segmentation of the heart and great vessels in CT images using a model-based adaptation framework. *Medical Image Analysis*, *15*(6), 863–876. https://doi.org/10.1016/j.media.2011.06.004

Elmariah, S., Palacios, I. F., McAndrew, T., Hueter, I., Inglessis, I., Baker, J. N., Kodali, S., Leon, M. B., Svensson, L., Pibarot, P., Douglas, P. S., Fearon, W. F., Kirtane, A. J., Maniar, H. S., & Passeri, J. J. (2013). Outcomes of transcatheter and surgical aortic valve replacement in high-risk patients with aortic stenosis and left ventricular dysfunction: Results from the placement of aortic transcatheter valves (PARTNER) Trial (Cohort A). *Circulation: Cardiovascular Interventions*, *6*(6), 604–614. https://doi.org/10.1161/CIRCINTERVENTIONS.113.000650

Emad, O., Yassine, I. A., & Fahmy, A. S. (2015). Automatic localization of the left ventricle in cardiac MRI images using deep learning. *2015 37th Annual International Conference of the IEEE Engineering in Medicine and Biology Society (EMBC)*, 683–686. https://doi.org/10.1109/EMBC.2015.7318454

Fahmy, A. S., El-Rewaidy, H., Nezafat, M., Nakamori, S., & Nezafat, R. (2019). Automated analysis of cardiovascular magnetic resonance myocardial native T1 mapping images using fully convolutional neural networks. *Journal of Cardiovascular Magnetic Resonance*, *21*(1), 7. https://doi.org/10.1186/s12968-018-0516-1

Feng, C., Li, C., Zhao, D., Davatzikos, C., & Litt, H. (2013). Segmentation of the Left ventricle using distance regularized two-layer level set approach. In C. Salinesi, M. C. Norrie, & Ó. Pastor (Eds.), *Advanced Information Systems Engineering* (Vol. 7908, pp. 477–484). Springer Berlin Heidelberg. https://doi.org/10.1007/978-3-642-40811-3_60

Fleischmann, D., Liang, D. H., & Herfkens, R. J. (2008). Technical advances in cardiovascular imaging. *Seminars in Thoracic and Cardiovascular Surgery*, *20*(4), 333–339. https://doi.org/10.1053/j.semtcvs.2008.11.015

Frangi, A. F., Niessen, W. J., & Viergever, M. A. (2001). Three-dimensional modeling for functional analysis of cardiac images, a review. *IEEE Transactions on Medical Imaging*, *20*(1), 2–5. https://doi.org/10.1109/42.906421

Généreux, P., Head, S. J., Hahn, R., Daneault, B., Kodali, S., Williams, M. R., van Mieghem, N. M., Alu, M. C., Serruys, P. W., Kappetein, A. P., & Leon, M. B. (2013). Paravalvular leak after transcatheter aortic valve replacement. *Journal of the American College of Cardiology, 61*(11), 1125–1136. https://doi.org/10.1016/j.jacc.2012.08.1039

Ghesu, F. C., Krubasik, E., Georgescu, B., Singh, V., Zheng, Y., Hornegger, J., & Comaniciu, D. (2016). Marginal space deep learning: Efficient architecture for volumetric image parsing. *IEEE Transactions on Medical Imaging, 35*(5), 1217–1228. https://doi.org/10.1109/TMI.2016.2538802

Gotardo, P. F. U., Boyer, K. L., Saltz, J., & Raman, S. V. (2006). A new deformable model for boundary tracking in cardiac MRI and its application to the detection of intraventricular dyssynchrony. *2006 IEEE Computer Society Conference on Computer Vision and Pattern Recognition – Volume 1 (CVPR'06), 1*, 736–743. https://doi.org/1 0.1109/CVPR.2006.34

Guo, Y., Liu, Y., Oerlemans, A., Lao, S., Wu, S., & Lew, M. S. (2016). Deep learning for visual understanding: A review. *Neurocomputing, 187*, 27–48.

Hajiaghayi, M., Groves, E. M., Jafarkhani, H., & Kheradvar, A. (2017). A 3-D active contour method for automated segmentation of the left ventricle from magnetic resonance images. *IEEE Transactions on Biomedical Engineering, 64*(1), 134–144. https://doi.org/10.1109/TBME.2016.2542243

Heiberg, E., Wigstrom, L., Carlsson, M., Bolger, A. F., & Karlsson, M. (2005). Time resolved three-dimensional automated segmentation of the left ventricle. *Computers in Cardiology, 2005*, 599–602. https://doi.org/10.1109/CIC.2005.1588172

Heimann, T., & Meinzer, H.-P. (2009). Statistical shape models for 3D medical image segmentation: A review. *Medical Image Analysis, 13*(4), 543–563. https://doi.org/10.1016/j.media.2009.05.004

Hesamian, M. H., Jia, W., He, X., & Kennedy, P. (2019). Deep learning techniques for medical image segmentation: Achievements and challenges. *Journal of Digital Imaging, 32*(4), 582–596. https://doi.org/10.1007/s10278-019-00227-x

Hinton, G. E., Srivastava, N., Krizhevsky, A., Sutskever, I., & Salakhutdinov, R. R. (2012). Improving neural networks by preventing co-adaptation of feature detectors. *ArXiv:1207.0580 [Cs]*. http://arxiv.org/abs/1207.0580

Isensee, F., Jaeger, P., Full, P. M., Wolf, I., Engelhardt, S., & Maier-Hein, K. H. (2018). Automatic cardiac disease assessment on cine-MRI via time-series segmentation and domain specific features. *ArXiv:1707.00587 [Cs], 10663*. https://doi.org/10.1007/978-3-319-75541-0

Jain, A. K., Zhong, Y., & Dubuisson-Jolly, M.-P. (1998). Deformable template models: A review. *Signal Processing, 71*(2), 109–129. https://doi.org/10.1016/S0165-1684(98)00139-X

Jang, Y., Hong, Y., Ha, S., Kim, S., & Chang, H.-J. (2018). Automatic segmentation of LV and RV in cardiac MRI. In M. Pop, M. Sermesant, P.-M. Jodoin, A. Lalande, X. Zhuang, G. Yang, A. Young, & O. Bernard (Eds.), *Statistical Atlases and Computational Models of the Heart. ACDC and MMWHS Challenges* (Vol. 10663, pp. 161–169). Springer International Publishing. https://doi.org/10.1007/978-3-319-75541-0_17

Keshavarz-Motamed, Z. (2020). A diagnostic, monitoring, and predictive tool for patients with complex valvular, vascular and ventricular diseases. *Scientific Reports, 10*(1), 6905. https://doi.org/10.1038/s41598-020-63728-8

Keshavarz-Motamed, Z., Garcia, J., Gaillard, E., Capoulade, R., Le Ven, F., Cloutier, G., Kadem, L., & Pibarot, P. (2014). Non-Invasive determination of left ventricular workload in patients with aortic stenosis using magnetic resonance imaging and doppler echocardiography. *PLoS ONE, 9*(1), e86793. https://doi.org/10.1371/journal.pone.0086793

Kimmel, R., Amir, A., & Bruckstein, A.M. (1995). Finding shortest paths on surfaces using level sets propagation. *IEEE Transactions on Pattern Analysis and Machine Intelligence*, *17*, 635–640. 10.1109/34.387512.

Keshavarz-Motamed, Z., Khodaei, S., Nezami, F. R., Amrute, J. M., Lee, S. J., Brown, J., Ben-Assa, E., Camarero, T. G., Calvo, J. R., Sellers, S., Blanke, P., & Leipsic, J. (2020). Mixed valvular disease following transcatheter aortic valve replacement: quantification and systematic differentiation using clinical measurements and image-based patient-specific in silico modeling., *9*(5), e015063.

Khened, M., Kollerathu, V. A., & Krishnamurthi, G. (2018). Fully convolutional multi-scale residual DenseNets for cardiac segmentation and automated cardiac diagnosis using ensemble of classifiers. *ArXiv:1801.05173 [Cs]*. http://arxiv.org/abs/1801.05173

Krizhevsky, A., Sutskever, I., & Hinton, G. E. (2017). ImageNet classification with deep convolutional neural networks. *Communications of the ACM*, *60*(6), 84–90. https://doi.org/10.1145/3065386

Kimia, Benjamin B., Tannenbaum, Allen R., & Zucker, Steven W. (1995). Shapes, shocks, and deformations I: The components of two-dimensional shape and the reaction-diffusion space. *International Journal of Computer Vision*, *15*, 189–224. 10.1007/bf01451741.

Leclerc, S., Smistad, E., Pedrosa, J., Østvik, A., Cervenansky, F., Espinosa, F., Espeland, T., Berg, E. A. R., Jodoin, P.-M., Grenier, T., Lartizien, C., D'hooge, J., Lovstakken, L., & Bernard, O. (2019). Deep learning for segmentation using an open large-scale dataset in 2D echocardiography. *IEEE Transactions on Medical Imaging*, *38*(9), 2198–2210. https://doi.org/10.1109/TMI.2019.2900516

Lee, H.-Y., Codella, N. C. F., Cham, M. D., Weinsaft, J. W., & Wang, Y. (2010). Automatic left ventricle segmentation using iterative thresholding and an active contour model with adaptation on short-axis cardiac MRI. *IEEE Transactions on Biomedical Engineering*, *57*(4), 905–913. https://doi.org/10.1109/TBME.2009.2014545

Lessmann, N., Išgum, I., Setio, A. A. A., de Vos, B. D., Ciompi, F., de Jong, P. A., Oudkerk, M., Mali, W. P. Th. M., Viergever, M. A., & van Ginneken, B. (2016). *Deep convolutional neural networks for automatic coronary calcium scoring in a screening study with low-dose chest CT* (G. D. Tourassi & S. G. Armato, Eds.; p. 978511). https://doi.org/10.1117/12.2216978.

Lieman-Sifry, J., Le, M., Lau, F., Sall, S., & Golden, D. (2017). FastVentricle: Cardiac segmentation with ENet. *ArXiv:1704.04296 [Cs]*. http://arxiv.org/abs/1704.04296

Lin, E., & Alessio, A. (2009). What are the basic concepts of temporal, contrast, and spatial resolution in cardiac CT? *Journal of Cardiovascular Computed Tomography*, *3*(6), 403–408. https://doi.org/10.1016/j.jcct.2009.07.003

Litjens, G., Kooi, T., Bejnordi, B. E., Setio, A. A. A., Ciompi, F., Ghafoorian, M., van der Laak, J. A. W. M., van Ginneken, B., & Sánchez, C. I. (2017). A survey on deep learning in medical image analysis. *Medical Image Analysis*, *42*, 60–88. https://doi.org/10.1016/j.media.2017.07.005

Long, J., Shelhamer, E., & Darrell, T. (2015). Fully convolutional networks for semantic segmentation. *Proceedings of the IEEE Conference on Computer Vision and Pattern Recognition (CVPR)*. https://doi.org/10.1109/TPAMI.2016.2572683

Lynch, M., Ghita, O., & Whelan, P. F. (2006). Left-ventricle myocardium segmentation using a coupled level-set with a priori knowledge. *Computerized Medical Imaging and Graphics*, *30*(4), 255–262. https://doi.org/10.1016/j.compmedimag.2006.03.009

Lynch, M., Ghita, O., & Whelan, P. F. (2008). Segmentation of the left ventricle of the heart in 3-D+t MRI data using an optimized nonrigid temporal model. *IEEE Transactions on Medical Imaging*, *27*(2), 195–203. https://doi.org/10.1109/TMI.2007.904681

McAllester, D. (2013). A PAC-Bayesian tutorial with a dropout bound. *ArXiv:1307.2118 [Cs]*. http://arxiv.org/abs/1307.2118

McInerney, T., & Terzopoulos, D. (1996). Deformable models in medical image analysis: A survey. *Medical Image Analysis, 1*(2), 91–108. https://doi.org/10.1016/S1361-8415 (96)80007-7

Milletari, F., Navab, N., & Ahmadi, S.-A. (2016). V-Net: Fully convolutional neural networks for volumetric medical image segmentation. *ArXiv:1606.04797 [Cs]*. http://arxiv.org/abs/1606.04797

Molaei, S., Shiri, M., Horan, K., Kahrobaei, D., Nallamothu, B., & Najarian, K. (2017). Deep convolutional neural networks for left ventricle segmentation. *2017 39th Annual International Conference of the IEEE Engineering in Medicine and Biology Society (EMBC)*, 668–671. https://doi.org/10.1109/EMBC.2017.8036913

Montagnat, J., Delingette, H., & Ayache, N. (2001). A review of deformable surfaces: Topology, geometry and deformation. *Image and Vision Computing, 19*(14), 1023–1040. https://doi.org/10.1016/S0262-8856(01)00064-6

Moradi, M., Gur, Y., Wang, H., Prasanna, P., & Syeda-Mahmood, T. (2016). A hybrid learning approach for semantic labeling of cardiac CT slices and recognition of body position. *2016 IEEE 13th International Symposium on Biomedical Imaging (ISBI)*, 1418–1421. https://doi.org/10.1109/ISBI.2016.7493533

Mozaffarian, Dariush, Benjamin, Emelia J., Go, Alan S., Arnett, Donna K., Blaha, Michael J., Cushman, Mary, de Ferranti, Sarah, Després, Jean-Pierre, Fullerton, Heather J., Howard, Virginia J., Huffman, Mark D., Judd, Suzanne E., Kissela, Brett M., Lackland, Daniel T., Lichtman, Judith H., Lisabeth, Lynda D., Liu, Simin, Mackey, Rachel H., Matchar, David B., McGuire, Darren K., Mohler, Emile R., Moy, Claudia S., Muntner, Paul, Mussolino, Michael E., Nasir, Khurram, Neumar, Robert W., Nichol, Graham, Palaniappan, Latha, Pandey, Dilip K., Reeves, Mathew J., Rodriguez, Carlos J., Sorlie, Paul D., Stein, Joel, Towfighi, Amytis, Turan, Tanya N., Virani, Salim S., Willey, Joshua Z., Woo, Daniel, Yeh, Robert W., & Turner, Melanie B. (2015). Heart Disease and Stroke Statistics—2015 Update. Circulation, 13110.1161/cir. 0000000000000152.

Moreno, R. A., de Sá Rebelo, M. F. S., Carvalho, T., Assunção, A. N., Dantas, R. N., do Val, R., Marin, A. S., Bordignom, A., Nomura, C. H., & Gutierrez, M. A. (2019). A combined deep-learning approach to fully automatic left ventricle segmentation in cardiac magnetic resonance imaging. In B. Gimi & A. Krol (Eds.), *Medical Imaging 2019: Biomedical Applications in Molecular, Structural, and Functional Imaging* (p. 68). SPIE. https://doi.org/10.1117/12.2512895

Mortazi, A., Burt, J., & Bagci, U. (2017). Multi-planar deep segmentation networks for cardiac substructures from MRI and CT. *ArXiv:1708.00983 [Cs, Stat]*. http://arxiv.org/ abs/1708.00983

Ngo, T. A., Lu, Z., & Carneiro, G. (2017). Combining deep learning and level set for the automated segmentation of the left ventricle of the heart from cardiac cine magnetic resonance. *Medical Image Analysis, 35*, 159–171. https://doi.org/10.1016/j.media. 2016.05.009

Noble, J. A., & Boukerroui, D. (2006). Ultrasound image segmentation: A survey. *IEEE Transactions on Medical Imaging, 25*(8), 987–1010. https://doi.org/10.1109/TMI.2 006.877092

Nombela-Franco, L., Ribeiro, H. B., Urena, M., Allende, R., Amat-Santos, I., DeLarochellière, R., Dumont, E., Doyle, D., DeLarochellière, H., Laflamme, J., Laflamme, L., García, E., Macaya, C., Jiménez-Quevedo, P., Côté, M., Bergeron, S., Beaudoin, J., Pibarot, P., & Rodés-Cabau, J. (2014). Significant mitral regurgitation left untreated at the time of aortic valve replacement. *Journal of the American College of Cardiology, 63*(24), 2643–2658. https://doi.org/10.1016/j.jacc.2014.02.573

Oktay, O., Ferrante, E., Kamnitsas, K., Heinrich, M., Bai, W., Caballero, J., Cook, S. A., de Marvao, A., Dawes, T., O'Regan, D. P., Kainz, B., Glocker, B., & Rueckert, D. (2018). Anatomically constrained neural networks (ACNNs): Application to cardiac image enhancement and segmentation. *IEEE Transactions on Medical Imaging, 37*(2), 384–395. https://doi.org/10.1109/TMI.2017.2743464

Orwat, S., Diller, G.-P., & Baumgartner, H. (2014). Imaging of congenital heart disease in adults: Choice of modalities. *European Heart Journal – Cardiovascular Imaging, 15*(1), 6–17. https://doi.org/10.1093/ehjci/jet124

O'Shea, K., & Nash, R. (2015). An introduction to convolutional neural networks. *ArXiv:1511.08458 [Cs].* http://arxiv.org/abs/1511.08458

Osher, S., & Sethian, J. A. (1988). Fronts propagating with curvature-dependent speed: Algorithms based on Hamilton-Jacobi formulations. *Journal of Computational Physics, 79*(1), 12–49. https://doi.org/10.1016/0021-9991(88)90002-2

Pace, D. F., Dalca, A. V., Geva, T., Powell, A. J., Moghari, M. H., & Golland, P. (2015). Interactive whole-heart segmentation in congenital heart disease. In N. Navab, J. Hornegger, W. M. Wells, & A. F. Frangi (Eds.), *Medical Image Computing and Computer-Assisted Intervention – MICCAI 2015* (Vol. 9351, pp. 80–88). Springer International Publishing. https://doi.org/10.1007/978-3-319-24574-4_10

Papolos, A., Narula, J., Bavishi, C., Chaudhry, F. A., & Sengupta, P. P. (2016). U.S. hospital use of echocardiography. *Journal of the American College of Cardiology, 67*(5), 502–511. https://doi.org/10.1016/j.jacc.2015.10.090

Paragios, N., Rousson, M., & Ramesh, V. (2002). Knowledge-based registration & segmentation of the left ventricle: A level set approach. *Sixth IEEE Workshop on Applications of Computer Vision, 2002. (WACV 2002). Proceedings.*, 37–42. https://doi.org/10.1109/ACV.2002.1182152

Park, S., Park, J., Shin, S.-J., & Moon, I.-C. (2018). Adversarial dropout for supervised and semi-supervised learning 8. The Thirty-Second AAAI Conference on Artificial Intelligence (AAAI-18). Deajeon, South Korea.

Patravali, J., Jain, S., & Chilamkurthy, S. (2017). 2D-3D fully convolutional neural networks for cardiac MR segmentation. *ArXiv:1707.09813 [Cs].* http://arxiv.org/abs/1707.09813

Payer, C., Štern, D., Bischof, H., & Urschler, M. (2018). Multi-label whole heart segmentation using CNNs and anatomical label configurations. In M. Pop, M. Sermesant, P.-M. Jodoin, A. Lalande, X. Zhuang, G. Yang, A. Young, & O. Bernard (Eds.), *Statistical Atlases and Computational Models of the Heart. ACDC and MMWHS Challenges* (Vol. 10663, pp. 190–198). Springer International Publishing. https://doi.org/10.1007/978-3-319-75541-0_20

Peng, P., Lekadir, K., Gooya, A., Shao, L., Petersen, S. E., & Frangi, A. F. (2016). A review of heart chamber segmentation for structural and functional analysis using cardiac magnetic resonance imaging. *Magnetic Resonance Materials in Physics, Biology and Medicine, 29*(2), 155–195. https://doi.org/10.1007/s10334-015-0521-4

Peters, J., Ecabert, O., Meyer, C., Kneser, R., & Weese, J. (2010). Optimizing boundary detection via Simulated Search with applications to multi-modal heart segmentation. *Medical Image Analysis, 14*(1), 70–84. https://doi.org/10.1016/j.media.2009.10.004

Peters, J., Ecabert, O., Meyer, C., Schramm, H., Kneser, R., Groth, A., & Weese, J. (2007). Automatic whole heart segmentation in static magnetic resonance image volumes. In N. Ayache, S. Ourselin, & A. Maeder (Eds.), *Medical Image Computing and Computer-Assisted Intervention – MICCAI 2007* (Vol. 4792, pp. 402–410). Springer Berlin Heidelberg. https://doi.org/10.1007/978-3-540-75759-7_49

Petitjean, C., & Dacher, J.-N. (2011). A review of segmentation methods in short axis cardiac MR images. *Medical Image Analysis, 15*(2), 169–184. https://doi.org/10.1016/j.media.2010.12.004

Petitjean, C., Zuluaga, M. A., Bai, W., Dacher, J.-N., Grosgeorge, D., Caudron, J., Ruan, S., Ayed, I. B., Cardoso, M. J., Chen, H.-C., Jimenez-Carretero, D., Ledesma-Carbayo, M. J., Davatzikos, C., Doshi, J., Erus, G., Maier, O. M. O., Nambakhsh, C. M. S., Ou, Y., Ourselin, S., ... Yuan, J. (2015). Right ventricle segmentation from cardiac MRI: A collation study. *Medical Image Analysis*, *19*(1), 187–202. https://doi.org/10.1016/j.media.2014.10.004

Pham, D. L., Xu, C., & Prince, J. L. (2000). A survey of current methods in medical image segmentation. *Image Segmentation*, 27.

Pham, Q. C., Vincent, F., Clarysse, P., Croisille, P., & Magnin, I. E. (2001). A FEM-based deformable model for the 3D segmentation and tracking of the heart in cardiac MRI. *ISPA 2001. Proceedings of the 2nd International Symposium on Image and Signal Processing and Analysis. In Conjunction with 23rd International Conference on Information Technology Interfaces (IEEE Cat. No.01EX480)*, 250–254. https://doi.org/10.1109/ISPA.2001.938636

Picano, E. (2005). Economic and biological costs of cardiac imaging. *Cardiovascular Ultrasound*, *3*(1), 13. https://doi.org/10.1186/1476-7120-3-13

Pluempitiwiriyawej, C., Moura, J. M. F., Yi-Jen Lin Wu, & Chien Ho. (2005). STACS: New active contour scheme for cardiac MR image segmentation. *IEEE Transactions on Medical Imaging*, *24*(5), 593–603. https://doi.org/10.1109/TMI.2005.843740

Ranganath, S. (1995). Contour extraction from cardiac MRI studies using snakes. *IEEE Transactions on Medical Imaging*, *14*(2), 328–338. https://doi.org/10.1109/42.387714

Redmon, J., Divvala, S., Girshick, R., & Farhadi, A. (2016). You only look once: Unified, real-time object detection. *2016 IEEE Conference on Computer Vision and Pattern Recognition (CVPR)*, 779–788. https://doi.org/10.1109/CVPR.2016.91

Rohé, M.-M., Sermesant, M., & Pennec, X. (2018). Automatic multi-atlas segmentation of myocardium with SVF-Net. In M. Pop, M. Sermesant, P.-M. Jodoin, A. Lalande, X. Zhuang, G. Yang, A. Young, & O. Bernard (Eds.), *Statistical Atlases and Computational Models of the Heart. ACDC and MMWHS Challenges* (Vol. 10663, pp. 170–177). Springer International Publishing. https://doi.org/10.1007/978-3-319-75541-0_18

Ronneberger, O., Fischer, P., & Brox, T. (2015). U-Net: Convolutional networks for biomedical image segmentation. *ArXiv:1505.04597 [Cs]*. http://arxiv.org/abs/1505.04597

Rupprecht, C., Huaroc, E., Baust, M., & Navab, N. (2016). Deep active contours. *ArXiv:1607.05074 [Cs]*. http://arxiv.org/abs/1607.05074

Santarelli, M. F., Positano, V., Michelassi, C., Lombardi, M., & Landini, L. (2003). Automated cardiac MR image segmentation: Theory and measurement evaluation. *Medical Engineering & Physics*, *25*(2), 149–159. https://doi.org/10.1016/S1350-4533(02)00144-3

Sarti, A., Corsi, C., Mazzini, E., & Lamberti, C. (2005). Maximum likelihood segmentation of ultrasound images with Rayleigh distribution. *IEEE Transactions on Ultrasonics, Ferroelectrics and Frequency Control*, *52*(6), 947–960. https://doi.org/10.1109/TUFFC.2005.1504017

Savarese, G., & Lund, L. H.(2017). Global public health burden of heart failure. *Cardiac Failure Review*, *3*(01), 7. https://doi.org/10.15420/cfr.2016:25:2

Shen, Dinggang, Wu, Guorong, & Suk, Heung-Il (2017). Deep Learning in Medical Image Analysis. *Annual Review of Biomedical Engineering*, 19, 221–24810.1146/annurev-bioeng-071516-044442.

Smistad, E., Ostvik, A., Haugen, B. O., & Lovstakken, L. (2017). 2D left ventricle segmentation using deep learning. *2017 IEEE International Ultrasonics Symposium (IUS)*, 1–4. https://doi.org/10.1109/ULTSYM.2017.8092573

Sapiro, Guillermo, & Tannenbaum, Allen (1993). Affine invariant scale-space. *International Journal of Computer Vision*, 11, 25–4410.1007/bf01420591.

Srivastava, N., Hinton, G., Krizhevsky, A., Sutskever, I., & Salakhutdinov, R. (2014). Dropout: A Simple way to prevent neural networks from overfitting. *Journal of Machine Learning Research.* 15, 1929–1958.

Suinesiaputra, A., Cowan, B. R., Al-Agamy, A. O., Elattar, M. A., Ayache, N., Fahmy, A. S., Khalifa, A. M., Medrano-Gracia, P., Jolly, M.-P., Kadish, A. H., Lee, D. C., Margeta, J., Warfield, S. K., & Young, A. A. (2014). A collaborative resource to build consensus for automated left ventricular segmentation of cardiac MR images. *Medical Image Analysis, 18*(1), 50–62. https://doi.org/10.1016/j.media.2013.09.001

Suri, J. S., Kecheng Liu, Singh, S., Laxminarayan, S. N., Xiaolan Zeng, & Reden, L. (2002). Shape recovery algorithms using level sets in 2-D/3-D medical imagery: A state-of-the-art review. *IEEE Transactions on Information Technology in Biomedicine, 6*(1), 8–28. https://doi.org/10.1109/4233.992158

Tan, L. K. (2017). Convolutional neural network regression for short-axis left ventricle segmentation in cardiac cine MR sequences. *Medical Image Analysis, 39,* 78–86.

Tan, L. K., Liew, Y. M., Lim, E., & McLaughlin, R. A. (2016). Cardiac left ventricle segmentation using convolutional neural network regression. *2016 IEEE EMBS Conference on Biomedical Engineering and Sciences (IECBES),* 490–493. https://doi.org/10.1109/IECBES.2016.7843499

Tao, Q., Yan, W., Wang, Y., Paiman, E. H. M., Shamonin, D. P., Garg, P., Plein, S., Huang, L., Xia, L., Sramko, M., Tintera, J., de Roos, A., Lamb, H. J., & van der Geest, R. J. (2019). Deep learning–based method for fully automatic quantification of left ventricle function from cine MR images: A multivendor, multicenter study. *Radiology, 290*(1), 81–88. https://doi.org/10.1148/radiol.2018180513

Tavakoli, V., & Amini, A. A. (2013). A survey of shaped-based registration and segmentation techniques for cardiac images. *Computer Vision and Image Understanding, 117*(9), 966–989. https://doi.org/10.1016/j.cviu.2012.11.017

Tong, Q., Ning, M., Si, W., Liao, X., & Qin, J. (2018). 3D deeply-supervised U-Net based whole heart segmentation. In M. Pop, M. Sermesant, P.-M. Jodoin, A. Lalande, X. Zhuang, G. Yang, A. Young, & O. Bernard (Eds.), *Statistical Atlases and Computational Models of the Heart. ACDC and MMWHS Challenges* (Vol. 10663, pp. 224–232). Springer International Publishing. https://doi.org/10.1007/978-3-319-75541-0_24

Tran, P. V. (2017). A fully convolutional neural network for cardiac segmentation in short-axis MRI. *ArXiv:1604.00494 [Cs].* http://arxiv.org/abs/1604.00494

Veni, G., Moradi, M., Bulu, H., Narayan, G., & Syeda-Mahmood, T. (2018). Echocardiography segmentation based on a shape-guided deformable model driven by a fully convolutional network prior. *2018 IEEE 15th International Symposium on Biomedical Imaging (ISBI 2018),* 898–902. https://doi.org/10.1109/ISBI.2018.8363716

Villarraga-Gómez, H., Lee, C., & Smith, S. T. (2018). Dimensional metrology with X-ray CT: A comparison with CMM measurements on internal features and compliant structures. *Precision Engineering, 51,* 291–307. https://doi.org/10.1016/j.precisioneng.2017.08.021

Wager, S., Wang, S., & Liang, P. (2013). Dropout training as adaptive regularization. *ArXiv:1307.1493 [Cs, Stat].* http://arxiv.org/abs/1307.1493

Wan, L., Zeiler, M., Zhang, S., LeCun, Y., & Fergus, R. (2013). Regularization of neural networks using DropConnect. *Proceedings of the 30th International Conference on Machine Learning, PMLR, 28*(3), 1058–1066.

Wang, C., MacGillivray, T., Macnaught, G., Yang, G., & Newby, D. (2018). A two-stage 3D Unet framework for multi-class segmentation on full resolution image. *ArXiv:1804.04341 [Cs].* http://arxiv.org/abs/1804.04341

Wang, C., & Smedby, Ö. (2018). Automatic whole heart segmentation using deep learning and shape context. In M. Pop, M. Sermesant, P.-M. Jodoin, A. Lalande, X. Zhuang, G. Yang, A. Young, & O. Bernard (Eds.), *Statistical Atlases and Computational Models of the Heart. ACDC and MMWHS Challenges* (Vol. 10663, pp. 242–249). Springer International Publishing. https://doi.org/10.1007/978-3-319-75541-0_26

Warde-Farley, D., Goodfellow, I. J., Courville, A., & Bengio, Y. (2014). An empirical analysis of dropout in piecewise linear networks. *ArXiv:1312.6197 [Cs, Stat].* http://arxiv.org/abs/1312.6197

Watson, S. R., Dormer, J. D., & Fei, B. (2018). Imaging technologies for cardiac fiber and heart failure: A review. *Heart Failure Reviews, 23*(2), 273–289. https://doi.org/10.1007/s10741-018-9684-1

Withey, D. J., & Koles, Z. J. (2008). A review of medical image segmentation: Methods and available software. *International Journal of Bioelectromagnetism,* 10(3), 125–148.

Wolterink, J. M., Leiner, T., de Vos, B. D., van Hamersvelt, R. W., Viergever, M. A., & Išgum, I. (2016). Automatic coronary artery calcium scoring in cardiac CT angiography using paired convolutional neural networks. *Medical Image Analysis, 34,* 123–136. https://doi.org/10.1016/j.media.2016.04.004

Wolterink, J. M., Leiner, T., Viergever, M. A., & Išgum, I. (2017). Dilated convolutional neural networks for cardiovascular MR segmentation in congenital heart disease. *ArXiv:1704.03669 [Cs], 10129,* 95–102. https://doi.org/10.1007/978-3-319-52280-7_9

Xu, Z., Wu, Z., & Feng, J. (2018). CFUN: Combining faster R-CNN and U-net network for efficient whole heart segmentation. *ArXiv:1812.04914 [Cs].* http://arxiv.org/abs/1812.04914

Yang, X., Bian, C., Yu, L., Ni, D., & Heng, P.-A. (2018a). 3D convolutional networks for fully automatic fine-grained whole heart partition. In M. Pop, M. Sermesant, P.-M. Jodoin, A. Lalande, X. Zhuang, G. Yang, A. Young, & O. Bernard (Eds.), *Statistical Atlases and Computational Models of the Heart. ACDC and MMWHS Challenges* (Vol. 10663, pp. 181–189). Springer International Publishing. https://doi.org/10.1007/978-3-319-75541-0_19

Yang, X., Bian, C., Yu, L., Ni, D., & Heng, P.-A. (2018b). Class-balanced deep neural network for automatic ventricular structure segmentation. In M. Pop, M. Sermesant, P.-M. Jodoin, A. Lalande, X. Zhuang, G. Yang, A. Young, & O. Bernard (Eds.), *Statistical Atlases and Computational Models of the Heart. ACDC and MMWHS Challenges* (Vol. 10663, pp. 152–160). Springer International Publishing. https://doi.org/10.1007/978-3-319-75541-0_16

Yang, D., Huang, Q., Axel, L., & Metaxas, D. (2018). Multi-component deformable models coupled with 2D-3D U-Net for automated probabilistic segmentation of cardiac walls and blood. *2018 IEEE 15th International Symposium on Biomedical Imaging (ISBI 2018),* 479–483. https://doi.org/10.1109/ISBI.2018.8363620

Ye, C., Wang, W., Zhang, S., & Wang, K. (2019). Multi-depth fusion network for whole-heart CT image segmentation. *IEEE Access, 7,* 23421–23429. https://doi.org/10.1109/ACCESS.2019.2899635

Yu, L., Cheng, J.-Z., Dou, Q., Yang, X., Chen, H., Qin, J., & Heng, P.-A. (2017). Automatic 3D cardiovascular MR segmentation with densely-connected volumetric ConvNets. *ArXiv:1708.00573 [Cs].* http://arxiv.org/abs/1708.00573

Zheng, Q., Delingette, H., Duchateau, N., & Ayache, N. (2018). 3D consistent & robust segmentation of cardiac images by deep learning with spatial propagation. *ArXiv:1804.09400 [Cs, Stat].* http://arxiv.org/abs/1804.09400

Zhou, Z., Rahman Siddiquee, M. M., Tajbakhsh, N., & Liang, J. (2018). UNet++: A Nested U-Net Architecture for Medical Image Segmentation. In D. Stoyanov, Z. Taylor, G. Carneiro, T. Syeda-Mahmood, A. Martel, L. Maier-Hein, J. M. R. S. Tavares, A. Bradley, J. P. Papa, V. Belagiannis, J. C. Nascimento, Z. Lu, S. Conjeti, M. Moradi, H. Greenspan, & A. Madabhushi (Eds.), *Deep Learning in Medical Image Analysis and Multimodal Learning for Clinical Decision Support* (Vol. 11045, pp. 3–11). Springer International Publishing. https://doi.org/10.1007/978-3-030-00889-5_1

Zhuang, X. (2013). Challenges and methodologies of fully automatic whole heart segmentation: A review. *Journal of Healthcare Engineering, 4*(3), 371–408. https://doi.org/
10.1260/2040-2295.4.3.371

Zhuang, X., Li, L., Payer, C., Štern, D., Urschler, M., Heinrich, M. P., Oster, J., Wang, C.,
Smedby, Ö., Bian, C., Yang, X., Heng, P.-A., Mortazi, A., Bagci, U., Yang, G.,
Sun, C., Galisot, G., Ramel, J.-Y., Brouard, T., … Yang, G. (2019). Evaluation of
algorithms for multi-modality whole heart segmentation: An open-access grand challenge. *Medical Image Analysis, 58*, 101537. https://doi.org/10.1016/j.media.2019.
101537

Zhukov, L., Bao, Z., Gusikov, I., Wood, J., & Breen, D. E. (2002). Dynamic deformable
models for 3D MRI heart segmentation. *Proceedings of SPIE – The International
Society for Optical Engineering.* https://doi.org/10.1117/12.467105

Zotti, C., Luo, Z., Lalande, A., Humbert, O., & Jodoin, P.-M. (2017). GridNet with automatic
shape prior registration for automatic MRI cardiac segmentation. *ArXiv:1705.08943
[Cs].* http://arxiv.org/abs/1705.08943

Zreik, M., Leiner, T., de Vos, B. D., van Hamersvelt, R. W., Viergever, M. A., & Isgum, I.
(2016). Automatic segmentation of the left ventricle in cardiac CT angiography
using convolutional neural networks. *2016 IEEE 13th International Symposium on
Biomedical Imaging (ISBI)*, 40–43. https://doi.org/10.1109/ISBI.2016.7493206

Zyuzin, V., & Chumarnaya, T. (2019). Comparison of UNet architectures for segmentation
of the left ventricle endocardial border on two-dimensional ultrasound images. *2019
Ural Symposium on Biomedical Engineering, Radioelectronics and Information
Technology (USBEREIT)*, 110–113. https://doi.org/10.1109/USBEREIT.2019.873
6616

3 Advances in Artificial Intelligence Applied to Heart Failure

Jose M. García-Pinilla and Francisco Lopez Valverde

CONTENTS

3.1 INTRODUCTION

Today we are immersed in a technological revolution led by Artificial Intelligence (AI) that is changing many aspects of our daily lives but is also altering many professions and medicine is one of them. If we analyze the common characteristics in the technological revolutions of the past, we can observe that in all of them there was a turning point, and the society that existed before was very different from the

DOI: 10.1201/9781003120902-3

one that followed. Undoubtedly, progress in medicine and specifically in heart failure will have a high impact on the society to come, improving their well-being notably in the near future.

AI is a branch of computing that focuses on solving problems by imitating certain cognitive abilities that are considered unique to the human species. As in the biological process there is a learning stage and subsequently a validation of that learning, in the computer engineering process a training stage is also carried out and the results are subsequently validated. The importance of data in this context must be emphasized, as it is seen as the "oil" that enables AI engines to run. Advances in the development of biometric sensors (Mohsin et al., 2018) have made it possible to have very useful information for AI. For example, in the case of patient monitoring, the parameters that may fluctuate rapidly need to be measured continuously and not intermittently as is currently done. Glucose is one of these parameters. In the most recent stage of the evolution of AI, the most significant advances have occurred in the area of Machine Learning and Deep Learning.

Machine Learning (ML) is a subarea of AI and consists of pattern-recognition algorithms used to define relationships and make predictions after training on selected data sets (Figure 3.1). The power of ML algorithms comes from the ability to automatically learn these patterns from large data sets for predictive analytics, allowing the user to obtain knowledge from past data and apply it to future predictions.

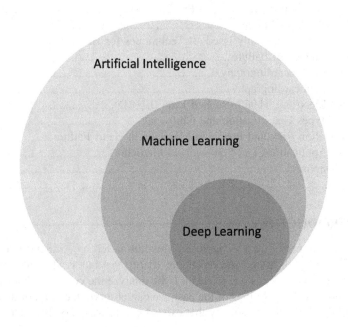

FIGURE 3.1   AI and related concepts.

Deep learning (DL) is a branch of ML (Figure 3.1) showing increasing promise in medicine, to assist in data classification, novel disease phenotyping, and complex decision making. DL is a form of ML typically implemented via multi-layered neural networks. DL has accelerated by recent advances in computer hardware and algorithms and is increasingly applied in e-commerce, finance, and voice and image recognition to learn and classify complex datasets. Strengths of DL include its ability to automate medical image interpretation, enhance clinical decision-making, identify novel phenotypes, and select better treatment pathways in complex diseases.

ML and AI have also a great potential in medicine for personalizing and improving patient care, which includes the diagnosis, management, and risk stratification of heart failure (HF).

HF is an epidemiologic disturbance because of its high and growing incidence and prevalence in developed countries, with a prevalence of 2–3% in the whole population, increasing with the elderly. HF is a serious condition associated with high morbidity and mortality rates. Twenty-six million adults globally are diagnosed with HF, while 3.6 million are newly diagnosed every year. Thirty percent of the patients suffering from HF die within the first year and the remaining die within 5 years. The related to HF management costs are approximately 1–2% of all healthcare expenditure, with most of them linked with recurrent hospital admissions (Virani et al., 2020). High prevalence and incidence are related to improving the prognosis of patients with acute myocardial infarction thanks to early coronary revascularization and better prognosis of HF with left ventricular reduced ejection fraction (HFrEF) (as a consequence of advances in pharmacologic and devices treatment in the last 30 years) and it is also related to aging of developed countries (Dunlay et al., 2017).

HF is a complex clinical syndrome associated with a heavy burden of symptoms and a wide range of therapeutic options. Despite improvements in HF management we still need approaches that might reduce the effects of this disease, preventing its insidious onset and worsening of symptoms, such as acute decompensations which is a challenge to the health providers as it generates high costs, mainly related to hospitalizations. However, the assessment of clinical safety and the evaluation of the potential benefits is still a matter of debate.

Rapid diagnosis and risk assessment of HF are essential to providing timely, cost-effective care. Traditional risk prediction tools have modest predictive value and the complexity of HF pathophysiology and treatment produces an array of information that is challenging for clinicians and researchers to process. AI and ML techniques offer a potential solution to organizing and interpreting these complex and increasing data.

ML has already demonstrated medical successes on image-based diagnoses, as a result of supervised 'deep learning,' a term referring to a neural network with many computational layers between the input data layer and output layer predicting the label. This approach requires very large datasets and significant computational power. The more traditional forms of supervised ML, which include decision trees and random forest ensemble affiliate, regularized linear models and boosted models, have also been employed in the development of algorithms in HF, offering the advantage of utility when only small datasets are available.

The goal of applying ML to HF is to improve the detection and classification of disease, make better predictions, and improve the personalization of medicine (Sanders et al., 2021).

3.2   UNDERSTANDING ARTIFICIAL INTELLIGENCE

One of the most widely used definitions of AI states that when a machine mimics the cognitive functions that humans associate with other human minds, such as perceiving, reasoning, learning and solving problems, it is considered intelligent. Another definition that is more applicable to many professions of the human being was formulated by Andreas Kaplan and Michael Haenlein (Kaplan & Haenlein, 2010) where they define artificial intelligence as "the ability of a system to correctly interpret specific data, to learn from said data and use that knowledge to achieve tasks and concrete goals through flexible adaptation". This is similar to what doctors do in some of their daily tasks.

3.2.1   TYPES OF AI

From the point of view of what the human can observe (Russel & Norvig, 2012) classified the systems using AI into:

- **Systems that act like** humans. Perform tasks in a similar way as humans do. For example, robots.
- **Systems that think like** humans. They are able to do complex tasks such as decision-making, problem-solving, and learning ability. For example, machine learning systems
- **Systems that act** rationally. They are able to obtain information about the environment, think like a human and act accordingly. For example, smart agents.

From the point of view of how they achieve the ability to think and develop learning, that is, how they work from the inside, we have two possibilities: rule-based systems and machine learning systems.

3.2.1.1   Rule-Based Systems

Different rules are programmed that can be applied under certain known circumstances. The system has to decide when it is presented with similar circumstances, but not the same as the ones programmed, which rules to apply. Based on the result obtained (hit or miss) the system develops a strategy (combination of rules) that leads to success. To program such a system we need a set of rules and a set of facts. A fact is an objective and verifiable data. Rules are the way to program decision-making when applying a strategy and have to be developed by a group of experts in the field (Wagner, 2017).

Rule-based systems are considered the simplest form of artificial intelligence and are very well suited for developing strategies in games like chess or Go. Their main drawback is that they cannot adapt to changes in the environment. If the environment changes then programmers and experts have to modify or add new rules.

3.2.1.2 Systems Based on Machine Learning

It is an AI technique with another approach but improved that helps eliminate the problems of rule-based systems. In this case, we have algorithms with generic learning capacity applicable to any field and that are trained with the data of a certain problem. In the next section, we will describe this concept in more detail.

3.2.2 UNDERSTANDING MACHINE LEARNING

Machine learning is the subfield of computer science and a branch of AI (Figure 3.1), whose goal is to develop techniques that allow computers to learn (Huang et al., 2015). An agent is said to learn when his performance improves with experience; that is, when the ability was not present in their genotype or birth traits. More specifically, machine learning researchers look for algorithms and heuristics to convert data samples into computer programs, without having to write the latter explicitly. The resulting models or programs must be able to generalize behaviors and inferences to a larger set of data.

It is the system that generates its own decision-making mechanism and is capable of learning based on the results obtained (success or failure). This technique is called "reinforcement learning" and consists of system learning by observing the environment that surrounds it. In other words, the system learns through trial and error, obtaining rewards or penalties in order to determine which is the best strategy. The strength of machine learning is that it can be adapted to the environment without having to program again. If circumstances change then the system adapts automatically. The system is constantly analyzing the data and testing to learn continuously (Arulkumaran et al., 2017).

3.2.3 THE IMPORTANCE OF DATA PRE-PROCESSING

Data is the lifeblood of AI's success today. Unfortunately, the data does not always come with the quality necessary or required by AI computational models and they need a prior process to improve them (Ramírez-Gallego et al., 2017).

- **Structuring**: It is often necessary to join data from different sources and they have a different structure. It is also possible that the data has no structure and we need for example to express it as an array.
- **Debugging**: Very often a dataset can have minor imperfections. Incomplete, repeated, or out-of-range data are examples of these situations.
- **Validation**: Sometimes data that appears to be correct may be technically improbable, so they need a review to be able to verify it and make sure that there have been no errors in the information gathering process.

3.2.4 TYPES OF LEARNING

Once we have the data, we are ready to begin the learning process. This process, carried out by an algorithm, tries to analyze and explore the data in search of hidden patterns. The result of this learning, sometimes, is nothing more than a function that

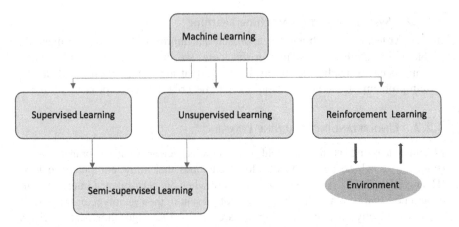

FIGURE 3.2 Types of learning.

operates on the data to calculate a certain prediction. Depending on the characteristics of the available data and the task we want to tackle, we can choose
between different types of learning. These are supervised learning, unsupervised
learning, semi-supervised learning, and reinforcement learning (Figure 3.2).

- **Supervised learning** needs labeled data sets, that is, we tell the model
  what we want it to learn (Singh et al., 2016). Suppose we have a database
  of electrocardiograms of patients who suffered an episode of heart failure
  in the 30-day period after this test was performed. And on the other hand,
  we have a database of healthy electrocardiograms without episodes of
  heart failure (Figure 3.3).
- **Unsupervised learning** works with data that has not been labeled. These
  algorithms are used mainly in tasks where it is necessary to analyze the
  data to extract new knowledge or group entities by affinity (Kiran et al.,
  2018). As an example of unsupervised learning, we have the clustering
  algorithms, which could be applied to search for patients with similar or
  similar characteristics in a clinical database and find causal relationships.
- **Semi-supervised learning** is a technique that combines the previous two.
  Sometimes you have a large set of data, but we only have a part of it
  labeled (Kiran et al., 2018). In this case, we have to use unsupervised
  learning techniques, but we take advantage of the data set that is labeled.
  On the one hand, an agent is trained with the tagged data and on the other
  hand, another agent is trained with the unlabeled data to find similarities.
  Going back to the EKG example, suppose we have a part of the data set
  labeled with patients who have had an episode of HF and another part a set
  of EKGs that we do not really know if they have had an episode of HF.
  First, we used unsupervised learning to find similarities in unlabeled
  electrons and supervised learning in labeled electrocardiograms.
  Subsequently, an algorithm evaluates the similarities found in the labeled
  data set to be able to associate a meaning or a causal relationship with it.

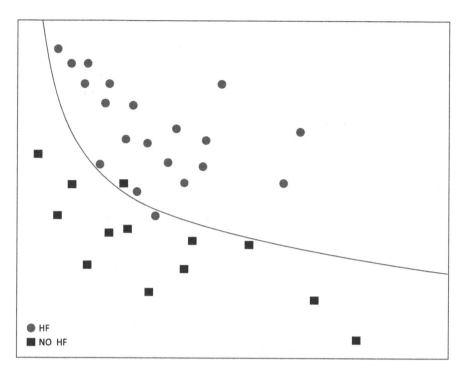

FIGURE 3.3   Unsupervised learning for HF detection.

• **Reinforcement learning** is inspired by behavioral psychology and determines what actions an agent must choose in their environment to obtain
  a reward or prize. Each individual action is associated with a reward or a
  penalty. The agent looks for a strategy that leads to the maximum number
  of rewards and minimum penalties. The problem to be solved must be
  formulated as a Markov chain decision process (Arulkumaran et al., 2017).
  This technique is usually very effective, but it is only possible to use it
  when the cost of experimentation is very low, and it is widely used in
  applications of marketing recommendations and game theory.

3.2.5   Most Commonly Used AI Techniques for Heart Failure

In this section we will briefly describe the AI techniques most used in HF applications and that are: decision trees, random forests, neural networks, deep learning,
support vector machines, clustering algorithms, and Bayesian networks.

A **decision tree** is a classic prediction model in which diagrams of logical
constructions are built, very similar to rule-based prediction systems, which serve to
represent and categorize a series of conditions that occur in succession, for the
resolution of a problem. It has the great advantage that they are transparent in the
description of the resolution of the problem, and therefore it is perfectly understood
how the solution is reached. Its main drawback is that its resolution capacity is low
(Saeys et al., 2007).

The **random forest** consists of a combination of decision trees to enhance their resolution capacity. It is a variant of the "bagging" technique in which a long collection of uncorrelated trees is built and subsequently averaged, in which each tree depends on its assigned value in a random vector (Breiman, 2001).

Artificial neural networks are a machine learning paradigm inspired by the neurons of biological nervous systems. It is a system of neuron links that collaborate with each other to produce an output stimulus. The connections have numerical weights that are adapted according to experience. In this way, neural networks adapt to an impulse and are capable of learning (Ripley, 2014).

Support vector machines are a combination of a set of supervised learning algorithms and are mainly applied to classification and regression problems. An SVM builds a set of hyperplanes in a space with high dimensionality in order to achieve an optimal separation between the different classes (Burges, 1998).

Deep Learning consists of a combination of learning algorithms through a cascading layered architecture. This combination manages to significantly increase the resolution capacity of the different learning algorithms in isolation or independently (Schmidhuber, 2015). Each layer is characterized by:

- Obtain the input information from the output of the previous layer
- Be able to use supervised or unsupervised learning.
- Be associated with a certain level of the extraction of characteristics or representation of the data.

This technique is often used for data modeling and pattern recognition. Its essence is fractional learning at multiple levels of features or data representations. These different levels are arranged in a hierarchy. The system learns from the multiple levels of representation that correspond to different levels of abstraction and finally a hierarchy of concepts is obtained. For example, if the input information is an electrocardiogram, the first level of representation would be in charge of extracting the wave and QRS parameters. The next level with these data could make a description of anomalies, and finally the third level a probability relationship with possible pathologies.

Clustering is the classification of observations into subgroups so that the observations in each group resemble each other according to certain criteria. Clustering techniques make different inferences about the structure of the data; they are usually guided by a specific similarity measure and by a level of internal compaction (similarity between members of a group) and the separation between different groups (Jain et al., 1999). Clustering is an unsupervised learning method and is a very popular technique for statistical data analysis.

A **Bayesian network** is a probabilistic model that represents a series of random variables and their conditional independencies through a directed acyclic graph (Friedman et al., 1997). A Bayesian network can represent, for example, the probabilistic relationships between diseases and symptoms. Given certain symptoms, the network can be used to calculate the probabilities that certain diseases are present in an organism. There are efficient algorithms that infer and learn using this type of representation.

3.3 DIAGNOSING HEART FAILURE

The diagnosis of HF requires symptoms and signs collected through patient's history and physical examination, laboratory data as well as objective demonstration of the structural or functional cardiac alterations as the primary underlying cause (mainly with laboratory and imaging data). ML-based methods aim to improve diagnosis through leveraging data found from each of these areas, including electrocardiography, echocardiography, electronic health record data, heart sounds, biomarkers, and other sources.

3.3.1 ELECTROCARDIOGRAPHY

The importance of electrocardiography (ECG) in the workup of HF is well established: 99% of patients with HF have abnormal ECG (Yancy et al., 2013; Ponikowski et al., 2016). As changes in electrical activity are due to physiologic changes, several groups have hypothesized that applying ML to these tracings will lead to more accurate detection of HF, and even more accuracy in the early stages of the disease (i.e. asymptomatic carriers of pathogenic mutations related to familial cardiopathies).

ECG have shown significant promise for risk stratification and diagnosis of HF. Some of the earlier studies using ECGs created algorithms to differentiate between healthy patients and those with HF, most using publicly available online ECG databases. These databases consist of a small set of patients that were either healthy and in normal sinus rhythm or had a known diagnosis of HF (Acharya et al., 2019; Wang and Zhou, 2019). The focus of these studies was the improvement and validation of ML methods, but the small size of these databases limits their application to clinical uses.

Other groups have evaluated larger and more varied data sets for the detection of HF: Attia et al. trained and validated a convolutional neural network on ECG data from 44,959 ECG–transthoracic echocardiogram (TTE) pairs to screen patients with asymptomatic left ventricular dysfunction and subsequently tested it on a separate set of 52,870 ECG-TTE pairs. The ML model predicted the presence of ventricular dysfunction with an area under the curve (AUC) of 0.93 and with a sensitivity, specificity, and accuracy of 86.3%, 85.7%, and 85.7%, respectively. In addition, patients who were identified as having an abnormal ECG were shown to have a four-fold risk of developing dysfunction in the future as compared to those who had a normal ECG (Attia et al., 2019).

Kwon et al. reported using a similar algorithm for detecting either HF with reduced ejection fraction (HFrEF) (Kwon et al., 2019). They incorporated a total of 55,163 ECG-TTE pairs from two separate hospitals, using the first hospital for both training and internal validation and the second hospital for external validation. For detection of HFrEF, they report AUCs for the internal and external validation sets of 0.84 and 0.88.

More recently, Adedinsewo et al. retrospectively studied an ECG algorithm for the identification of HFrEF in a cohort presenting to the emergency department with breathlessness. This algorithm employed had an AUC 0.89 as well as sensitivity, specificity, negative predictive value, and positive predictive value of 74%, 87%, 97%, and 40%, respectively (Adedinsewo et al., 2020).

A recent work investigated the possibility of detecting HF by extracting multi-modal features and employing ML techniques analyzing heart rate variability from ECGs, because when this parameter is reduced, it is considered to be a predictor of negative cardiovascular outcomes in HF patients. Performance was evaluated by measuring specificity, sensitivity, positive predictive value, negative predictive value, and AUC. The proposed approach can provide an effective and efficient tool for the automatic detection of HF patients (Hussain et al., 2020).

ECG evaluation by ML is rapidly evolving with novel features being isolated from ECG wavelet transformation techniques, which use the wavelet energy in the peak of the ventricular depolarization complex consisting of Q, R, and S waves to scale the appearance of the energy in the T-wave display (Sengupta et al., 2018). This permits the extraction of features associated with reduced early diastolic myocardial relaxation velocity indicative of LV diastolic dysfunction. Such ML models may soon provide a cost-effective screening method for the early detection of HF with LV diastolic dysfunction.

3.3.2 ECHOCARDIOGRAPHY

Echocardiography is the standard tool for HF characterization and management. However, the interpretation of echocardiographic images has levels of subjectivity and inter-reader reliability, and in many cases, as HF with preserved ejection fraction, different parameters are required to be considered and its interpretation is not easy. ML techniques have shown excellent ability in medical image processing and analysis and there has been considerable interest in applying these same successes to echocardiography.

In 2018, Zhang et al. created a model that automatically evaluated >14,000 echocardiographs to measure cardiac structure and function using a convolutional neural network (Zhang et al., 2018). Among many features of their algorithm, the model automatically calculated left ventricular ejection fraction (LVEF) and longitudinal strain (LS), which showed respective median absolute differences of 6% and 1.4% when compared with manual tracings.

Asch et al., with a different deep learning approach, developed and trained an ML algorithm to estimate EF while circumventing standard border detection techniques, under a database of more than 50,000 echocardiograms (Asch et al., 2019). Automatically calculated EF showed a mean absolute deviation of 2.9% and excellent agreement with the expert-derived values (r = 0.95). In addition, the algorithm was tested to detect EF ≤ 35%, which it did so with a sensitivity and specificity of 0.93 and 0.87.

ML has also been applied to standard features in echocardiography reports for the detection and evaluation of heart failure, including heart failure with preserved ejection fraction (HFpEF). For example, Reddy et al. used ML techniques to identify clinical and echocardiographic variables that were the most predictive of having a diagnosis of HFpEF (Reddy et al., 2018). Echocardiographic variables under consideration included features such as the estimated pulmonary artery filling pressure and E/e′ ratio. Using six variables selected by their ML algorithm, they created a risk score that resulted in an AUC of 0.841 for the diagnosis of HFpEF. This was notably higher than the performance of the European Society of Cardiology's 2016 algorithm for HFpEF detection on the same set of patients (AUC of 0.672), whose algorithm relies on variables such as BNP and other echocardiographic or imaging features. Creating a risk score based on ML-selected features allows for easier incorporation into clinical use, as these scores often are based on easy-to-obtain features.

However, they face similar requirements for external validation and generally have lower performance than a full-feature algorithm.

In a recent work, a machine learning (ML) algorithm utilizing both complex echocardiographic data and clinical parameters could be used to phenogroup an HF cohort and identify patients with beneficial response to cardiac resynchronization therapy (Cikes et al., 2019).

Combining clinic, biochemical, and echocardiographic data from 654 patients included in the TOPCAT clinical trial, Segar et al. got phenomapping patients with HF and preserved EF using ML-based unsupervised cluster analysis. They described three different phenotypes. Phenogroup 1 had higher burden of co-morbidities, natriuretic peptides, and abnormalities in left ventricular structure and function; phenogroup 2 had lower prevalence of cardiovascular and non-cardiac co-morbidities but higher burden of diastolic dysfunction; and phenogroup 3 had lower natriuretic peptide levels, intermediate co-morbidity burden, and the most favorable diastolic function profile. In adjusted Cox models, participants in phenogroup 1 (vs. phenogroup 3) had a significantly higher risk for all adverse clinical events including the primary composite endpoint, all-cause mortality, and HF hospitalization. Phenogroup 2 (vs. phenogroup 3) was significantly associated with a higher risk of HF hospitalization but a lower risk of atherosclerotic event (myocardial infarction, stroke, or cardiovascular death), and a comparable risk of mortality. These conclusions were validated in the non-echocardiographic cohort or the TOPCAT study and in the external cohort of the RELAX trial (Segar et al., 2020).

3.3.3   Electronic Health-Related Data (EHRD)

The use of EHRD is available in widespread hospital systems. Patient care has resulted in the assimilation of a vast amount of available medical data, which have repeatedly proven valuable in the development of diagnostic algorithms and care pathways. The early identification of hospitalized patients with HF, particularly those previously unrecognized, has the potential to markedly improve care and aid in overall cost reduction. This has also prompted the evaluation of the minimum amount and type of data needed to train effective disease onset predictive ML models using longitudinal EHRD.

In 2016, Blecker et al compared several supervised ML algorithms in identifying patients currently hospitalized with HF using patient demographics, laboratory tests, and other clinical data (Blecker et al., 2016). Their ML model outperformed the ability of multiple non-ML algorithms, which included the current problem list, having a BNP of 500 pg/mL or higher, and an LR model of routine clinical variables. This analysis included over 47,000 hospitalizations, 13% (6549) carried the diagnosis of HF. This group compared ML results to multiple traditional algorithms utilizing problem lists, medications, brain natriuretic peptide, and a logistic regression model of the clinical variable. They reported an AUC of 0.974, with a sensitivity of 83% and a positive predictive value of 90% for the identification of hospitalized HF patients.

Other groups have incorporated natural-language processing (NLP) algorithms for the identification of hospitalized HF patients, where NLP is a subfield of AI that often relies on supervised and DL techniques to process free text (such as the text in unstructured medical reports). By incorporating NLP techniques into their identification algorithm, Evans et al improved detection of hospitalized HF patients, leading to a sensitivity of 95.3% and PPV of 97.5%, an increase of greater than 10% sensitivity when compared with their previous algorithm using structured clinical data alone (Evans et al., 2016).

A recent study checked the feasibility of the combined approach of untargeted metabolomics and ML to create a simple and potentially useful diagnostic panel for HFrEF. The study included 67 chronic HFrEF patients and 39 controls. Fasting serum samples were fingerprinted by liquid chromatography-mass spectrometry. Feature selection based on random-forest models fitted to resampled data and followed by linear modeling, resulted in the selection of eight metabolites (uric acid, two isomers of LPC 18:2, LPC 20:1, deoxycholic acid, docosahexaenoic acid and one unknown metabolite), demonstrating their predictive value. The accuracy of a model based on metabolites panel was comparable to BNP (0.85 vs 0.82) (Marcinkiewicz-Siemion et al., 2020).

3.4   HF SUBTYPES CLASSIFICATION AND CLUSTERING

HF manifests at least two major subtypes, which are commonly distinguished based on the measurement of the left ventricular ejection fraction (LVEF). Patients with LVEF larger or equal to 50% are characterized as patients with HF with preserved ejection fraction (HFpEF), while patients with LVEF lower than 40% are characterized as patients with HF with reduced ejection fraction (HFrEF). Machine learning techniques have been applied to classify HF subtypes.

Austin et al. classified HF patients according to two disease subtypes (HFpEF vs. HFrEF) using different classification methods. The training of the classifiers was performed using the EFFECT-1 sample of Enhanced Feedback for Cardiac Treatment (EFFECT) study, while for the validation of the classifiers the EFFECT-2 sample was used. Removing subjects with missing values and subjects whom ejection fraction could not be determined, 3.697 patients for training and 4.515 patients for testing were finally employed. For each patient, 34 variables were recorded expressing information regarding demographic characteristics, vital signs, presenting signs and symptoms, laboratory data, and previous medical history. The results indicate that patients can be classified into one of the two mutually exclusive subtypes with a 69.6% positive predictive value using the Random Forests classifier (Austin et al., 2013).

The gray zone ambiguity arises whenever the left ventricular ejection fraction is between 40 and 50%. This ambiguity hinders the classification of heart failure as HFpEF or HFrEF. Alonso-Betanzos et al. used both the metric EF and the basic variables that define the EF, systolic volume (ESV), and end-diastolic volume (EDV). The authors employed two datasets for the evaluation of the above-mentioned ML techniques. The first dataset included data from 48 real patients (35 belong to the class HFpEF and 13 to the class HFrEF), while the second dataset includes simulated data, generated using Monte Carlo simulation approach, that correspond to 63 instances (34 from class HFpEF and 29 from class HFrEF). The results of the unsupervised methods revealed interesting dividing patterns of the two subtypes. Both supervised and unsupervised machine learning algorithms were used and demonstrated that selected machine learning models offer promise for classifying heart failure patients (including the gray zone) whenever driven by ventricular volume data (Alonso-Betanzos et al., 2015).

Shah et al. focused on the distinction of HFpEF subtypes. They employed 397 HFpEF patients and performed detailed clinical, laboratory, electrocardiographic phenotyping of the participating patients. The extracted 67 continuous variables were given as input to statistical learning algorithms and penalized model-based clustering. The analysis revealed three distinct pheno-groups in terms of clinical characteristics, cardiac structure and function, hemodynamics and outcomes (Shah et al., 2015).

HFpEF is a heterogeneous syndrome. A recently published work, based on clinical and echocardiogram data using ML, compared characteristics, proteomics and outcomes across phenogroups, applying a model-based clustering to 32 echocardiogram and 11 clinical and laboratory variables collected in stable conditions from 320 HFpEF Swedish outpatients. Six phenogroups were identified, for which significant differences in the prevalence o concomitant atrial fibrillation, anemia, and kidney disease were observed. Fifteen out of 86 plasma proteins differed between phenogroups, including biomarkers of HF, atrial fibrillation, and kidney function. The prognosis was significantly different between phenogroups at short-term, mid-term, and long-term follow-up (Hedman et al., 2020).

In a similar way, Kagiyama et al. recently published the results of a new work that demonstrated the value of ML to assess the presence of left ventricular dysfunction based on ECG features. A multicenter prospective study was conducted at four institutions enrolling a total of 1,202 patients. Patients from three institutions (814) formed an internal cohort and were randomly divided into training and internal test (80:20). ML models were developed using signal-processed ECG, traditional ECG, and clinical features. Data from the fourth institution was reserved as an external test set (388) to evaluate the model generalizability. They demonstrated that a quantitative prediction of myocardial relaxation can be performed using easily obtained clinical and ECG features, which could be a cost-effective strategy as a first clinical step for assessing the presence of HFpEF (Kagiyama et al., 2020).

3.5 STRATIFICATION RISK AND PRECISION MEDICINE IN HEART FAILURE

In addition to improving the diagnosis of HF, ML techniques would be able to tailor therapy to individual patients for improved outcomes. Areas in which HF research seeks to improve precision include stratifying patients and identifying those who will benefit from advanced treatments, such as cardiac resynchronization therapy (CRT) and left ventricular assist device (LVAD) therapy.

A major component of precision medicine includes making accurate predictions for patients based on personal characteristics and clinical data. Predicting the incidence and outcome of patients with HF is difficult, notwithstanding many risk scores reported in recent decades.

Sax et al. used a retrospective cohort design to identify all emergency department visits for acute HF during 2 years in 21 hospitals. They tested whether predictive accuracy in a different population could be enhanced with additional electronic health-record-based variables or ML approaches. Among 26,189 total encounters, the base model had an AUC of 0.76. Incorporating additional variables led to improved accuracy with logistic regression to 0.80 and 0.85 with ML. They found that using the ML model improved 30-day-risk prediction compared with the conventional approach: 11.1%, 25.7%, and 49.9% of the study population had predicted serious adverse events risk of less than 3%, 5%, and 10%, respectively; and 28% of those with less than 3% were admitted at 30 days (Sax et al., 2021).

In other recent work, data from 299 patients with HF were collected in 2015. Several ML classifiers were applied to predict the patient's survival. An alternative feature ranking analysis by traditional biostatistics test was performed and compared these results with those provided by the ML algorithms. Both feature ranking approaches clearly identify serum creatinine and ejection fraction as the two most relevant features, and ML survival prediction models on these two factors alone were built, and demonstrated that serum creatinine and ejection fraction were sufficient to predict survival (Chicco & Jurman, 2020).

Not all the ML approaches are similar: An Australian study developed a multilayer perceptron-based approach to predict 30 days HF readmission or death and compared the predictive performances using area under the precision-recall curve, sensitivity and specificity with other ML and regressions models. 10,757 patients were included, and the novel approach produced the highest area under the curve (0.62) and area under the precision-recall curve (0.46) with 48% sensitivity and 70% specificity (Awan et al., 2019).

Other recent work included 312 consecutive patients referred for transthoracic echocardiography and subsequently diagnosed with HF. Two ML techniques, multilayer perception, and multi-task learning were compared with logistic regression for their ability to predict all-cause mortality. Both ML approaches showed better prediction performance than logistic regression; so ML will provide additional value for improving outcome prediction (Tse et al., 2020).

However, a recent meta-analysis that included 686,842 patients with different ML methods (random forest, decision trees, regression trees, support vector machines, neural networks, and Bayesian techniques, suggested that an external validation of ML-based studies of prediction modeling is needed, and that ML-based studies should also be evaluated using clinical quality standards for prognosis research (Shin et al., 2021).

Adler et al. used an ML algorithm (by training a boosted decision tree algorithm to relate a subset of the patient data with a very high or very low mortality risk) to capture correlations between patient characteristics and mortality. 5,822 hospitalized and ambulatory patients with HF were included. From this model a risk score was derived; and it was able to discriminate between low and high risk of death by identifying eight variables (diastolic blood pressure, creatinine, blood urea nitrogen, hemoglobin, white blood cell count, platelets, albumin, and red blood cell distribution width) (Adler et al., 2020).

One of the hopes of ML development is the identification of responders to particular HF therapies. CRT is a therapy that has shown improving survival of selected HFrEF patients. However, a significant percentage of the patients are non-responders to the therapy. In a recent study performed in 1,106 HF patients included in the MADIT-CRT trial, an unsupervised ML algorithm was used to categorize subjects by similarities in clinical parameters, and left ventricular volume and deformation traces at baseline into mutually exclusive groups. The analysis identified two phenogroups (based on clinical features, biomarker values, measures of left and right ventricular structure and function) with a higher response to CRT (Cikes et al., 2019). In a similar way, ML has been shown to predict long-term mortality and graft failure of heart transplant recipients (Agasthi et al., 2020).

3.6 FUTURE APPLICATIONS AND PERSONALIZED MEDICINE

A higher proportion of patients with HF have benefitted from a wide and expanding variety of sensor-enabled implantable devices. These patients can now also take advantage of the ever-increasing availability and affordability of consumer electronics. Wearable, on- and near-body sensor technologies, much like implantable devices, generate massive amounts of data. The connectivity of all these devices has created opportunities for pooling data from multiple sensors and for AI to provide new diagnostic, triage, risk-stratification, and disease management insights for the delivery of better, more personalized, and cost-effective healthcare. Implantable cardiac sensors have shown promise in reducing rehospitalization for HF, but the efficacy of noninvasive approaches has not been determined.

The addition of further sensors to devices that are already being implanted poses a significant opportunity. Cardiac pacemakers or defibrillators provide such a platform. However, clinically unutilized data could also be captured from existing pacemaker sensors. As established, physical activity is highly predictive of cardiovascular outcomes. Internal accelerometers, incorporated for the primary purpose of rate-responsive pacing, passively generate low-cost data that can serve as a surrogate of physical activity, highlighting an as-yet untapped clinical and ML opportunity that could inform interventions (Bachtiger et al., 2020).

CRT devices have already demonstrated capacity for additional sensing features, including intrathoracic impedance. Rising impedance can be used to stratify patients into varying mortality risks and serves as a superior measure by preceding weight gain by 2 weeks, predicting an increased risk of HF hospitalization sooner.

Several newer implantable sensor technologies include measuring pulmonary artery pressure that is directly deployed into the pulmonary artery (CardioMEMs, Abbot). Using home transmission with an implanted pressure sensor, long-term hospital admission rates showed significant improvements, with the reduction in rates of admissions to hospital by 33% compared with a control group (Abraham et al., 2016).

Raised left atrial pressure is the most specific and earliest sign of impending HF exacerbation, long before clinical symptoms occur. V-LAP (Vectorious Medical Technologies) is a miniature, wireless, and battery-free microcomputer for cardiac monitoring in HF, able to directly monitor the left atrial pressure. Ex-vivo and animal findings have been published and the first human study of the device is underway (Perl et al., 2019). The device is designed to be inserted in the interatrial septum and aims to permit data collection daily. The sensor is not powered by batteries and can work for the life of the patient. It is charged remotely and transmits the patient's cardiac activity information wirelessly to doctors and to the hospital, where it is analyzed by cardiologists.

The LINK-HF study (Multisensor Non-invasive Remote Monitoring for Prediction of Heart Failure Exacerbation) examined the performance of a personalized analytical platform using continuous data streams to predict rehospitalization after HF admission. Study subjects were monitored for up to 3 months using a disposable multisensor patch placed on the chest that recorded physiological data. Data were uploaded continuously via smartphone to a cloud analytics platform. Machine learning was used to design a prognostic algorithm to detect HF exacerbation. A hundred subjects were enrolled. After discharge, the analytical platform derived a personalized baseline model of expected physiological values. Differences between baseline model estimated vital signs and actual monitored values were used to trigger a clinical alert.

The platform was able to detect precursors of hospitalization for HF exacerbation with 76–88% sensitivity and 85% specificity. The median time between initial alert and readmission was 6.5 days (Stehlik et al., 2020).

A new algorithm to monitor HF patients was tested in the Multisensor Chronic Assessment in Ambulatory Heart Failure study (MultiSENSE) study. The HeartLogic index combines data from multiple sensors (first and third accelerometer-based heart sounds, intrathoracic impedance, respiratory rate, the ratio between respiratory rate and tidal volume, nocturnal heart rate, and patient activity) integrated with the implantable defibrillator and has proved to be a sensitive, timely and efficient predictor of HF decompensation. This device calculates daily a composite index by integrating inputs received from the sensors. The activation of the associated alert can enable early detection of clinical deterioration and suggest an action to be taken in patients who are not yet critical, potentially preventing adverse events by stratifying their risk of occurrence (Boehmer et al., 2017).

The risk of potential dehumanization of medicine because of the increased distance between the physician and the patient has to be evaluated with caution. Telehealth should be used as a tool to improve patient care and serve as an adjunct to – rather than a replacement for – face-to-face contact. Telehealth, in particular, may exert a difference for patients who have physical disabilities or those with financial difficulties in traveling long distances. Telemedicine platforms might offer healthcare providers the opportunity to contact patients across long distances, reaching those in the most isolated locations. The topic of the evolving relationship between the physician and patient has been also recently addressed.

3.7 CONCLUSIONS

ML, AI, and Telemedicine allow a complete assessment of cardiac function, which is critical in HF patients and will become a basic tool to optimize the treatment of these patients. The combination of AI, big data, and massively parallel computing offer the ability to create a way of practicing evidence-based, cost-effective, and personalized medicine.

AI represents a different way to yield evidence for clinical practice, and deep learning challenges the usual ways by which the medical community has achieved scientific consensus to date. Because of the epidemic proportion of HF as a clinical syndrome, the ability to process big data from public health registries represents an ongoing opportunity for uncovering the evolving scope of the disease.

REFERENCES

Abraham, W. T., Stevenson, L. W., Bourge, R. C., Lindenfeld, J. A., Bauman, J. G., & Adamson, P. B. (2016). Sustained Efficacy of Pulmonary Artery Pressure to Guide Adjustment of Chronic Heart Failure Therapy: Complete Follow-up Results from the CHAMPION Randomised Trial. *The Lancet*, *387*(10017). doi:10.1016/S0140-6736(15)00723-0.

Acharya, U. R., Fujita, H., Oh, S. L., et al. (2019). Deep Convolutional Neural Network for the Automated Diagnosis of Congestive Heart Failure Using ECG Signals. *Applied Intelligence*, *49*, 16–27.

Adedinsewo, D., Carter, R. E., Attia, Z., Johnson, P., Kashou, A. H., Dugan, J. L., Albus, M., et al. (2020). Artificial Intelligence-Enabled ECG Algorithm to Identify Patients With Left Ventricular Systolic Dysfunction Presenting to the Emergency Department With Dyspnea. *Circulation: Arrhythmia and Electrophysiology*, *13*(8). doi:10.1161/CIRCEP. 120.008437.

Adler, E. D., Voors, A. A., Klein, L., Macheret, F., Braun, O. O., Urey, M. A., Zhu, W., et al. (2020). Improving Risk Prediction in Heart Failure Using Machine Learning. *European Journal of Heart Failure*, *22*(1), 139–147. doi:10.1002/ejhf.1628.

Agasthi, P., Buras, M. R., Smith, S. D., Golafshar, M. A., Mookadam, F., Anand, S., Rosenthal, J. L., Hardaway, B. W., DeValeria, P., & Arsanjani, R. (2020). Machine Learning Helps Predict Long-Term Mortality and Graft Failure in Patients Undergoing Heart Transplant. *General Thoracic and Cardiovascular Surgery*, *68*(12). doi:10.1007/s11748-020-01375-6.

Alonso-Betanzos, A., Bolón-Canedo, V., Heyndrickx, G. R., & Kerkhof, P. L.M.. (2015). Exploring Guidelines for Classification of Major Heart Failure Subtypes by Using Machine Learning. *Clinical Medicine Insights: Cardiology*, *9s1*(January). doi:10.4137/CMC.S18746.

Arulkumaran, K., Deisenroth, M. P., Brundage, M., & Bharath, A. A. (2017). Deep Reinforcement Learning: A Brief Survey. *IEEE Signal Processing Magazine*, *34*(6). doi:10.1109/MSP.2017.2743240.

Asch, F. M., Poilvert, N., Abraham, T., Jankowski, M., Cleve, J., Adams, M., Romano, N., et al. (2019). Automated Echocardiographic Quantification of Left Ventricular Ejection Fraction Without Volume Measurements Using a Machine Learning Algorithm Mimicking a Human Expert. *Circulation: Cardiovascular Imaging*, *12*(9). doi:10.11 61/CIRCIMAGING.119.009303.

Attia, Z. I., Kapa, S., Lopez-Jimenez, F., et al. (2019). Screening for Cardiac Contractile Dysfunction Using an Artificial Intelligence–Enabled Electrocardiogram. *Nat Med*, *25*(1), 70.

Austin, P. C., Tu, J. V., Ho, J. E., Levy, D., & Lee, D. S. (2013). Using Methods from the Data-Mining and Machine-Learning Literature for Disease Classification and Prediction: A Case Study Examining Classification of Heart Failure Subtypes. *Journal of Clinical Epidemiology*, *66*(4). doi:10.1016/j.jclinepi.2012.11.008.

Awan, S. E., Bennamoun, M., Sohel, F., Sanfilippo, F. M., & Dwivedi, G. (2019). Machine Learning-based Prediction of Heart Failure Readmission or Death: Implications of Choosing the Right Model and the Right Metrics. *ESC Heart Failure*, *6*(2). doi:10. 1002/ehf2.12419.

Bachtiger, P., Plymen, C. M., Pabari, P. A., Howard, J. P., Whinnett, Z. I., Opoku, F., Janering, S., Faisal, A. A., Francis, D. P., & Peters, N. S. (2020). Artificial Intelligence, Data Sensors and Interconnectivity: Future Opportunities for Heart Failure. *Cardiac Failure Review*, *6*(May). doi:10.15420/cfr.2019.14.

Blecker, S., Katz, S. D., Horwitz, L. I., Kuperman, G., Park, H., Gold, A., & Sontag, D.. (2016). Comparison of Approaches for Heart Failure Case Identification From Electronic Health Record Data. *JAMA Cardiology*, *1*(9). doi:10.1001/jamacardio.201 6.3236.

Boehmer, J. P., Hariharan, R., Devecchi, F. G., Smith, A. L., Molon, G., Capucci, A., An, Q., et al. (2017). A Multisensor Algorithm Predicts Heart Failure Events in Patients With Implanted Devices. *JACC: Heart Failure*, *5*(3). doi:10.1016/j.jchf.2016.12.011.

Breiman, L. (2001). Random Forests. *Machine Learning*, *45*(1), 5–32. doi:10.1023/A:101 0933404324.

Burges, C. J. C. (1998). A Tutorial on Support Vector Machines for Pattern Recognition. *Data Mining and Knowledge Discovery*, *2*(2), 121–167. doi:10.1023/A:1009715 923555.

Chicco, D., & Jurman, G. (2020). Machine Learning Can Predict Survival of Patients with Heart Failure from Serum Creatinine and Ejection Fraction Alone. *BMC Medical Informatics and Decision Making*, *20*(1). doi:10.1186/s12911-020-1023-5.

Cikes, M., Sanchez-Martinez, S., Claggett, B., Duchateau, N, Piella, G., Butakoff, C., Pouleur, A. C., et al. (2019). Machine Learning-Based Phenogrouping in Heart Failure to Identify Responders to Cardiac Resynchronization Therapy. *European Journal of Heart Failure*, *21*(1). doi:10.1002/ejhf.1333.

Dunlay, S. M., Roger, V. L., & Redfield, M. M. (2017). Epidemiology of Heart Failure with Preserved Ejection Fraction. *Nature Reviews Cardiology*, *14*, 591–602.

Evans, R. S., Benuzillo, J., Horne, B. D., Lloyd, J. F., Bradshaw, A., Budge, D., Rasmusson, K. D., et al. (2016). Automated Identification and Predictive Tools to Help Identify High-Risk Heart Failure Patients: Pilot Evaluation. *Journal of the American Medical Informatics Association*, *23*(5). doi:10.1093/jamia/ocv197.

Friedman, N., Geiger, D., & Goldszmidt, M. (1997). Bayesian Network Classifiers. *Machine Learning*, *29*(2–3). doi:10.1023/a:1007465528199.

Hedman, Å. K., Hage, C., Sharma, A., Brosnan, M. J., Buckbinder, L., Gan, L.-M., Shah, S. J., et al. (2020). Identification of Novel Pheno-Groups in Heart Failure with Preserved Ejection Fraction Using Machine Learning. *Heart*, *106*(5). doi:10.1136/heartjnl-2019-315481.

Huang, G., Huang, G. B., Song, S., & You, K. (2015). Trends in Extreme Learning Machines: A Review. *Neural Networks*, 61, 32–48. doi:10.1016/j.neunet.2014.10.001.

Hussain, L., Awan, I. A., Aziz, W., Saeed, S., Ali, A., Zeeshan, F., & Kwak, K. S. (2020). Detecting Congestive Heart Failure by Extracting Multimodal Features and Employing Machine Learning Techniques. *BioMed Research International*, *2020* (February). doi:10.1155/2020/4281243.

Jain, A. K., Murty, M. N., & Flynn, P. J. (1999). Data Clustering: A Review. In *ACM Computing Surveys*, *31*, 264–323. doi:10.1145/331499.331504.

Kagiyama, N., Piccirilli, M., Yanamala, N., Shrestha, S., Farjo, P. D., Casaclang-Verzosa, G., Tarhuni, W. M., et al. (2020). Machine Learning Assessment of Left Ventricular Diastolic Function Based on Electrocardiographic Features. *Journal of the American College of Cardiology*, *76*(8). doi:10.1016/j.jacc.2020.06.061.

Kaplan, A. M., & Haenlein, M. (2010). Users of the World, Unite! The Challenges and Opportunities of Social Media. *Business Horizons*, *53*(1), 59–68. doi:10.1016/j.bushor.2009.09.003.

Kiran, B. R., Thomas, D. M., & Parakkal, R. (2018). An Overview of Deep Learning Based Methods for Unsupervised and Semi-Supervised Anomaly Detection in Videos. *Journal of Imaging*, *4*(2). doi:10.3390/jimaging4020036.

Kwon, J. M., Kim, K. H., Jeon, K. H., et al. (2019). Development and Validation of Deep-Learning Algorithm for Electrocardiographybased Heart Failure Identification. *Korean Circulation Journal*, *49*(7), 629–639.

Marcinkiewicz-Siemion, M., Kaminski, M., Ciborowski, M., Ptaszynska-Kopczynska, K., Szpakowicz, A., Lisowska, A., Jasiewicz, M., et al. (2020). Machine-Learning Facilitates Selection of a Novel Diagnostic Panel of Metabolites for the Detection of Heart Failure. *Scientific Reports*, *10*(1). doi:10.1038/s41598-019-56889-8.

Mohsin, A. H., Zaidan, A. A., Zaidan, B. B., Albahri, A. S., Albahri, O. S., Alsalem, M. A., & Mohammed, K. I. (2018). Real-Time Remote Health Monitoring Systems Using Body Sensor Information and Finger Vein Biometric Verification: A Multi-Layer Systematic Review. *Journal of Medical Systems*, *42*(12). doi:10.1007/s10916-018-1104-5.

Perl, L., Soifer, E., Bartunek, J., Erdheim, D., Köhler, F., Abraham, W. T., & Meerkin, D. (2019). A Novel Wireless Left Atrial Pressure Monitoring System for Patients with Heart Failure, First Ex-Vivo and Animal Experience. *Journal of Cardiovascular Translational Research*, *12*(4). doi:10.1007/s12265-018-9856-3.

Ponikowski, P., Voors, A. A., Anker, S. D., et al. (2016). 2016 ESC Guidelines for the Diagnosis and Treatment of Acute and Chronic Heart Failure: The Task Force for the Diagnosis and Treatment of Acute and Chronic Heart Failure of the European Society of Cardiology (ESC). Developed with the Special Contribution of the Heart Failure Association (HFA) of the ESC. *European Journal of Heart*, *18*, 891–975.

Ramírez-Gallego, S., Krawczyk, B., García, S., Woźniak, M., & Herrera, F. (2017). A Survey on Data Preprocessing for Data Stream Mining: Current Status and Future Directions. *Neurocomputing*, *239* (May), 39–57. doi:10.1016/j.neucom.2017.01.078.

Reddy, Y. N. V., Carter, R. E., Obokata, M., Redfield, M. M., & Borlaug, B. A. (2018). A Simple, Evidence-Based Approach to Help Guide Diagnosis of Heart Failure With Preserved Ejection Fraction. *Circulation*, *138*(9). doi:10.1161/CIRCULATIONAHA.118.034646.

Ripley, B. D. (2014). *Pattern Recognition and Neural Networks*. Cambridge University Press. doi:10.1017/CBO9780511812651.

Russel, S., & Norvig, P. (2012). *Artificial Intelligence—a Modern Approach 3rd Edition. The Knowledge Engineering Review*. Pearson Education Limited.

Saeys, Y., Inza, I., & Larrañaga, P. (2007). A Review of Feature Selection Techniques in Bioinformatics. *Bioinformatics*, 23(19), 2524–2507. doi:10.1093/bioinformatics/btm344.

Sanders, W. E., Burton, T., Khosousi, A., & Ramchandani, S. (2021). Machine Learning: At the Heart of Failure Diagnosis. *Current Opinion in Cardiology*, *36*(2), 227–233. doi:10.1097/HCO.0000000000000833.

Sax, D. R., Mark, D. G., Huang, J., Sofrygin, O., Rana, J. S., Collins, S. P., Storrow, A. B., Liu, D., & Reed, M. E. (2021). Use of Machine Learning to Develop a Risk-Stratification Tool for Emergency Department Patients With Acute Heart Failure. *Annals of Emergency Medicine*, *77*(2). doi:10.1016/j.annemergmed.2020.09.436.

Schmidhuber, J. (2015). Deep Learning in Neural Networks: An Overview. *Neural Networks*, *61*, 85–117. doi:10.1016/j.neunet.2014.09.003.

Segar, M. W., Patel, K. V., Ayers, C., Basit, M., Tang, W.H. W., Willett, D., Berry, J., Grodin, J. L., & Pandey, A. (2020). Phenomapping of Patients with Heart Failure with Preserved Ejection Fraction Using Machine Learning-based Unsupervised Cluster Analysis. *European Journal of Heart Failure*, *22*(1). doi:10.1002/ejhf.1621.

Sengupta, P. P., Kulkarni, H., & Narula, J. (2018). Prediction of Abnormal Myocardial Relaxation From Signal Processed Surface ECG. *Journal of the American College of Cardiology*, *71*(15). doi:10.1016/j.jacc.2018.02.024.

Shah, S. J., Katz, D. H., Selvaraj, S., Burke, M. A., Yancy, C. W., Gheorghiade, M., Bonow, R. O., Huang, Chiang-Ching, & Deo, R. C. (2015). Phenomapping for Novel Classification of Heart Failure With Preserved Ejection Fraction. *Circulation*, *131*(3). doi:10.1161/CIRCULATIONAHA.114.010637.

Shin, S., Austin, P. C., Ross, H. J., Abdel-Qadir, H., Freitas, C., Tomlinson, G., Chicco, D., et al. (2021). Machine Learning vs. Conventional Statistical Models for Predicting Heart Failure Readmission and Mortality. *ESC Heart Failure*, *8*(1). doi:10.1002/ehf2.13073.

Singh, A., Thakur, N., & Sharma, A. (2016). A Review of Supervised Machine Learning Algorithms. *Proceedings of the 10th INDIACom; 2016 3rd International Conference on Computing for Sustainable Global Development, INDIACom 2016*, 1310–1315. Institute of Electrical and Electronics Engineers Inc.

Stehlik, J., Schmalfuss, C., Bozkurt, B., Nativi-Nicolau, J., Wohlfahrt, P., Wegerich, S., Rose, K., et al. (2020). Continuous Wearable Monitoring Analytics Predict Heart Failure Hospitalization. *Circulation: Heart Failure*, *13*(3). doi:10.1161/CIRCHEARTFAILURE.119.006513.

Tse, G., Zhou, J., Woo, S. W. D., Ko, C. H., Lai, R. W. C., Liu, T., Liu, Y., et al. (2020). Multi-modality Machine Learning Approach for Risk Stratification in Heart Failure with Left Ventricular Ejection Fraction ≤ 45%. *ESC Heart Failure*, *7*(6). doi:10.1002/ehf2.12929.

Virani, S. S., Alonso, A., Benjamin, E. J., et al. (2020). Heart Disease and Stroke Statistics—2020 Update: A Report from the American Heart Association. *Circulation*, *141*, e139–e596.

Wagner, W. P. (2017). Trends in Expert System Development: A Longitudinal Content Analysis of over Thirty Years of Expert System Case Studies. *Expert Systems with Applications*, *76*(June), 85–96. doi:10.1016/j.eswa.2017.01.028.

Wang, L, & Zhou, X. (2019). Detection of Congestive Heart Failure Based on LSTM-Based Deep Network via Short-Term RR Intervals. *Sensors*, *19*(7), 1502.

Yancy, C. W., Jessup, M., Bozkurt, B., et al. (2013). 2013 ACCF/AHA Guideline for the Management of Heart Failure: A Report of the American College of Cardiology Foundation/American Heart Association Task Force on Practice Guidelines. *Journal of the American College of Cardiology*, *62*, e147–e239.

Zhang, J., Gajjala, S., Agrawal, P., Tison, G. H., Hallock, L. A., Beussink-Nelson, L., Lassen, M. H., et al. (2018). Fully Automated Echocardiogram Interpretation in Clinical Practice. *Circulation*, *138*(16). doi:10.1161/CIRCULATIONAHA.118.034338.

4 A Combination of Dilated Adversarial Convolutional Neural Network and Guided Active Contour Model for Left Ventricle Segmentation

Heming Yao, Jonathan Gryak, and
Kayvan Najarian

CONTENTS

DOI: 10.1201/9781003120902-4

4.1   INTRODUCTION

Cardiovascular disease is the number one cause of death globally, taking the lives of 17.7 million people every year, and comprises 31% of all global deaths (Organization & Unit, 2014). Early diagnosis and treatment of cardiovascular disease can significantly improve patient prognosis and quality of life. One of the critical diagnostic tools is Cardiovascular Magnetic Resonance (CMR) imaging. Due to its diagnostic accuracy and reproducibility, CMR is currently the standard used to quantitatively evaluate global and regional cardiac function. Based on CMR images, quantitative analysis can be performed to derive numerical physiological parameters such as end-systolic volume (ESV), end-diastolic volume (EDV), ejection fraction (EF), and left ventricle (LV) mass. These parameters can facilitate the diagnosis and management of a variety of cardiac diseases, including ischemic and non-ischemic cardiomyopathies (Schulz-Menger et al., 2013; Yusuf et al., 2003). Studies have also shown that these parameters are significant predictors of prognosis and can be used to guide the treatment of heart disease (Bluemke et al., 2008; Knauth et al., 2008; Sachdeva et al., 2015).

In practice, measurements of these parameters depend upon manual or semi-automatic delineation of endocardial and epicardial contours of the LV. However, manual endocardium and epicardium wall segmentation is very time-consuming and prone to high intra- and inter-observer variability, which ultimately leads to variability in subsequent assessments. A fast, accurate, and fully automated LV segmentation framework can overcome these issues while providing the following advantages. First, it can decrease medical costs by reducing the time for image analysis. Second, unlike manual delineations that are often performed only during end-diastole (ED) and end-systole (ES) (Schulz-Menger et al., 2013) due to the effort involved, fully automated methods can quickly generate the segmentation throughout the cardiac cycle to calculate more sophisticated physiological parameters such as peak ejection rate (Bacharach et al., 1979) for LV dyssynchrony detection. Third, in population analysis, automatic segmentation can generate vast amounts of quantitative data from images, while its integration with clinical data can facilitate a more extensive analysis of cardiovascular disease. For example, the segmentation of endocardial and epicardial contours can help derive LV wall thickness. An accurate LV wall thickness estimation can help monitor changes in thickness over the cardiac cycle or longitudinally. This can enhance the diagnostic and prognostic utility of estimation in patients with LV hypertrophy, cardiac arrhythmia, or coronary artery disease (Pouleur et al., 1984).

In this chapter, we first discuss the existing LV segmentation techniques and their limitations. After that, a novel LV segmentation method is proposed, which combines deep learning techniques and deformable models to improve the generalization and accuracy of LV segmentation. The chapter is organized as follows. Section 4.2 describes the existing methods for LV segmentation and the motivation for the proposed method. Section 4.3 gives a detailed description of the proposed segmentation framework and evaluation metrics. Section 4.4 provides the experimental setup, including hyper-parameter tuning, dataset information, and the validation process. Section 4.5 discusses evaluation results, with the performance of the proposed method compared with other published work. Finally, Section 4.6 concludes with the overall performance and efficacy of the proposed method. Potential limitations and future directions are also discussed.

4.2   LITERATURE REVIEW

Numerous techniques have been proposed for automated LV segmentation in CMR images. In this section, we review and summarize these techniques into several categories, with the common characteristics of methods within each category discussed.

4.2.1   IMAGE PROCESSING TECHNIQUES

Popular image processing methods include thresholding (Katouzian et al., 2006), intensity-based clustering (Lynch et al., 2006), dynamic programming-based border detection (Yeh et al., 2005), graph cut techniques (Cousty et al., 2009), and random walks (Eslami et al., 2013) have been applied to LV segmentation. In (Katouzian et al., 2006), a region of interest was selected semi-automatically, after which the endocardial and epicardial contours were identified by thresholding and morphological operations. In Pednekar et al. (2006), the difference in temporal intensity through a sequence of CMR images was calculated to identify the circular region near the LV boundaries. Then, an initial myocardial region estimation was obtained using a Gaussian mixture model calculated via expectation-maximization. After that, the estimated myocardial region was propagated along the temporal sequence with the help of an intensity-based fuzzy affinity map and a dynamic programming approach. A new formulation of the random walk was proposed in Eslami et al. (2013), which retrieves the closest sample in the training set to guide the segmentation. With this technique, rare cases in the training set can equally contribute to the segmentation of new cases.

Image processing techniques utilize intensity distribution variations to segment the LV cavity and myocardial region from other anatomical structures. Additionally, spatiotemporal information from the sequence of image slices are used to improve the robustness of region segmentation. Due to their dependency on the intensity distribution, these methods have difficulty dealing with noise, grayscale variations, or low gradients arising from technical issues, different imaging cohorts, or pathological tissues.

4.2.2 DEFORMABLE MODELS

Snakes or active contour model is an algorithm for contour detection. A contour is defined by a set of points. The contour will be iteratively updated to minimize the image energy. The energy function is usually defined as a weighted combination of internal forces such as rigidity, elasticity, and external edge-based or region-based forces (Kass et al., 1988; Lei et al., 2017; Li et al., 2010). The method in Yang et al. (2004) was motivated by the observation that neighboring anatomical structures have consistent locations and shapes, which can facilitate myocardial segmentation. An extended deformable model was proposed in Heiberg et al. (2012), which incorporates temporal information and a priori anatomical information for edge detection. A joint probability distribution over the variations of the neighboring objects is defined, and the contours can evolve according to the information from both neighboring objects and the image intensity. In Lee et al. (2010), iterative thresholding was first applied to segment the LV endocardium. After that, the epicardial contour was extracted using the active contour model guided by the estimated endocardial contour.

In deformable models, an optimal contour can be detected autonomously and adaptively by evolving the contour to minimize the energy function. However, the initialization of the contour and the convergence policy is crucial to success. Also, these methods have difficulty segmenting images with low contrast.

4.2.3 ATLAS AND REGISTRATION

In this class of segmentation techniques, an atlas that describes the structure of the target object is first generated from one or more manually delineated images. Image registration is a process of mapping the coordinates of one image to those of another image. Given an atlas, a new image can be segmented by image registration between the new image and the atlas. A three-step registration framework was proposed in Zhuang et al. (2010). A global affine registration was first applied to localize the heart, then a locally affine registration method was employed to initialize the substructures. Finally, a free-form deformation with adaptive control point status was proposed to refine the local details. The method achieved good performance on heart images with diverse morphology and pathology. In Lorenzo-Valdés et al. (2004), a 4D probabilistic heart atlas was constructed from 14 healthy cases, which encodes both spatial anatomical information and temporal information. With the 4D probabilistic atlas, the expectation-maximization algorithm was applied to perform the segmentation.

For atlas-guided segmentation techniques, constructing an atlas that captures the variety of anatomical structures is essential. A single atlas usually is insufficient, and enriching the atlas pool is an effective way to improve the segmentation performance. However, the multi-atlas segmentation method requires a considerable computational cost. In addition, the registration process does not impose anatomical constraints on the transformation. As a result, the transformation calculated from the registration process can be less anatomically valid.

4.2.4 DEEP LEARNING TECHNIQUES

More recently, many publications have successfully applied Convolutional Neural Networks (CNN) for automated LV segmentation, greatly improving segmentation accuracy and speed. An end-to-end fully convolutional network (FCN) model with 15 convolutional layers was proposed in Tran (2016). A transfer learning technique was applied to leverage the learned feature representation from a large dataset. In Poudel et al. (2016), a recurrent FCN incorporating the U-Net architecture and gated recurrent units modules was proposed to extract the spatial anatomical dependencies across adjacent slices. In Tan et al. (2017), the task of LV segmentation was parametrized using the radial distances between the LV center point and the endocardial and epicardial contours in polar space. Two CNNs were constructed – the first one localized the LV center point, while the second produced a regression model to infer the radial distances. This method achieved the highest published overall result in comparison to other automated algorithms on a public dataset (Suinesiaputra et al., 2014).

While these aforementioned deep learning methods have achieved the best results on various cardiac datasets, they are limited in two important ways. First, successive down-sampling layers have been used to enlarge the receptive field, and as a result, the resolution of the full-sized segmentation mask is impaired. Second, because segmentation is performed as pixel-wise classification, constraints from anatomical structures are not used, and the segmentation map may be less anatomically valid in low-contrast or pathological cases. In the context of LV segmentation, we have some prior anatomical knowledge, such as the LV wall being roughly circular in shape and the LV lumen comprising the interior of the LV and being brighter than other LV regions. Incorporating these constraints into LV segmentation can lead to a more accurate and efficient automatic segmentation.

4.2.5 HYBRID TECHNIQUES

There are several studies that combine methods from different categories to improve the robustness and accuracy of segmentation. In Hu et al. (2014), a local binary fitting model that combines edge and region information is first applied to detect the blood pool and thereby determine the endocardial contour. After that, a dynamic programming technique was used to detect the epicardial contour. In Avendi et al. (2016), an FCN was used to localize and segment the blood pool, after which a region-based deformable model that incorporates FCN output was applied to detect the epicardial and endocardial contours. The method uses the segmented blood pool from the FCN as an initialization, which provides a more robust and accurate segmentation, particularly for challenging basal and apical slices.

4.2.6 MOTIVATION OF THE PROPOSED METHOD

In this study, we present a fully automatic LV segmentation framework for CMR slices in the short axis. It aims to overcome the existing limitations in deep learning techniques. In the proposed framework, a dilated FCN (DFCN) is developed using dilated convolutions, where most of the up-sampling layers and down-sampling layers in a conventional FCN are removed to improve image resolution for pixel-wise prediction. In addition to the FCN, an adversarial network has been developed to extract global features by differentiating the entire automatic segmentation from the manual segmentation. The loss from the adversarial network is fused into the segmentation loss to integrate global information from segmented components and enforce higher-order consistency in shape inference. To further utilize intensity homogeneity characteristics, a guided active contour model (GACM) is proposed as a post-processing method for FCN output. While the combination of deep learning and deformable-model approaches is not new (Avendi et al., 2016), it is the first time that the active contour model has been extended for the segmentation of two contours simultaneously. The segmentation results of the proposed method were extensively evaluated with respect to contour quality as well as clinical parameter estimation. The proposed framework achieves the best performance on the LV2011 dataset and good performance in another independent dataset that is comparable to other extant methods. From our analysis, the proposed framework generalizes well for images from different vendors and with different patient pathologies.

4.3 METHODS

4.3.1 PREPROCESSING

An FCN performs pixel-wise prediction by extracting hierarchical features of receptive fields around the pixel. Thus, the scale of the LV in the image matters. To standardize CMR sequences from multiple sources, we resized the images using linear interpolation with the *Pixel Spacing* parameter in the DICOM metadata to make one pixel represent. As the original MRI slices have a large background region where all pixel values are zero and the chamber view is located at the center of the image, a 256×253 sub-region is cropped around the center of the image. Note that the described cropping is only used in the training phase expressly for the purpose of size standardization and reducing the training time. In the test phase, the whole resized image will be fed into the network to generate the segmentation. (Note: As no fully connected layer is involved in the proposed segmentation network, a fixed input size is not required during the test phase.) An illustration of the preprocessing procedure in both the training and test phases is given in Figure 4.1(a).

4.3.2 CONVOLUTIONAL NEURAL NETWORK

Figure 4.1 provides detailed information on the proposed networks. A dilated FCN (Figure 4.1(b)), which we call the *Segmentor*, performs segmentation, while an *adversarial network* (Figure 4.1(c)) is used to boost the segmentation performance by adding global information.

Segmentor: Conventional CNNs use down-sampling layers to enlarge the size of receptive fields and extract more complex patterns, but these down-sampling

(a)

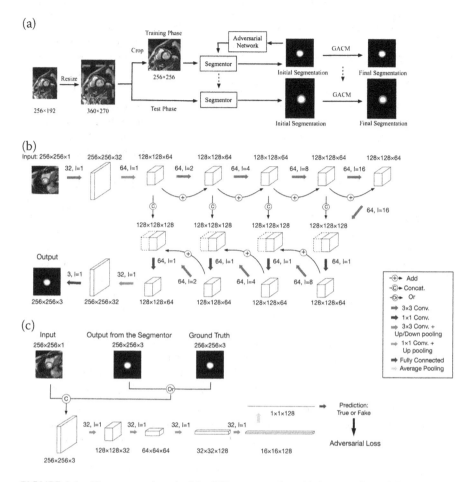

FIGURE 4.1 The proposed method for LV segmentation. (a) An overview of the proposed method in the training phase and test phase. (b) The architecture of the Segmentor. Each cube represents feature maps from the previous convolutional layer and the size of the feature map is given as height × width × the number of channels. The number of filters and the dilation factor used in each convolutional layer are provided as (the number of filters, l=dilation rate). All convolutional layers have a stride of 1 and pooling layers have a stride of 2. The Segmentor takes MRI slices as input and produces pixel-wise class predictions. (c) The architecture of the Adversarial Network. The input for the Adversarial Network is either the concatenation of MRI slices and the corresponding labels or the concatenation of MRI slices and the output from the Segmentor. The predicted class label is true or fake. The Adversarial Network is fused into the Segmentor by adding the adversarial loss.

layers will largely reduce the spatial size of feature maps. While conventional CNNs work well for image classification, dense prediction in segmentation tasks requires successive up-sampling layers to recover the resolution. To avoid information loss and additional work during the down-sampling and up-sampling processes, dilated convolution was proposed in Yu & Koltun (2015) for multi-scale context aggregation.

As shown in Figure 4.1b, instead of using successive max-pooling layers in the FCN, we only use one max-pooling layer and one up-pooling layer to reduce the memory storage and computational cost. For convolutional layers with dilation factors, both the number of channels and spatial size of feature maps are kept the same. Previous publications found lower layers extract general features, such as oriented edge and textural characteristics, while intermediate and higher layers encode more task-specific and highly-semantic features (Pinheiro et al., 2016). To enhance the integration of spatially rich features from lower levels with the coarse, object-specific features from higher layers, both long skip connections and shortcut connections are added in the Segmentor. The features from local receptive fields in lower layers can provide sufficient fine structure information for pixel-wise prediction. 1 × 1 convolution layers are added to reduce the computational burden induced by feature map concatenation.

The output layer consists of a 1 × 1 convolutional layer along with an element-wise soft-max function that generates the final probability map. In the last convolutional layer, the number of feature channels is the same as the number of classes K. In this study, we have three classes and thus $K = 3$, where class 1, 2, and 3 represent pixels in the background region, myocardium, and LV lumen, respectively.

We now introduce some mathematical notation that will be utilized throughout this section. Denote the input MRI slice of size $H{\times}W$ after preprocessing by $I: \Omega \rightarrow \mathbb{R}$, where $\Omega = \{1, 2, ...,H\} \times \{1, 2, ...,W\}$. The ground truth using one-hot encoding is a function $L: \Omega \rightarrow \{e_1, e_2, ..., e_K\}$, where $e_i = (0, \cdots, 0, 1, 0, \cdots, 0) \in \mathbb{B}^K$. Let $\Delta_K = \{(p_1, p_2, ..., p_K) \in [0, 1]^K : \Sigma_{i=1}^K p_i = 1\}$ be the set of all probability vectors of length K. Let $o(\cdot): \Omega \rightarrow \Delta_K$ be the function that generates the probability map from the output layer over K classes. The probability map is of size $H \times W \times K$.

Given the training image I and the ground truth L, the probability map P is generated as $o(I)$ and the Jaccard loss from the Segmentor can be written as:

$$loss_{seg} = -\Sigma_{k=1}^K \Sigma_{\mathbf{x} \in \Omega} \frac{L_k(\mathbf{x}) P_k(\mathbf{x})}{L_k(\mathbf{x}) + P_k(\mathbf{x}) - L_k(\mathbf{x}) P_k(\mathbf{x})}$$

where $P_k(\mathbf{x}): \Omega \rightarrow [0, 1]$ is the kth element of the probability vector, i.e., the scalar probability of the pixel at the position $\mathbf{x} \in \Omega$ belonging to the kth class.

Adversarial Network: Adversarial training has been successfully applied to image generation (Zhu et al., 2017). The motivation was to integrate higher-order potentials for image segmentation by training an adversarial classification model that can distinguish the predicted segmentation from the ground truth. Adversarial training can improve higher-order consistency by enforcing the Segmentor to generate results having less mismatch with the ground truth (Luc et al., 2016).

Given the training image I and ground truth L, the input of the Adversarial Network is the concatenation of an MRI slice with either the segmentation from the Segmentor $(I, o(I))$ or the ground truth (I, L). The Adversarial Network consists of four convolutional layers, three max-pooling layers, one global pooling layer, and two fully connected layers. The final classification is whether the input was generated from the Segmentor (Fake) or the ground truth (True). Since the Adversarial Network has a receptive field of the whole input, higher-order characteristics of the segmentation, such as the shape of a region labeled with a certain class or the spatial relationship between two regions labeled with a different class, can be extracted.

We denote $p_{true}(I, L)$ to be the scalar probability of the Adversarial Network predicting that L is the ground truth of I, and $p_{fake}(I, o(I))$ the probability of predicting that $o(I)$ is a predicted segmentation of I.

The binary classification loss for the Adversarial Network is defined as:

$$loss_{ad} = -\log(p_{true}(I, L)) - \log(p_{fake}(I, o(I)))$$

The first term in the adversarial loss depends only on the Adversarial Network. To force the Segmentor to generate segmentations less distinguishable from the ground truth of the Adversarial Network, the hybrid Segmentor loss is defined as:

$$loss'_{seg} = loss_{seg} + \log(p_{fake}(I, o(I)))$$

When the prediction from the Segmentor is less distinguishable from the ground truth, the second term in $loss'_{seg}$ becomes smaller. Training the Adversarial Network is equivalent to minimizing $loss_{ad}$ and training the Segmentor is equivalent to minimizing $loss'_{seg}$. The algorithm to train the Segmentor combined with the Adversarial Network is given as follows (Figure 4.2):

Algorithm 1 Training the proposed CNN. θ_{ad} denotes parameters in the Adversarial Network and θ_{seg} denotes the parameters in the Segmentor.

1: **for** *the number of training steps* **do**
2:    Feed in a mini-batch of training images $\{I_1, I_2, ..., I_m\}$ and labels $\{L_1, L_2, ..., L_m\}$
3:    Generate the segmentation from the Segmentor $\{o(I_1), o(I_2), ..., o(I_m)\}$
4:    Update the Adversarial Network via stochastic gradient descent: $\nabla_{\theta_{ad}} \frac{1}{m} \sum_{j=1}^m loss_{ad}(I_j, L_j, o(I_j))$
5:    Feed in another mini-batch of training images $\{I'_1, I'_2, ..., I'_m\}$ and labels $\{L'_1, L'_2, ..., L'_m\}$
6:    Generate the segmentation from the Segmentor $\{o(I'_1), o(I'_2), ..., o(I'_m)\}$
7:    Update the Segmentor via stochastic gradient descent: $\nabla_{\theta_{seg}} \frac{1}{m} \sum_{i=1}^m loss'_{seg}(L'_i, o(I'_i))$
8: **end for**

FIGURE 4.2 Algorithm of the proposed model training.

4.3.3   GUIDED ACTIVE CONTOUR MODEL

We propose a guided region-based active contour model (GACM) as a post-processing method to refine the endocardial and epicardial contour. In this work, the energy function is defined as a combination of a guided region-based term (E_{region}), a prior shape term (E_{shape}), and a regularization term (E_{reg}) (Li et al., 2010; Lei et al., 2017).

Let us define $\phi: \Omega \rightarrow \mathbb{R}$ to be the level set function of a Euclidean signed distance function. The contour is represented by $C = \{\mathbf{x} \in \Omega | \phi(\mathbf{x}) = 0\}$, where points inside the contour have $\phi(\mathbf{x}) > 0$ and points outside the contour have $\phi(\mathbf{x}) < 0$. Given an image I, C_{epi} and C_{endo} denote the predicted epicardial and endocardial contour from the Segmentor, respectively. Let ϕ_{epi} and ϕ_{endo} be the corresponding level set functions. ϕ_{endo} is used as an initial contour $\phi^{(0)}$ for curve evolution that is also integrated into the energy formulation as a prior shape. From our experimental results, we found that the prediction ϕ_{epi} from the Segmentor works well, but the predicted endocardial contour ϕ_{endo} sometimes needs refinement. For this reason, we propose the GACM to refine the endocardial contour, where ϕ_{epi} from the Segmentor is used as a guide contour to define a sub-region that improves the average intensity estimation in the foreground and background, and enhances the performance of region-based ACM.

The energy function can be written as

$$E(\phi) = \alpha_1 E_{region}(\phi) + \alpha_2 E_{shape}(\phi) + \alpha_3 E_{reg}(\phi)$$

$$
\begin{aligned}
E_{region}(\phi) \;=\; & \int_\Omega |I_s(\mathbf{x}) - c_1|^2 H(\phi) d\mathbf{x} \\
& + \int_\Omega |I_s(\mathbf{x}) - c_2|^2 (H(\phi_{epi}) - H(\phi)) d\mathbf{x} \\
E_{shape}(\phi) \;=\; & \int_\Omega (\phi - \phi_{endo})^2 d\mathbf{x} \\
E_{reg}(\phi) \;=\; & \int_\Omega p(|\nabla \phi|) d\mathbf{x},
\end{aligned}
$$

where $H(\cdot)$ is the Heaviside function, $\nabla(\cdot)$ is the gradient operation, and $p: [0, \infty) \rightarrow \mathbb{R}$ is the double-well potential function defined in (Li et al., 2010). The role of p is to maintain the signed distance property $|\nabla \phi| = 1$ within the vicinity of the zero level set.

Unlike in (Chan & Vese, 2001), in the above equation, I_s represents the sub-region inside C_{epi}. This is based on the prior knowledge that the endocardial contour is always inside the epicardial contour and helps avoid the influence of surrounding anatomical structures. c_1 and c_2 are calculated as

$$c_1(\phi) = \frac{\int_\Omega I(\mathbf{x}) H(\phi) d\mathbf{x}}{\int_\Omega H(\phi) d\mathbf{x}}$$

$$c_2(\phi) = \frac{\int_\Omega I(\mathbf{x})(H(\phi_{epi}) - H(\phi))d\mathbf{x}}{\int_\Omega (H(\phi_{epi}) - H(\phi))d\mathbf{x}}.$$

The final contour can be obtained by using the gradient descent algorithm to minimize the energy function over ϕ:

$$\phi^\star = \underset{\phi}{\arg\min}\, E(\phi).$$

By letting ϕ be parametrized by t, the Euler-Lagrange equation for minimizing the energy functional can be presented as

$$\frac{\partial \phi}{\partial t} = -\frac{\partial E}{\partial \phi} = \delta(\phi)[\alpha_1 \mathrm{div}(\frac{\nabla \phi}{|\nabla \phi|}) + \alpha_2(I_s - c_2)^2$$
$$- \alpha_2(I_s - c_1)^2 - \alpha_3(\phi - \phi_{epi})] + \alpha_4 \mathrm{div}(d_p(|\nabla \phi|)\nabla \phi),$$

where $\delta(\cdot)$ is the delta function.

The gradient descent starts with an initialization of $\phi^{(0)}$ and is updated iteratively

$$\phi^{(t+1)} = \phi^{(t)} + \mu \frac{\partial \phi}{\partial t},$$

where μ is the step size. The updating process continues until either the solution is stationary, i.e., $|\phi^{(t+1)} - \phi^{(t)}| < \epsilon$, where ϵ is a small value, or t exceeds a prescribed maximum number of iterations.

4.3.4 EVALUATION METRICS

Various evaluation methods were used to assess the proposed segmentation framework and compare its performance with other published results.

For myocardium segmentation, let TP and TN be the number of pixels that were correctly predicted as myocardium and non-myocardium and let FP and FN be the number of pixels that were misclassified as myocardium and non-myocardium. The sensitivity (SN), specificity (SP), positive predictive value (PPV), and negative predictive value (NPV) were used to evaluate the pixel-level accuracy. Let a and b be the predicted and manually delineated epicardial/endocardial contours, with A and B the areas enclosed by a and b, respectively. To evaluate the similarity between A and B, the Dice coefficient and Jaccard index were calculated, which are defined as

$$D(A, B) = \frac{|A \cap B|}{|A| + |B|},$$

$$J(A, B) = \frac{|A \cap B|}{|A \cup B|},$$

where $|\cdot|$ denotes the cardinality of the set.

The distance between two contours is another measure by which to evaluate whether two contours are closely matched. Average perpendicular distance (APD) was calculated, as well as the Hausdorff distance (HD), defined as

$$HD(a, b) = \max(\max_{i \in a}(\min_{j \in b} \|i - j\|), \max_{i \in b}(\min_{j \in a} \|i - j\|)).$$

Clinical parameters including EDV, ESV, EF, and LV mass are calculated. LV volume is derived by stacking endocardial contours of the entire LV at the same phase and summing all pixels in the LV lumen multiplied by slice spacing. Myocardium volume is calculated similarly. In this study, papillary muscles were assigned to the LV lumen. The LV end-diastolic/end-systolic phases are determined as the phases with the largest/smallest LV lumen volume. EF measures the percent of blood the left ventricle pumps out at each contraction and is defined as

$$EF = (EDV - ESV)/EDV \times 100\%.$$

LV mass is calculated by the myocardium volume at ED multiplied by the myocardial density ($1.05 \, \text{gcm}^{-3}$).

4.4   EXPERIMENTAL SETUP

4.4.1   DATASET

LV Segmentation Challenge 2011 (LV2011) (Suinesiaputra et al., 2014): The LV2011 dataset is derived from the 2011 LV Segmentation Challenge, part of the 2011 STACOM Workshop. LV2011 dataset contains 100 training cases (over 20,000 slices) and 100 validation cases (over 20,000 slices) from patients with coronary artery diseases. The cine images were obtained from multiple MR scanner systems using steady-state free precession (SSFP) during repeated breath-holds. A short-axis stack from the base to the apex of the LV was acquired at 1-cm intervals (either a slice thickness with gap or slice thickness with gap). The temporal resolution is between 19 and 30 frames. Manual segmentation was provided for the training set over the cardiac cycle, and the consensus myocardium segmentation (CS* consensus) was used as ground truth for the validation set. In this study, the training set in LV2011 was used to train and validate the proposed method.

Sunnybrook Cardiac Dataset (Radau et al., 2009): The Sunnybrook dataset is from the MICCAI 2009 LV Segmentation Challenge. It contains 45 cases (over 600 annotated slices in total) from patients in four groups: healthy, hypertrophy (HYP), heart failure with infarction (HF-I), and heart failure without infarction (HF-NI). The cine images were acquired using a 1.5T GE Signa MRI during 10–15 second breath-holds. A short-axis stack from the atrioventricular ring to apex was obtained with slice thickness and gap. The manual endocardial and epicardial contours were provided for slices at the ED and ES phases as the ground truth. In the 2009 LV Segmentation Challenge, the dataset was randomly divided into the training set (15 cases) and test set (30 cases). In this study, all 45 cases in Sunnybrook dataset were used to test the proposed method trained on LV2011 training set.

4.4.2 CONFIGURATIONS

Based on the manual delineation of endocardial and epicardial contours in the datasets, the label map was created by assigning 0 (background) to regions outside epicardial contours, 1 (myocardium) to regions between endocardial and epicardial contours, and 2 (LV lumen) to regions inside the endocardial contours. After that, the label map was converted to a one-hot encoding for subsequent network training.

The proposed CNN was implemented using the TensorFlow library and trained on an NVIDIA GTX Titan X GPU. Random rotation was used to augment the training data. The Adam optimizer was employed with a learning rate of 10^{-3} to minimize the loss during both Segmentor and Adversarial Network training. The model was trained for 20,000 steps with a mini-batch size of ten. A five-fold cross-validation was performed on the LV2011 training set for the network selection before post-processing.

The output from five-fold cross-validation was used to tune the parameters for the GACM algorithm. Based on a sensitivity analysis, the values of α_1, α_2, and α_3 in the guided active contour model were set to 1, 3, and 0.1, respectively, with μ set to 0.2 and the number of iterations capped at 50. For some slices where no LV lumen region was segmented by the Segmentor, the center point of the region enclosed by the epicardial border was used as the initial contour for evolution and α_2 and α_3 were set to zero.

The finalized model (DFCN-AD) was trained using all cases in the LV2011 training set and tested using the LV2011 validation set and the entire Sunnybrook dataset. The proposed GACM algorithm was used for post-processing. The final segmentation results were compared with other published results.

4.4.3 EVALUATION PROCESS

Three different sets of evaluation metrics were used in this study. The first one is a customized method for evaluating the performance of various network architectures. To evaluate the predicted segmentation more extensively and have a better understanding of each network, the Jaccard index and Hausdorff distance were calculated for endocardial and epicardial contours, while the Jaccard index, SN, and PPV were calculated for myocardium segmentation, i.e., the region enclosed by the endocardial and epicardial contours. These values were calculated per image and the average values from five-fold cross-validation were used to compare the performance of different networks.

After the final model was trained, the predicted segmentation on the LV2011 validation set was evaluated using the public criteria proposed by creators of the LV2011 dataset, which includes the Jaccard index, SN, SP, TPV, and NPV.

To avoid bias due to large background areas, regions of interest defined by the "assign consensus voxels" option (Suinesiaputra et al., 2014) were used to calculate SP and NPV. The ground truth for LV function including EDV, ESV, EF, and LV mass were provided by the dataset creators. Correlation and Bland-Altman plots to visualize the agreement between estimated clinical parameters and the ground truth are presented. To ensure a fair comparison, the same LV coverage range was taken.

The predicted segmentation on the Sunnybrook dataset was evaluated using the criteria proposed by the creators of the dataset. An official evaluation code was used to calculate the Dice, APD, and percentage of "good contours" for epicardial and endocardial contours. A contour with APD smaller than was regarded as a "good contour" by the dataset creators. Only good contours were included for calculating patient-wise average Dice and APD. No images from the Sunnybrook dataset were used to train the proposed model – the entire Sunnybrook dataset was used solely as a disjoint validation dataset.

4.5   RESULTS AND DISCUSSION

4.5.1   MODEL SELECTION

Five-fold cross-validation was used to evaluate the proposed FCN models. In Table 4.1, the FCN with dilated convolution (DFCN) and the combination of DFCN and adversarial network (DFCN-AD) were compared with U-Net as proposed in (Ronneberger et al., 2015) and the FCN proposed in (Tran, 2016). For U-Net and FCN, to achieve their best performance, we directly used the

TABLE 4.1

Performance Metrics for Different Models. The Proposed Models (DFCN, DFCN-AD) were Compared with U-Net (Ronneberger et al., 2015) and FCN (Tran, 2016). Values Are Provided as Mean (Standard Deviation) Over the Image

Network	Epi		Endo		Myocardium			Total Parameters
	Jaccard	HD (mm)	Jaccard	HD (mm)	Jaccard	SN	PPV	
U-Net	0.83 (0.19)	6.50 (10.67)	0.76 (0.28)	5.40 (7.55)	0.68 (0.20)	0.88 (0.13)	0.77 (0.28)	31.4
FCN	0.81 (0.25)	6.78 (10.98)	0.74 (0.30)	5.46 (7.62)	0.67 (0.19)	0.86 (0.13)	0.77 (0.19)	11.2
DFCN	0.85 (0.14)	5.52 (7.58)	0.76 (0.28)	5.22 (7.13)	0.74 (0.15)	0.88 (0.12)	0.81 (0.17)	**2.0**
DFCN-AD	**0.86 (0.12)**	**4.40 (5.12)**	**0.78 (0.25)**	**4.14 (5.21)**	**0.76 (0.13)**	**0.89 (0.11)**	**0.84 (0.15)**	2.4

hyper-parameters adopted in the publications. From Table 4.1, the integration of dilated convolution into FCN can efficiently improve segmentation performance by greatly reducing the number of parameters. Generally, in the design of CNN and FCN, the application of one down-sampling layer will reduce the resolution of feature maps, which requires an increase in the number of filters to compensate. Dilated convolution does not suffer from this loss of resolution, negating the need to increase the number of filters and thereby reducing the redundancy in higher layers.

Comparing results from DFCN and DFCN-AD, the integration of the adversarial network significantly improves the segmentation performance for the myocardium. This indicates that global features extracted from the Adversarial Network can provide the Segmentor with information from a different point of view and improve the overall performance. When comparing the predicted segmentation from DFCN with that of DFCN-AD, we find that DFCN-AD works especially better for slices whose epicardial and/or endocardial contours are fuzzy. Figure 4.3 presents four examples, for which the CMR image, manually labeled segmentation (ground truth), and automatic segmentations from the DFCN and DFCN-AD are given. The Jaccard coefficients for the automatic segmentation are provided under the image. For the two cases in the first row, both the location and the boundary of the myocardium are hard to detect. The segmentation from DFCN has a correct LV localization but fails in inferring the shape of the LV wall while DFCN-AD successfully delineated the myocardium boundary with a circular shape. Regarding the two cases in the second row, the DFCN-AD has a better prediction on LV wall thickness when the surrounding structures are less distinguishable, or the area of LV lumen is small. Our results show that the introduction of the Adversarial Network makes the final segmentation have greater consistency with the ground truth. Figure 4.4 compares the average Jaccard values and the standard derivations of the segmentations from DFCN and DFCN-AD for apical, mid-ventricle, and basal slices, respectively. As expected, DFCN-AD achieved greater gains in Jaccard value for apical slices and basal slices, where the segmentation tends to be less accurate compared with mid-ventricle slices because of fuzzy edges and intensity heterogeneity.

4.5.2 GACM

The dense prediction from the FCN is based on the receptive field centered on the target pixel while the GACM considers the intensity difference of the region inside the contour with that outside the contour, as well as the boundary smoothness and curvature. The last rows in Table 4.1 give the performance improvement using GACM on the LV2011 validation set. From these experimental results, GACM helps refine the boundary of the predicted endocardial border, especially for those with poor quality. Figure 4.5 provides some examples of the performance improvement resulting from GACM. For the first example shown in Figure 4.5, the LV lumen was missed by the Segmentor while the GACM was able to segment the

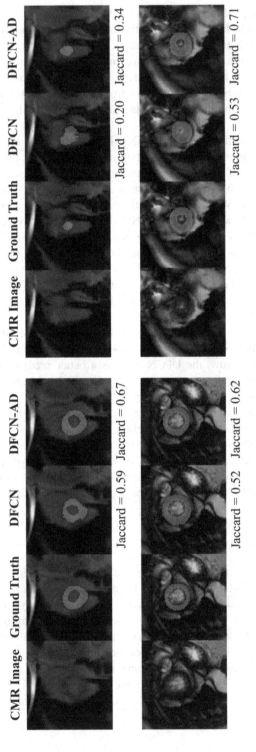

FIGURE 4.3 Comparison among the ground truth, outputs from DFCN, and outputs from DFCN-AD.

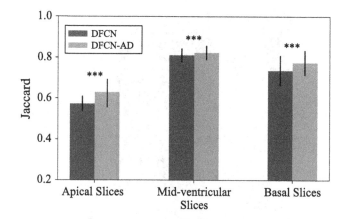

FIGURE 4.4 Comparison of Jaccard index between the segmentation from DFCN and DFCN-AD. Basal, mid-ventricle, and apical slices were defined using the method described in (Suinesiaputra et al., 2014). The standard deviations over cases are shown. The significance is calculated using paired t-test.

LV lumen by initializing the seed as the center of the region enclosed by the predicted epicardial border.

4.5.3 LV2011 DATASET

The final segmentation on the LV2011 validation dataset was generated by the DFCN-AD model, and followed with post-processing using GACM. The first two rows in Figure 4.6 depict the predicted epicardial and endocardial contours for one case in the LV2011 validation dataset. Slices are selected both from the ES to ED phases and from the apex to the base. The results show that the segmentation is robust against changes in both slice location and thickness during the cardiac cycle. The papillary muscles can also be correctly distinguished from the myocardium.

In Table 4.2, we compared the performance of the proposed framework on the LV2011 validation dataset with other published results. Our proposed method achieves the best results as compared to other fully automatic methods. The derived clinical parameters were evaluated using correlation and Bland-Altman plots in Figure 4.7. The correlations between the estimated EDV, ESV, EF, LV mass with ground truths are 0.99, 0.99, 0.90, and 0.99, respectively. The average differences shown in the Bland-Altman plots indicate that the proposed method slightly over-segments the myocardium and under-segments the LV lumen. Overall, the bias and variance in the estimations are small, and the average signed difference between the estimated EF and the ground truth is $0.57\% \pm 5.32\%$. In addition, the absolute differences between the estimations and the ground truths for EDV, ESV, EF, and LV mass are 5.30 cm^3 $\pm$ 5.34 cm^3, 5.96 cm^3 $\pm$ 5.26 cm^3, 3.82% $\pm$ 3.73%, and 7.52g $\pm$ 6.04g, respectively. From these results, we can conclude our automatic segmentation achieves accurate estimations for clinical parameters, evincing the high quality of automatic contour delineation.

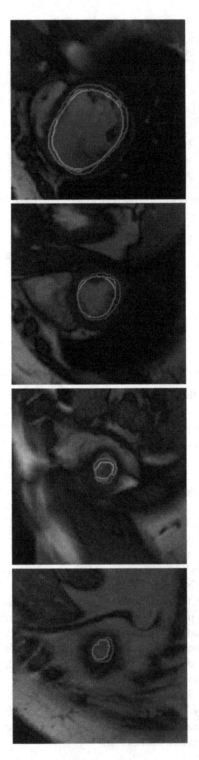

FIGURE 4.5 Comparison between the ground truth (blue lines), the output from Segmentor (red lines), and the output from GACM (green lines).

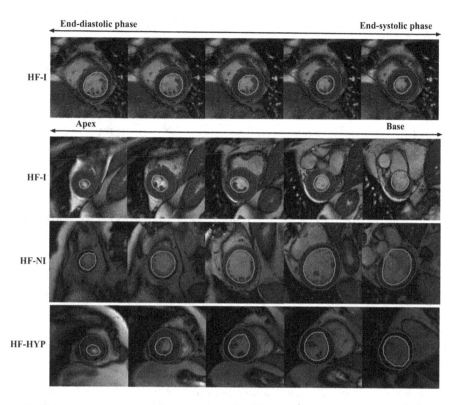

FIGURE 4.6 Automatic segmentations on the LV2011 and SunnyBrook datasets. Blue lines are predicted epicardial contours and red lines are predicted endocardialcontours. The first row shows delineations of both contours for slices during the cardiac cycle and the next three rows show automatic delineations for slices from the apex to the base. The cases on the first and the second rows are from the LV2011 validation dataset. The case in the third row is from the HF-NIgroup in the Sunnybrook dataset and the case in the fourth row is from the HYP group in the Sunnybrook dataset.

4.5.4 SUNNYBROOK DATASET

To further evaluate the proposed method, the Sunnybrook dataset was used as a disjoint validation dataset. The predicted epicardial and endocardial contours were generated using the DFCN-AD that was trained on LV2011 training data and applied to the Sunnybrook dataset for evaluation. The third row in Figure 4.6 presents one case in the HF-NI group and the fourth row presents one case in the HYP group, showing that accurate epicardial and endocardial contours can be generated on the disjoint dataset from apical to basal slices. In Table 4.3, we compared the purposed method with other

TABLE 4.2

Comparison of Myocardium Segmentations with Other Published Results on LV2011 Validation Dataset Based on CS* Consensus. The Methods Are Listed in Descending Order by the Jaccard Index. The Value Is Formatted as Mean (Standard Deviation) Over Images

Method	Jaccard	SN	SP	TPV	NPV
Proposed (with GACM)	0.76 (0.11)	0.88 (0.10)	0.95 (0.04)	0.86 (0.10)	0.96 (0.03)
Proposed (w/o GACM)	0.76 (0.12)	0.87 (0.11)	0.95 (0.04)	0.84 (0.11)	0.96 (0.03)
M. Khened et al.	0.74 (0.15)	0.84 (0.16)	0.96 (0.03)	0.87 (0.10)	0.95 (0.03)
P.V. Tran	0.74 (0.13)	0.83 (0.12)	0.96 (0.03)	0.86 (0.10)	0.95 (0.03)
SCR	0.69 (0.23)	0.74 (0.23)	0.96 (0.05)	0.87 (0.16)	0.89 (0.09)
INR	0.43 (0.10)	0.89 (0.17)	0.56 (0.15)	0.50 (0.10)	0.93 (0.09)

published methods on the Sunnybrook dataset. Contrary to other published methods in Table 4.3, no case in the Sunnybrook dataset was used for training or parameters tuning. Methods from were tested on 30 cases because the remaining 15 cases were used to train a CNN model. Although we did not use any cases in Sunnybrook dataset to design the proposed method, its performance is comparable with other methods specifically designed for Sunnybrook dataset. The correlations between the estimated EDV, ESV, EF, and LV mass with ground truth are 0.99, 0.99, 0.93, and 0.96, respectively. These results demonstrate that our model has good generalization. To further understand the model, we evaluated its performance on four pathological groups, the results of which are listed in Table 4.4. Different from the official evaluation methods described earlier, all contours were used to calculate Dice and HD to evaluate the performance comprehensively. Generally, the performances are similar among different groups. The proposed method worked best on the HF-I group, which is reasonable as the model was trained on cases from the DETERMINE study, where patients were diagnosed with coronary artery disease with an infarct mass $\geq 10\%$ of the LV mass.

4.5.5 QUALITY ANALYSIS

Figure 4.8 shows some challenging examples where the proposed method still performed accurate segmentations. In Figure 4.8(a), the surrounding structures have a very low contrast with myocardium; in Figure 4.8(b), the CMR slice has a low quality with noise and artifacts from imaging; in Figure 4.8(c), the edge of the endocardial contour is very fuzzy and heterogeneities exist in the brightness of the ventricle cavity due to increased trabeculations; in Figure 4.8(d), part of the LV wall is very thin. We further evaluated the

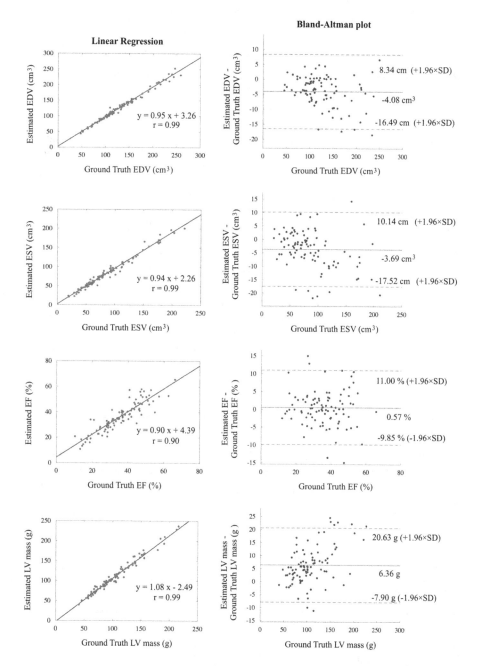

FIGURE 4.7 Correlation graph (left) and Bland–Altman plot (right) for estimated clinical parameters including EDV, ESV, EF, and LV mass. In the correlationgraph, the fitted linear regression model is given. *r* is the Pearson correlation. In the Bland–Altman plot, the mean difference and the standard deviation (SD) of the differences are shown.

TABLE 4.3

Comparison of the Epicardial and Endocardial Contour Segmentation with Other Published Results on the Sunnybrook Dataset. The Value Is Formatted as Mean (Standard Deviation) Over Cases

Method	The number of Cases	Epi			Endo		
		Dice	APD (mm)	Good Contours (%)	Dice	APD (mm)	Good Contours (%)
Proposed	45	0.95 (0.01)	1.75 (0.47)	96.56 (6.06)	0.91 (0.03)	2.25 (0.45)	96.02 (5.47)
Proposed (w/o GACM)	45	0.95 (0.01)	1.75 (0.47)	96.56 (6.06)	0.91 (0.03)	2.25 (0.45)	96.02 (5.47)
P.V. Tran	30	0.96 (0.01)	1.65 (0.31)	99.17 (2.20)	0.92 (0.03)	1.73 (0.35)	98.48 (4.06)
Ngo *et al.*	45	0.93 (0.02)	1.99 (0.46)	98.52 (5.74)	0.88 (0.04)	2.22 (0.46)	94.55 (9.31)
H. Hu *et al.*	45	0.94 (0.02)	2.21 (0.45)	91.21 (8.52)	0.89 (0.03)	2.24 (0.40)	91.06 (0.42)
Queirâs *et al.*	45	0.94 (0.02)	1.80 (0.41)	92.70 (9.5)	0.90 (0.05)	1.76 (0.45)	95.40 (9.6)

TABLE 4.4

Segmentation Evaluation on Four Pathological Groups from the Sunnybrook Dataset. HF-1: Heart failure with Infarction; HF-NI: Heart Failure Without Infarction; HYP: LV Hypertrophy. Values Are Provided as Mean (Standard Deviation) Over Cases

Group	Epi		Endo		Myocardium		
	Dice	HD (mm)	Dice	HD (mm)	Dice	SN	PPV
Normal (n = 9)	0.93 (0.07)	5.11 (5.38)	0.89 (0.13)	4.64 (1.76)	0.76 (0.09)	0.90 (0.08)	0.67 (0.11)
HF-I (n = 12)	0.95 (0.04)	5.14 (2.62)	0.93 (0.05)	4.85 (2.04)	0.77 (0.08)	0.88 (0.08)	0.70 (0.10)
HF-NI (n = 12)	0.93 (0.07)	5.61 (3.82)	0.91 (0.07)	4.87 (2.66)	0.73 (0.10)	0.89 (0.08)	0.63 (0.12)
HYP (n = 12)	0.92 (0.06)	5.83 (5.78)	0.87 (0.12)	5.07 (1.73)	0.75 (0.10)	0.93 (0.06)	0.65 (0.13)
Mean (std)	0.93 (0.01)	5.42 (0.31)	0.90 (0.02)	4.86 (0.15)	0.75 (0.01)	0.90 (0.02)	0.66 (0.03)

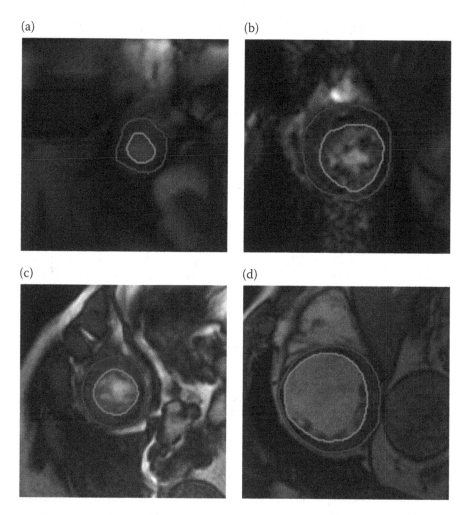

FIGURE 4.8 Challenging examples. Blue line: epicardial contour; Red line: endocardial contour.

spatiotemporal variation among segmentations from the proposed method as exhibited in Figure 4.9. In general, the Jaccard values for segmented mid-ventricle slices are higher and less scattered, while the segmented apical slices are less accurate and have a higher variance. The second row lists the average Jaccard value and standard deviation for slices over the cardiac cycle. The Jaccard values for slices near the ES phase are slightly lower and more scattered.

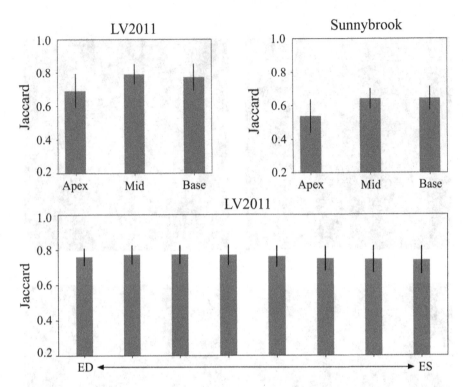

FIGURE 4.9 Spatiotemporal variances in Jaccard index of the automatic segmentations. Basal, mid-ventricle and apical slices were defined using the method described in (Suinesiaputra et al., 2014). The standard deviations over cases are shown.

4.6 CONCLUSION

In this chapter, we present a fully convolutional neural network, combined with an adversarial network, to perform LV segmentation. The global feature extracted by the adversarial network, and boundary smoothness and regional information by GACM can improve segmentation performance. Comprehensive evaluations demonstrate the proposed method can achieve accurate and robust segmentation over images from various institutions, scanners, and anatomical structures of high variability.

The proposed method can be further improved. The results show that the segmentation for apical slices and basal slices is less accurate than that for mid-ventricle slices. Thus, integrating information from spatially-adjacent MRI slices has the potential to facilitate more accurate localization and position estimation. Also, slight differences exist in segmentation performance for patients in different pathological groups (shown in Table 4.4 in part because the training cases are all from patients with coronary artery disease and previous infarct. It indicates that further training with cases from different pathological conditions is vital to building a robust segmentation model. Testing and refining the segmentation model for a variety of cardiac disease states including abnormally shaped, deformed, and hyper-trabeculated ventricles are likely to improve the accuracy of the method and enhance clinical applications. Future work should also investigate comparisons of manual segmentation and automatic segmentation derived parameters on CMR to clinical data, traditional measurements such as those obtained by echocardiography, and with patient outcomes.

REFERENCES

Avendi, M., Kheradvar, A., & Jafarkhani, H. (2016). A combined deep-learning and deformable-model approach to fully automatic segmentation of the left ventricle in cardiac MRI. *Medical Image Analysis, 30*, 108–119.

Bacharach, S. L., Green, M. V., Borer, J. S., Hyde, J. E., Farkas, S. P., & Johnston, G. S. (1979). Left-ventricular peak ejection rate, filling rate, and ejection fraction–frame rate requirements at rest and exercise: Concise communication. *Journal of Nuclear Medicine, Society of Nuclear Medicine, 20*(3), 189–193.

Bluemke, D. A., Kronmal, R. A., Lima, J. A., Liu, K., Olson, J., Burke, G. L., & Folsom, A. R. (2008). The relationship of left ventricular mass and geometry to incident cardiovascular events: The MESA (Multi-Ethnic Study of Atherosclerosis) study. *Journal of the American College of Cardiology, 52*(25), 2148–2155.

Chan, T. F., & Vese, L. A. (2001). Active contours without edges. *IEEE Transactions on Image Processing, 10*(2), 266–277.

Cousty, J., Bertrand, G., Najman, L., & Couprie, M. (2009). Watershed cuts: Minimum spanning forests and the drop of water principle. *IEEE Transactions on Pattern Analysis and Machine Intelligence, 31*(8), 1362–1374.

Eslami, A., Karamalis, A., Katouzian, A., & Navab, N. (2013). Segmentation by retrieval with guided random walks: Application to left ventricle segmentation in MRI. *Medical Image Analysis, 17*(2), 236–253.

Heiberg, E., Wigström, L., Carlsson, M., Bolger, A. F., & Karlsson, M. (2012). Time resolved three-dimensional segmentation of the left ventricle in multimodality cardiac imaging. 2013 8th International Symposium on Image and Signal Processing and Analysis (ISPA). https://doi.org/10.1109/ISPA.2013.6703796

Hu, H., Gao, Z., Liu, L., Liu, H., Gao, J., Xu, S., Li, W., & Huang, L. (2014). Automatic segmentation of the left ventricle in cardiac MRI using local binary fitting model and dynamic programming techniques. *PloS One, 9*(12), e114760.

Kass, M., Witkin, A., & Terzopoulos, D. (1988). Snakes: Active contour models. *International Journal of Computer Vision, 1*(4), 321–331.

Katouzian, A., Prakash, A., & Konofagou, E. (2006). A new automated technique for left-and right-ventricular segmentation in magnetic resonance imaging. *2006 International Conference of the IEEE Engineering in Medicine and Biology Society*, 3074–3077.

Knauth, A. L., Gauvreau, K., Powell, A. J., Landzberg, M. J., Walsh, E. P., Lock, J. E., del Nido, P. J., & Geva, T. (2008). Ventricular size and function assessed by cardiac MRI predict major adverse clinical outcomes late after tetralogy of Fallot repair. *Heart, 94*(2), 211–216.

Lee, H.-Y., Codella, N. C., Cham, M. D., Weinsaft, J. W., & Wang, Y. (2010). Automatic left ventricle segmentation using iterative thresholding and an active contour model with adaptation on short-axis cardiac MRI. *IEEE Transactions on Biomedical Engineering, 57*(4), 905–913.

Lei, Y., Shi, J., & Wu, J. (2017). Region-driven distance regularized level set evolution for change detection in remote sensing images. *Multimedia Tools and Applications, 76*(23), 24707–24722.

Li, C., Xu, C., Gui, C., & Fox, M. D. (2010). Distance regularized level set evolution and its application to image segmentation. *IEEE Transactions on Image Processing, 19*(12), 3243–3254.

Lorenzo-Valdés, M., Sanchez-Ortiz, G. I., Elkington, A. G., Mohiaddin, R. H., & Rueckert, D. (2004). Segmentation of 4D cardiac MR images using a probabilistic atlas and the EM algorithm. *Medical Image Analysis, 8*(3), 255–265.

Luc, P., Couprie, C., Chintala, S., & Verbeek, J. (2016). Semantic segmentation using adversarial networks. *ArXiv Preprint ArXiv:1611.08408.*

Lynch, M., Ghita, O., & Whelan, P. F. (2006). Automatic segmentation of the left ventricle cavity and myocardium in MRI data. *Computers in Biology and Medicine, 36*(4), 389–407.

Organization, W. H., & Unit, W. H. O. M. of S. A. (2014). *Global status report on alcohol and health, 2014.* World Health Organization.

Pednekar, A., Kurkure, U., Muthupillai, R., Flamm, S., & Kakadiaris, I. A. (2006). Automated left ventricular segmentation in cardiac MRI. *IEEE Transactions on Biomedical Engineering, 53*(7), 1425–1428.

Pinheiro, P. O., Lin, T.-Y., Collobert, R., & Dollár, P. (2016). Learning to refine object segments. *European Conference on Computer Vision*, 75–91.

Poudel, R. P., Lamata, P., & Montana, G. (2016). Recurrent fully convolutional neural networks for multi-slice MRI cardiac segmentation, in Zuluaga M., Bhatia K., Kainz B, Moghari M, & Pace D. (Eds), *Reconstruction, Segmentation, and Analysis of Medical Images* (pp. 83–94). Springer.

Pouleur, H., Rousseau, M., van Eyll, C., & Charlier, A. (1984). Assessment of regional left ventricular relaxation in patients with coronary artery disease: Importance of geometric factors and changes in wall thickness. *Circulation, 69*(4), 696–702.

Radau, P., Lu, Y., Connelly, K., Paul, G., Dick, A., & Wright, G. (2009). Evaluation framework for algorithms segmenting short axis cardiac MRI. *The MIDAS Journal-Cardiac MR Left Ventricle Segmentation Challenge*, 49.

Ronneberger, O., Fischer, P., & Brox, T. (2015). U-Net: Convolutional networks for biomedical image segmentation. *International Conference on Medical Image Computing and Computer-Assisted Intervention*, 234–241.

Sachdeva, S., Song, X., Dham, N., Heath, D. M., & DeBiasi, R. L. (2015). Analysis of clinical parameters and cardiac magnetic resonance imaging as predictors of outcome in pediatric myocarditis. *American Journal of Cardiology, 115*(4), 499–504.

Schulz-Menger, J., Bluemke, D. A., Bremerich, J., Flamm, S. D., Fogel, M. A., Friedrich, M. G., Kim, R. J., von Knobelsdorff-Brenkenhoff, F., Kramer, C. M., Pennell, D. J., & others. (2013). Standardized image interpretation and post processing in cardiovascular magnetic resonance: Society for Cardiovascular Magnetic Resonance (SCMR) board of trustees task force on standardized post processing. *Journal of Cardiovascular Magnetic Resonance, 15*(1), 35.

Suinesiaputra, A., Cowan, B. R., Al-Agamy, A. O., Elattar, M. A., Ayache, N., Fahmy, A. S., Khalifa, A. M., Medrano-Gracia, P., Jolly, M.-P., Kadish, A. H., & others. (2014). A collaborative resource to build consensus for automated left ventricular segmentation of cardiac MR images. *Medical Image Analysis, 18*(1), 50–62.

Tan, L. K., Liew, Y. M., Lim, E., & McLaughlin, R. A. (2017). Convolutional neural network regression for short-axis left ventricle segmentation in cardiac cine MR sequences. *Medical Image Analysis, 39*, 78–86. https://doi.org/10.1016/j.media.2017.04.002

Tran, P. V. (2016). A fully convolutional neural network for cardiac segmentation in short-axis MRI. *ArXiv Preprint ArXiv:1604.00494.*

Yang, J., Staib, L. H., & Duncan, J. S. (2004). Neighbor-constrained segmentation with level set based 3-D deformable models. *IEEE Transactions on Medical Imaging, 23*(8), 940–948.

Yeh, J.-Y., Fu, J., Wu, C., Lin, H., & Chai, J. (2005). Myocardial border detection by branch-and-bound dynamic programming in magnetic resonance images. *Computer Methods and Programs in Biomedicine, 79*(1), 19–29.

Yu, F., & Koltun, V. (2015). Multi-scale context aggregation by dilated convolutions. ICLR 2016. ArXiv Preprint ArXiv:1511.07122

Yusuf, S., Pfeffer, M. A., Swedberg, K., Granger, C. B., Held, P., McMurray, J. J., Michelson, E. L., Olofsson, B., Östergren, J., Investigators, C., Committees, & others. (2003). Effects of candesartan in patients with chronic heart failure and preserved left-ventricular ejection fraction: The CHARM-Preserved Trial. *The Lancet, 362*(9386), 777–781.

Zhu, J.-Y., Park, T., Isola, P., & Efros, A. A. (2017). Unpaired image-to-image translation using cycle-consistent adversarial networks. *Proceedings of the IEEE International Conference on Computer Vision*, 2223–2232.

Zhuang, X., Rhode, K. S., Razavi, R. S., Hawkes, D. J., & Ourselin, S. (2010). A registration-based propagation framework for automatic whole heart segmentation of cardiac MRI. *IEEE Transactions on Medical Imaging, 29*(9), 1612–1625.

5 Automated Methods for Vessel Segmentation in X-ray Coronary Angiography and Geometric Modeling of Coronary Angiographic Image Sequences: A Survey

Zijun Gao, Kritika Iyer, Lu Wang, Jonathan Gryak,
C. Alberto Figueroa, Kayvan Najarian,
Brahmajee K. Nallamothu, and
Reza Soroushmehr

CONTENTS

DOI: 10.1201/9781003120902-5

5.1 INTRODUCTION

Coronary artery disease (CAD) is the most common type of heart disease and a leading cause of mortality globally (Anon n.d.). In 2016, it was reported to affect 330 million people and was responsible for 17.9 million deaths around the world with an upward trend (Nowbar et al. 2019). CAD occurs when plaque builds up in the coronary artery, thereby restricting blood flow to the heart (Libby & Theroux 2005). The narrowing is clinically termed stenosis and can lead to deprivation of oxygen and nutrients in the cardiac muscle cells, eventually causing myocardial infarction, or heart attack (Thygesen et al. 2007). X-ray coronary angiography (XCA), computed tomography angiography (CTA), and magnetic resonance angiography are the main techniques for imaging coronary arteries. As the gold standard of CAD diagnosis and a crucial part of the interventional process, XCA locates the blockage by releasing radio-opaque contrast agents into the coronary arteries through a catheter in a procedure known as heart (cardiac) catheterization (Grossman 1986). Trained cardiologists then assess the severity of blockage in arteries based on select frames from XCA videos – typically via visual estimation.

In the past few years, artificial intelligence (AI) has rapidly seen an increase in many application domains, including medicine. Many automated models and methods have been developed for healthcare systems with the goal of enabling faster and more accurate diagnostic decisions, thus saving patients' lives and reducing costs. One of the applications of AI in CAD diagnosis/prognosis is to develop automated stenosis evaluation systems that could help clinicians in avoiding unnecessary tests and consequently reduce healthcare costs. 3D reconstruction of cardiac anatomies is another application of AI which provides patient-specific heart models that could be beneficial in complex surgeries. In this paper, we review state-of-the-art methods developed for CAD diagnosis with emphasis on coronary artery segmentation, 3D vessel modeling, and stenosis detection techniques using XCA. In terms of approaches used for vessel segmentation, methods could be categorized into machine learning and non-machine learning (non-learning). In non-learning models/methods, rules and relations between inputs and outputs are usually human-generated while the machine learning models are typically configured automatically. Whether machine learning or non-learning, automatic segmentation of coronary arteries is not an easy task given the heterogeneity of the background contrast and inherent noise in images. The presence of catheters, pacemakers, and organs may lead to artifacts in segmentation. Heart motion and camera motion can reduce the quality of images and influence the accurate assessment of coronary arteries. Moreover, since XCA videos are taken from different angles, an automatic information fusion to capture vessel structure is challenging due to overlap between different structures such as catheters and bones, intensity heterogeneity, and motion artifacts (e.g., respiratory and cardiac motions). In non-learning approaches, extracting rules to deal with the image challenges is time-consuming and requires substantial trial-and-error. On the other hand, deep learning and machine learning methods require many manually labeled samples (e.g., annotated/ground-truth images) to automatically learn the relationships between the inputs and outputs.

Therefore, achieving robust, scalable, and accurate models that could be deployed in clinical settings is a real challenge.

Most of the methods reviewed in this chapter have been published in the past six years. The main focus of this review is on automated techniques, using both machine learning and non-learning methods, for artery segmentation. Additionally, we summarize several applications of coronary segmentation including stenosis characterization and 3D coronary geometric modeling. We first review in Section 2, a number of methods used as pre-processing techniques applied in machine learning or non-learning methods to improve the quality of images or increase their size and diversity. In Section 3 we review state-of-the-art learning and non-learning methods used for coronary artery segmentation. In this section, we also analyze the methods in terms of the validation strategy. In Section 4, we review a number of metrics used for evaluating the performance of the methods. In Section 5, we focus on methods used for stenosis measurement and 3D geometric modeling and reconstruction as two applications of coronary artery segmentation. We summarize the chapter by providing a brief conclusion in Section 6.

5.2 PRE-PROCESSING TECHNIQUES

To improve the quality of the images or to enhance the vessel-like (or tubular) structures, a pre-processing step for the XCA images is often included as part of the segmentation pipeline. Pre-processing encompasses filter-based image noise reduction and foreground and background contrast adjustment (Cervantes-Sanchez et al. 2019; Jo et al. 2019).

Samuel (Samuel & Veeramalai 2021) applied contrast limited adaptive histogram equalization and adaptive fractional differential approach as a pre-processing step for image enhancement. To better suit model training, converting images to grayscale, resizing images to fixed input size, applying normalization or standardization are usually employed. For those handling XCA image sequences, the pre-processing step is more complex and specific. For example, Fan et al. (2018) registered one image to the others in the multichannel inputs for catheter removal. Xia et al. (2020) employed matrix decomposition to separate foreground and background components. They solved an optimization problem to decompose a feature matrix (consisting of a frame column) into a static part and a moving part for foreground detection. Then, an energy function with the input of the hierarchical vessel was used in vessel detection and vessel segmentation. It is worth mentioning that the use of different methods makes the definition of pre-processing step ambiguous. Some preprocessing steps used in machine learning or deep learning models could be part of the XCA handling pipeline in non-learning methods.

A more general discussion on pre-processing techniques could be found in a review paper (Krig 2016).

While the pre-processing techniques described above are used for both learning and non-learning methods, data augmentation is widely applied in deep learning to alleviate overfitting problems resulting from limited data. Most of the literature in this review used data augmentation techniques such as random scaling, cropping, flipping and rotation with some of them applied shear and elastic deformation

(Au et al. 2018; Cervantes-Sanchez et al. 2019; Fan et al. 2018; Iyer et al. 2021; Jo et al. 2019; Samuel and Veeramalai 2021; Shin et al. 2018; Yang et al. 2019). For more information on this aspect, the survey (Shorten & Khoshgoftaar 2019) is a helpful reference.

5.3 SEGMENTATION METHODS

5.3.1 Non-Learning Methods

In this section, we review state-of-the-art methods developed for artery segmentation using non-learning approaches. These methods generally employ image processing techniques that do not require a training process. For instance, region growing, thresholding, active contour/snake or level-set modeling, Fuzzy C-means could be applied as non-learning methods for coronary artery segmentation. After performing pre-processing and artery segmentation, a refinement step, usually called post-processing, may be necessary to improve the segmentation performance. In the following, we review coronary artery segmentation techniques and summarize them in Table 5.1.

Habijan et al. (2020) proposed a semi-automatic method for centerline extraction and tracking of the single coronary artery based on ridge edge detection (using Frangi's method) and thresholding. After finding the skeleton of the vessel, they removed the smaller branches and manually selected one point at the start of a single coronary artery and another point at its end. Dijkstra's algorithm was then applied on the branch to find the shortest path between selected points and then motion tracking was performed using template matching of 16-neighborhood around the selected points. Another semi-automated method based on centerline detection was proposed in Lv et al. (2019) where the centerline was extracted using the minimal path propagation with backtracking (MPP-BT) algorithm with manual start points (Chen et al. 2015). The centerline was then used in constraining the Chan-Vese active contour (level-set) algorithm (Chan & Vese 2001).

Kerkeni et al. (2016) developed a multiscale region growing for segmenting coronary arteries from 2D XCAs. They used pixels with the highest Frangi's filter response as seed points in each scale. In addition to vesselness information, they used direction information in the growing rule that could result in avoiding blockage caused by low vesselness values in vascular regions. Ma et al. (2020) also proposed a coronary artery segmentation by finding seed points from Frangi's filter response and then applying a region growing approach. Tache and Glotsos (2020) proposed a semi-automated method for coronary artery segmentation and stenosis detection. They applied Frangi's vesselness filtering, region growing, morphological operators, manual delimitation, and Canny edge detector. They deployed Dijkstra's algorithm for contour tracking and assessed the quality of the segmented images visually with the quantitative coronary arteriography (QCA) software. One of the drawbacks of non-learning methods is their lack of generalizability as they usually develop using a small dataset and might not consider a variety of challenges with XCAs such as artifacts and noise.

TABLE 5.1

Summary of Non-Learning Methods

Ref	Pre-processing	Technique	Method Inputs	Dataset (Training/Test)	Validation Technique	Evaluation Metrics	Branch	Method Type
(Cruz-Aceves et al. 2016)	Multiscale Gabor filters	Global thresholding method based on multi-objective optimization	Single image	80 images of 27 patients.	Training: Testing = 1:1	AUROC, Sensitivity, Specificity, Accuracy, Positive predictive value (PPV)	Entire	Filter-based
(Kerkeni et al. 2016)	Frangi filter	Multiscale region growing algorithm	Single image	50 images of 25 patients.	Training: Testing = 7:43	Dice, Sensitivity, Precision	Entire	Model-based
(Ma et al. 2017)	N/A	Robust PCA	Image sequences	Evaluated on 42 angiography sequences from 21 patients	N/A	contrast-to-noise ratio	Entire	Model-based
(Sameh et al. 2017)	Threshold decomposition driven adaptive filter and Frangi Filter	Region growing algorithm for segmentation; K-Nearest Neighbor for classification	Single image	Evaluated on angiography images from 75 patients	N/A	True Positive (TP), True Negative (TN), False Positive (FP), and False Negative (FN) rates	Entire	Model-based
(Cervantes-Sanchez et al. 2016)	Gabor filters with parameters tuned by differential evolution	Ostu's thresholding	Single image	Trained on 40 angiogram images; Tested on 40 angiogram images	N/A	AUROC, Accuracy	Entire	Filter-based
(Galassi et al. 2018)	Multiscale Hessian-based filters	The fast matching level set algorithm	Single image	24 stenotic segments from 12 patients	N/A	Reduction in diameter in stenosis regions	Entire	Model-based
(Wan et al. 2018)	Filter-based noise reduction, background homogenization, and vessel enhancement	Statistical region merging	Single image	Evaluated on 100 angiography images selected from 100 patients	N/A	Mean absolute difference, Hausdorff distance, Dice score, Sensitivity, Specificity, Accuracy	Entire	Model-based

(Continued)

TABLE 5.1 (Continued)
Summary of Non-Learning Methods

Ref	Pre-processing	Technique	Method Inputs	Dataset (Training/Test)	Validation Technique	Evaluation Metrics	Branch	Method Type
(Wan et al. 2018)	Mean filter, non-local mean filter, and Frangi filter	Ostu's thresholding	Single image	Evaluated on 143 angiogram sequences with 267 stenosis	N/A	Accuracy, Sensitivity, Specificity, F1-score	Entire	Filter-based
(Lv et al. 2019)	Centerline extraction with minimal path propagation with backtracking (MPP-BT) algorithm	Level-set Evolution	Single image	Evaluated on 4 2D angiogram image and 9 3D angiogram image	N/A	Precision, Dice, Recall	Entire	Model-based
(Xia et al. 2020)	Vessel enhancement by foreground and background separation with matrix decomposition	Energy optimization function	Image sequences	40 angiogram videos	N/A	MAE (mean absolute error) and F-measure	Entire	Model-based
(Qin et al. 2019)	Global logarithm transformation	Tensor Completion	Image sequences	Evaluated on 12 angiography sequences and 10 sequences synthetic images	N/A	Detection rate, Precision and F1-Score	Entire	Model-based
(Ma et al. 2020)	Frangi filter	Region growing	Single image	Evaluated on angiogram image from 6 patients	N/A	Dice, Sensitivity	Entire	Model-based

Robust Principal Component Analysis (PCA) approaches have also been used for layer separation to aid angiographic coronary segmentation. Ma et al. (2017) proposed an automatic online layer separation approach that robustly separates interventional X-ray angiograms into three layers: a breathing layer, a quasi-static layer, and a vessel layer that contains information of coronary arteries and medical instruments. The method used morphological closing and an online robust PCA algorithm to separate the three layers. The potential of the proposed approach was demonstrated by enhancing the contrast of vessels in X-ray images with low vessel contrast, which would facilitate the use of a reduced amount of contrast agent to prevent contrast-induced side effects such as itching and contrast-induced nephropathy. Qin et al. (2019) proposed a similar method using robust PCA known as VRBC-t-TNN (vessel region background completion with twist tensor nuclear norm). In this approach, a logarithmic mapping was used to create an X-ray attenuation sum model. A combination of robust PCA and spatially adaptive filtering was performed on the X-ray attenuation sum model to separate the vessel structures from the background. t-TNN was used to complete the background image where the vessels had been removed, and the vessels were recovered from the original X-ray angiography image by subtracting the t-TNN completed background image.

5.3.2 Learning Methods

In recent years, machine learning methods have become more commonly adopted for coronary segmentation in X-ray angiography images. The main advantage of machine learning methods is that they require less manual correction than non-learning methods such as edge-detection algorithms (Yang et al. 2019), and are less sensitive to structures such as the catheter or motion artifacts often present in angiography images (Fan et al. 2018; Iyer et al. 2021). Most machine learning algorithms for image segmentation use deep convolutional neural networks (CNNs), as they have demonstrated high accuracy in a wide range of segmentation applications (Badrinarayanan et al. 2017; Chen et al. 2018; Ronneberger et al. 2015; Zhao et al. 2017).

Several deep learning methods for angiographic segmentation have been developed. The first class of deep learning approaches utilizes additional stages on top of state-of-the-art segmentation CNNs. Samuel and Veeramalai (Samuel & Veeramalai 2021) developed VSSC Net (Vessel Specific Skip Chain Network), which added two stages on top of VGG16 (Visual Geometry Group network with 16 layers), a seminal CNN for image classification (Simonyan & Zisserman 2015). The additional stages were composed of Vessel Specific Convolutional (VSC) blocks, which identified the vessel regions, and skip chain convolutional layers, which allowed feature-rich information to propagate between VSC blocks. These stages refined the feature maps learned by VGG16 to improve vessel segmentation. Similar to Samuel and Veeramalai, Shin et al. (2018) developed a multi-stage network for coronary segmentation. In this approach, a graphical connectivity network was used to learn the tree structure of the vessels of interest. The DRIU (Deep Retinal Image Understanding) network (Maninis et al. 2016), a well-established network for retinal segmentation, was simultaneously employed to learn local pixel-level features of the vessels. The outputs of these two networks were then input into an inference CNN which generated a probability map of the vessel pixels. Iyer et al. (2021) trained a two-stage network for coronary segmentation called AngioNet. An Angiographic Processing Network, or APN, was designed to improve boundary sharpness and enhance local contrast of angiograms by learning the best possible pre-processing filter. The filtered angiography image from the APN was then input into Deeplabv3+ (Chen et al. 2018), a widely-adopted network for semantic segmentation. Jo et al. (2019) developed a two-stage CNN for the segmentation of a specific coronary branch: the left anterior descending artery (LAD). In the first stage, a filter pruning method known as Selective Feature Mapping (SFM) was used to detect convolutional layers which learned features most pertinent to the LAD. The resulting feature map and original input to the SFM stage were both passed to a segmentation CNN (U-Net, VGG16, or DenseNet), followed by a post-processing step using a thresholding method to generate a binary segmentation map.

The second type of deep learning method uses modified versions of state-of-the-art segmentation networks without a multi-stage approach (Shi et al. 2020; Yang et al. 2019). UE-Net, developed by Shi et al. (2020), is a generative adversarial network (GAN) where U-net was used as the generator, and a three-layer "E-shaped" pyramid structure was used as the discriminator. A GAN was used rather than a deep CNN to exploit the correlation between neighboring pixels and improve the connectivity of the vessel tree. Yang et al. (2019) tested several variations of U-Net on segmentation of the main coronary artery branches (right coronary, left anterior descending artery, and left circumflex). They compared the performance of the original U-Net against variations of U-Net where the encoder was replaced with ResNet, InceptionResNetv2, or DenseNet.

The third class of deep learning methods includes networks designed specifically for angiographic segmentation. MSN-A (Multichannel Segmentation Network – with Aligned inputs) is a multi-scale CNN method developed by Fan et al. (2018) for angiographic segmentation. The CNN, a multi-channel fully convolutional network (FCN) inspired by U-Net, takes two inputs: the angiographic frame of interest and its aligned mask image. The mask image was from the same angiographic series as the frame of interest, and captured the background and catheter before the dye is injected into the vessels. The mask image was aligned to match the frame of interest using hierarchical deformable dense matching, the idea being that the network could learn to subtract out the background using this image and better identify the vessels. In the work of Wang et al. (2020), both 3D and 2D convolutions were used to exploit the time-series information of an angiographic series as well as pixel-level features in individual frames. The 2D network was composed of a down-sampling encoder, an up-sampling decoder, bottle-neck modules, and skip connections, which accomplished the segmentation task, while the 3D network captured temporal information.

Others have sought to improve angiographic segmentation via the training process for deep learning methods (Zhang et al. 2020; Zhu 2020, 2017), rather than by modifying or creating a new network architecture. Zhu et al. (Xia et al. 2020) developed a transfer learning strategy for small data sets using regularization to prevent over-fitting. They implemented this strategy using PSPNet (Zhao et al. 2017), or Pyramid Scene Parsing Network, a deep convolutional network designed to incorporate features from multiple scales. The authors did not report any modifications to the original PSPNet architecture. Zhang et al. (2020) developed a weakly supervised deep learning framework to learn from automatically generated pseudo-labels rather than manual annotations. Pseudo-labels were generated using a combination of morphological operations, Robust Principal Component Analysis (RPCA) decomposition, and Otsu thresholding to separate vessel structures from the background. U-Net was then trained to refine the noise and systematic error in the pseudo-labels using a self-paced learning scheme, local manual refinement of difficult regions, and manual superpixel annotation.

Hybrid methods including both filters and deep-learning algorithms have also been employed for coronary segmentation. Nasr-Esfahani (et al. 2018) applied multiscale top hat filters to angiogram images to improve contrast enhancement. The angiograms were then subdivided into patches and fed into a CNN for local and global feature extraction. The learned features were input into a second CNN along with an edge map generated by a canny edge detector to refine the probability map of the vessel pixels from the first CNN. The probability map was converted to a binary segmentation using adaptive thresholding based on the largest connected component in the image. Another hybrid method was developed by Cervantes-Sanchez et al. (2019). In this method, multi-scale Gaussian Matched Filters (GMF) and Gabor filters were applied to coronary angiograms to enhance the vessel-like structures present in the images. The filtered images were then input into a multi-layer perceptron, and the output of the network was thresholded using a value determined by a soft computing procedure to segment the vessels.

The deep learning and hybrid approaches described here have demonstrated greater accuracy than many non-learning methods. These methods also require less user interaction or manual correction than non-learning methods, due to their robustness in avoiding structures that are of similar pixel intensity as the vessels, such as the catheter or background artifacts. A third advantage is their speed in generating a segmentation image, on the order of less than a second for some deep learning methods (Fan et al. 2018; Iyer et al. 2021; Yang et al. 2019), compared to several seconds for non-learning methods (Vukicevic et al. 2018; Xia et al. 2020). Although they can often achieve higher accuracy than non-learning methods, deep learning approaches require much larger computational resources such as GPUs and larger datasets to train them. As computational resources become more accessible, we expect the trend of recent developments in deep learning algorithms for coronary segmentation to continue. In Table 5.2, we summarize the machine learning methods reviewed in this chapter.

5.4 POST-PROCESSING TECHNIQUES

Post-processing in coronary artery segmentation usually aims to smooth the mask and remove small irrelevant pixel groups. Felfelian et al. (2016) used morphological opening operators to remove artifact pixels and kept the largest connected component for final segmentation output, whereas in Jo et al. (2019) the size of the

TABLE 5.2
Summary of Machine Learning Methods

Ref	Pre-processing	Technique	Method Inputs	Dataset (Training/Test)	Validation Technique	Evaluation Metrics	Branch	Method Type
(Nasr-Esfahani et al. 2018)	Multiscale Top-hat filters	Two shallow convolutional neural networks followed by adaptive thresholding	Image patches	44 angiogram images, 990,000 patches.	Training: Validation: Testing = 2:1:1	Dice score, Accuracy, Sensitivity, Specificity, Border Error	Entire	Filter-based features with shallow NN
(Fan et al. 2018)	Registrate images before and after dye injection	U-Net with multichannel inputs	Two registered images	148 angiogram sequences.	Training: Testing = 130:18	Dice score, Sensitivity, Specificity, Precision, Accuracy	Entire	Deep-learning
(Jo et al. 2019)	Frangi filter	CNN and selective feature mapping	Single image	1987 images of 1180 patients	Training: Testing = 200:1787	Accuracy, Precision, Recall, Specificity, F1-Score	Left anterior descending (LAD) artery	Deep-learning
(Vlontzos and Mikolajczyk, 2018)	N/A	Weakly supervised learning with optical flow and U-Net	Image sequences	365 angiogram sequences of 26 patients.	Training: Testing = 363:2	Dice Score	Entire	Weakly supervised deep-learning

Reference	Pre-processing	Method	Input	Dataset	Split	Metrics	Region	Category
(Au et al. 2018)	Standardization, resizing	YOLONet for localization, U-Net for segmentation	Single image for localization, patches for segmentation	RCA of 1024 study participants.	Training: Validation: Testing = 70:15:15	F1-Score	Branches with Lesion Area	Deep-learning
(Yang et al. 2019)	Normalization	U-Net with changed backbones	Single image	3302 angiogram images of 2042 patients.	Five-folds cross-validation and an external validation set with 181 images of 128 patients.	F1-Score	Major Branches	Deep-learning
(Cervantes-Sanchez et al. 2019)	Gaussian matched filters, multiscale Gabor filters	Multilayer perceptron followed with performance-based thresholding	Filtering output of multiple scales	130 X-ray angiograms. (Dataset available)	Training: Testing = 2:1	AUROC, Dice, Sensitivity, Specificity	Entire	Filter-based features with shallow NN
(Shin et al. 2019)	N/A	Graph neural network and U-Net	Single image	3137 angiogram images of 85 patients.	Training: Testing = 2958:179	AUROC, F1-Score	Entire	Deep-learning
(Hao et al. 2019)	Manually-selected salient frames	Modified U-Net	4 consecutive images	60 angiogram sequences	N/A	Sensitivity, Precision, F1-score	Entire	Deep-learning
(Wang et al. 2020)	Resize to fit input size	A combination of 3D convolutional layers and a 2D network based on U-Net	3 or 5 adjacent images	170 angiogram videos with 8835 images.	Training: Testing = 5:1	IOU (Intersection over Union), Sensitivity, Specificity and Accuracy	Entire	Deep-learning

(Continued)

TABLE 5.2 (Continued)
Summary of Machine Learning Methods

Ref	Pre-processing	Technique	Method Inputs	Dataset (Training/Test)	Validation Technique	Evaluation Metrics	Branch	Method Type
(Zhang et al. 2020)	Vessel enhancement by layer separation with Robust Principal Component Analysis, pseudo label generation	Annotation-refining self-paced learning with U-Net as segmentation model	Single image	171 angiogram images of 30 patients.	112 images of 17 patients for training, 25 images of 4 patients for validation and 54 images of 9 patients for testing	Recall, Precision, Dice	Entire	Weakly supervised deep-learning
(Shi et al. 2020)	N/A	Adapted generative adversarial networks with U-Net like structure in a generator	Blood vessel binary image	754 images from the cranial view and 3119 images from the Right view.	Training: Testing = 9:1	Mean pixel accuracy	Trunk and first-level branches	Deep-learning
(Zhu et al. 2020)	Normalization, convert to grayscale	Pyramid Scene Parsing Network	Single image	angiogram images from 109 patients.	Training: Testing = 2:1	Sensitivity, Specificity, Accuracy	Entire	Deep-learning
(Wu 2020)	Normalization	U-Net	Image sequences	148 angiogram sequences.	123 for 5 folds cross-validation, 25 for testing	Sensitivity, positive predictive value, F1-score	Entire	Deep-learning

					Training: Testing		Entire	Deep-learning
(Samuel and Veeramalai 2021)	Adaptive fractional differential approach, contrast limited adaptive histogram equalization, and Gaussian filter	Pre-trained VGG-16	XCA image, 3 channels from preprocessing results	160 angiogram images (Dataset available)	Training: Testing = 3:1	Contrast to Noise Ratio (CNR), Sensitivity, Accuracy, AUC	Entire	Deep-learning
(Iyer et al. 2021)	Normalization	Angiographic Processing Network with Deeplabv3+	Single image	462 angiogram images of 161 patients.	5-fold cross-validation and a separate testing set	Dice Score, Accuracy	Entire	Deep-learning

redundant region was determined by the average size of vessel regions in the mask. On the other hand, several (Nasr-Esfahani et al. 2018; Vukicevic et al. 2018) used the largest connected component to identify the vessel masks in the segmented image as post-processing. This method may compromise the subsequent stenosis detection step as severe stenosis sometimes results in discontinuity in the initial segmentation mask and selecting the largest connected components would ignore this case.

Here we also want to briefly mention different thresholding techniques for obtaining the final segmentation mask since it belongs to post-model processing and is important for segmentation and stenosis detection performance. Currently, Otsu thresholding (Ridler & Calvard 1978) and fix-value thresholding are the most used in the papers we reviewed. In the paper of Cervantes-Sanchez et al. (2019), they compared fourteen automatic thresholding tools with 13 frequently used in vessel-segmentation literature and 1 developed by themselves. In their previous paper (Cruz-Aceves et al. 2016), they proposed a global thresholding method based on multi-objective optimization.

5.5   EVALUATION METRICS

Several metrics could be used for evaluating the performance of vessel segmentation techniques. True positive (TP), true negative (TN), false negative (FN), and false positive (FP) are commonly used in evaluation metrics, where positive and negative refer to pixels belonging to vessels and background, respectively. The true or false characterization is determined by comparing the algorithm output to the Gold Standard (GS) segmentation. Segmentation performance measures are summarized in Table 5.1. Accuracy (Acc), sensitivity (Se), specificity (Sp), Dice/F1-score, positive and negative rate, area under receiver operating characteristic (AUROC/AUC), area under the precision-recall curve (AUPRC), and Cohen's Kappa coefficient (κ) (Moccia et al. 2018) are among popular metrics defined in this section (Table 5.3).

Acc measures how many observations, both true positive and negative, were correctly classified, among the total number of test cases (n). Sensitivity, also known as true positive rate, measures the proportion of positives that were correctly classified. Specificity measures the proportion of negatives, both TN and FP correctly classified. Although a high Se reflects the desirable algorithm inclination to detect vessels, a high Se with low Sp indicates that the segmentation includes many pixels that do not belong to vessels. Consequently, an algorithm that provides high Se and low Sp is acceptable if the post-processing step can remove possible FP.

Even though Acc, Se, and Sp are the most frequently adopted performance metrics, other derived metrics are also often employed. Examples include FP rate, which is equal to 1 − Sp, positive predictive value (PPV), which is the proportion of TP among TP+FP, and negative predictive value (NPV), which is the ratio between TN and TN+FN. PPV gives an estimation of how likely a pixel belongs to a vessel given that the algorithm classifies it as positive. NPV corresponds to the likelihood that a pixel does not belong to a vessel, given that the algorithm classifies it as negative class.

TABLE 5.3

Evaluation Metrics for Vessel Segmentation Algorithms

Index	Definition
Accuracy (Acc)	$\frac{TP+TN}{TP+TN+FP+FN}$
Sensitivity (Se)	$\frac{TP}{TP+FN}$
Specificity (Sp)	$\frac{TN}{TN+FP}$
False Positive rate ($FP\ rate$)	$1 - Sp$
Positive Predictive Value (PPV)	$\frac{TP}{TP+FP}$
Negative Predictive Value (NPV)	$\frac{TN}{TN+FN}$
AUROC	Area Under the ROC curve
AUPRC	Area Under the Precision-recall curve
F1-score/Dice Similarity Coefficient (DSC)	$\frac{2TP}{FP+FN+2TP}$
Cohen's Kappa coefficient (κ)	$\frac{Acc - Pe}{1 - Pe}$, where Pe is the hypothetical probability of chance agreement

The receiver operating characteristic (ROC) curve, which illustrates the performance of a binary classifier system as its discrimination threshold is varied, is also often reported. The area under the ROC (AUROC) is used as an evaluation metric, which indicates whether the model can discriminate between cases (positive examples) and non-cases (negative examples.). A perfect classifier is assumed to have the value of one for AUROC.

The precision-recall curve can be used, too. Precision corresponds to PPV, while recall to Se. The precision-recall curve compares TP with FN and FP, excluding TN, which is less relevant for the vessel segmentation performance evaluation since the proportion of TP (vessels) and TN (background) is highly skewed. Also in this case, the area under the precision-recall curve (AUPRC) can be exploited.

The most commonly used spatial overlapping index is the Dice similarity coefficient (DSC), also known as the F1 score, which is calculated from the precision and recall of the test. However, the F-measures do not take true negatives into account, hence measures such as the Matthews correlation coefficient, Informedness or Cohen's kappa may be preferred to assess the performance of a binary classifier.

5.6 APPLICATIONS

5.6.1 STENOSIS MEASUREMENT

Coronary artery segmentation is a crucial step in the quantitative coronary angiography (QCA) pipeline. Stenosis severity is typically visually estimated rather than measured in the clinical workflow. Designed to give a more reliable and objective measure of stenosis severity compared with visual stenosis diagnosis, QCA produces a score based on the percentage diameter stenosis (%DS). Coronary revascularization may be performed if %DS is greater than 50% (Montalescot et al., 2013), where a %DS of 70% or greater is considered clinically significant for cardiac lesions (Zhang et al. 2020). There are several different methods for calculating %DS (Hideo-Kajita et al. 2019). It could be calculated, for example, with the formula below by averaging the proximal and distal normal reference vessel diameter in the lesion segment.

$$\%DS = 1 - \frac{minimum\ lumen\ diameter}{average\ reference\ vessel\ diameter} * 100$$

Among the papers we reviewed that used non-learning methods for segmentation, some identify the stenosis region and determine %DS by calculating the diameters along the vessel structure when the segmentation is done, while others grade the stenosis severity with different %DS thresholds, with 25%, 50%, and 70% being the most commonly used for staging. Cervantes-Sanchez et al. (2016) used Gaussian matched filters whose parameters were estimated based on a genetic algorithm to detect coronary arteries and then obtained the segmentation by thresholding with Otsu's method. Candidate stenosis regions were marked as the vessel width local minima by calculating the second-order derivative. Cervantes-Sanchez et al. (2019) updated the aforementioned pipeline with Gaussian matched filters tuned by differential evolution for enhancement step and used iterative thresholding for segmentation. By replacing the second-order derivative methods with the naive Bayesian classifier applied on a 3D feature vector, they reported improved performance in stenosis detection. Wan et al. (2018) denoised the image with non-local mean filtering, enhanced vessel structure with Hessian-based filters, and used Otsu threshold to obtain the segmentation. They then extracted the vessel tree skeleton, estimated the diameter with a parametric model, and conducted stenosis detection by identifying the local minimum points with the backward difference method. Sameh et al. (2017) first enhanced the images with adaptive and Frangi filters, then used the region growing algorithm for segmentation, followed by the K-nearest neighbor algorithm for coronary artery blockage stage classification.

Recently, researchers applied deep learning to stenosis detection with XCA image patches. Antczak and Liberadzki (2018) created an artificial dataset with an algorithm that mainly employed Bézier curves to train convolutional neural networks of up to five layers from scratch and then used real XCA images patches for further training. Au et al. (2018) proposed a deep learning framework for handling XCA images to achieve real-time stenosis detection with three network modules that serve to localize, segment, and identify clinically significant stenosis regions. Inputs of the segmentation network module are small patches obtained from the localization module. Ovalle-Magallanes et al. (2020) focused on transfer learning for stenosis detection in XCA image patches by introducing a network-cut and fine-tuning hybrid method. Various setups and fine-tuning strategies were tested on different network structures, with both the classification results and activation maps for a visual explanation presented.

In 2019, researchers started to focus on end-to-end approaches where deep-learning networks received a full XCA sequence as input, selected the high-quality frames, and delivered the location (and severity) of possible blockage as output. C. Cong et al. (2019) employed three different network structures for angiographic projections classification, salient frame selection, and stenosis detection. The third network could output the normal/stenosis classification results, the activation map and stenosis localization information. Wu et al. (2020) used the Unet segmentation of vessel structure as the quality measure to select salient frames, detected stenosis with deconvolutional single-shot multibox detector network, and applied sequence false positive suppression to filter false positive stenosis regions with the temporal characterization of the XCA sequence. Moon et al. (2021) proposed a weakly-supervised deep learning approach. They applied frame selection by saving top-5 frames, denoted as keyframes, which had the highest number of vessel pixels in the binary map generated by vessel enhancement filtering, Fangi filtering, and Otsu thresholding. Keyframes were assigned to inception V3 for stenosis classification, with attention modules that covered both the spatial and channel attention added to enable the deep learning model to focus on the characteristics of stenosis in training. Gradient-weighted class activation mapping was applied to visualize stenosis-relevant discriminatory regions in the XCA image. Zhang et al. (2020) proposed a hierarchical attentive multi-view learning model that exploits the complementary information from different angulated views and provides direct quantification of coronary artery stenosis based on XCA sequences from the main-view, the support view and a XCA image from the keyframe. The model was composed of main-view, support-view, keyframe-view, and regression modules to handle different input. Spatio-temporal features were extracted from 3D convolution on the main and support view, with discriminative information obtained from attention block and stenosis representation enhanced from the keyframe-view module.

5.6.2 3D GEOMETRIC MODELING AND RECONSTRUCTION

Angiographic segmentation has been used as part of several workflows to develop 3D models of the coronary tree. These 3D coronary models can better characterize stenosis severity than 2D X-ray angiography images alone, which are subject to issues such as vessel overlap and foreshortening (Çimen et al. 2016). The two main classes of 3D reconstruction methods are those based on epipolar constraints and those based on deformable models (Çimen et al. 2016). Epipolar constraint methods determine corresponding points or segments in multiple images and project these corresponding structures to generate the 3D model (Banerjee et al. 2020; Blondel et al. 2006; Cardenes et al. 2012; Chen and Carroll 2000; Galassi et al. 2018; Hoffmann et al. 2000; Jandt et al. 2009; Liao et al. 2010; Movassaghi et al. 2004; Vukicevic et al. 2018). The deformable model approach uses energy functions and prescribed mechanical properties to deform a 3D snake such that its projections match a series of X-ray angiography images (Canero et al. 2002; Cong et al. 2015;

Yang et al. 2014; Zheng et al. 2010). Here, we review several methods that have been developed in the last decade.

In recent years, a subset of epipolar constraint algorithms has followed a Non-Uniform Rational B-Spline (NURBS) based approach to create 3D models of the coronary tree from single plane angiography images. NURBS surfaces and curves are commonly used to analytically represent shapes in computer vision (Piegl & Tiller 1996). In the first stage of their algorithm, Galassi et al. (2018) applied multiscale Hessian-based filters to pairs of angiogram images to generate vesselness probability maps. This was followed by a fast-marching level set algorithm to identify the vessel boundaries. The centerlines of the vessels were extracted from the segmented image using a subvoxel skeletonization algorithm and corresponding segments were labeled on each image of the pair. The 3D position of these vessels was determined by finding the projected intersection of the segment surfaces, and NURBS contours were used to define the luminal contours of the 3D vessel surface. Vukicevic et al. (2018) also used a NURBS-based semi-automatic approach to 3D reconstruction of the coronary tree. Their segmentation algorithm involved user specification of the start and endpoints of each vessel of interest, followed by a Frangi Filter to extract the vessel-like structures. Wave propagation via the Fast Marching method from the user-defined points was used to extract the centerlines of the vessels and split them into branches. The 2D centerlines were projected using partial matching to create a 3D centerline, followed by B-spline interpolation. The coronary branch surface was then reconstructed by projecting vessel border points from the 2D images to create 3D vessel cross-sectional patches. The 3D model was finally discretized using NURBS-based meshing. Banerjee et al. (2020) used a similar approach as Galassi et al. for the vessel segmentation portion of their method but implemented a fully automated 3D vessel reconstruction algorithm. Input images were initially processed using a multiscale Hessian-based filter, followed by an active contour region-growing algorithm to isolate the vessels. The 2D centerlines and vessel boundaries were extracted using a fast-marching level set algorithm, and centerlines were corrected for both rigid and nonrigid motion. Centerline points from corresponding branches were projected to form an outer hull point cloud containing many possible point correspondences. Several algorithms such as B-spline fitting, distance thresholding, and re-projection error were used to refine the point cloud into the true 3D centerline.

Another subset of epipolar constraint methods has used rotational X-ray angiography sequences to reconstruct the 3D coronary tree. Jandt et al. (2009) adopted a fully automatic reconstruction method that uses gated angiography projections from the same point in the cardiac cycle. Projections were filtered using multiscale Frangi filters to compute the vesselness response of each pixel. The 2D vesselness responses were projected into 3D and multiplied. The 3D vesselness response, projected angle differences, and penalties for motion or noise-induced response made up a 3D speed function, which guided a fast-marching algorithm to generate the 3D model. Finally, endpoints were detected using propagation speed and backtracing was used to identify the shortest path of the wavefront or 3D centerline. Liao et al. (2010) applied coherence enhanced diffusion, multiscale Frangi filters, and hysteresis thresholding to ECG-gated projections to segment the vessel.

Morphological thinning was used to extract the 2D centerlines, followed by fast-marching between user-specified seed points or B-spline smoothing to edit the centerlines if needed. 3D symbolic reconstruction was used to assign 3D depth to centerline pixels in multiple projections by minimizing an energy function based on the soft-epipolar constraint and reprojection error via α-expansion graph cuts. Finally, the coronary tree was recovered from the symbolic reconstruction points using Euclidean minimum spanning tree algorithm and 2D vessel radii were projected to reconstruct the 3D vessel lumen.

We now present several recent examples of deformable models and energy-based methods for coronary tree reconstruction. The method of Yang et al. (Cong et al. 2015; Yang et al. 2014) did not directly generate a segmentation image; however, active contour models were used to identify the vessel boundaries and find the centerline of the vessels in each 2D image plane. External energy terms derived from generalized gradient vector flow (Xu & Prince 1998) were used to deform a 2D parametric curve until it matched the centerline of the image. The endpoints of the curve were back-projected into 3D and a cylindrical vessel with assigned internal energy was initialized between these endpoints. The 2D external energy forces were then back-projected and applied to this 3D vessel while the internal energy governed its deformation; the 3D vessel was iteratively reprojected and deformed until convergence. Zheng et al. (2010) proposed a semi-automatic method to reconstruct the coronaries using deformable snakes. Similar to the method of Yang et al. (2014), this method did not directly segment the coronary vessels; however, pixel intensity and pixel intensity gradients were used to identify vessel boundaries and inform the energy function used to determine the vessel centerline. First, the user-determined homologous points along the vessel using epipolar constraints to initialize the snake. These points were back-projected to create an initial 3D snake. Energy functions relating to the 2D and 3D snakes were used to deform the 3D centerline and maintain continuity, smoothness, and accurate reprojection onto the 2D planes. The process was repeated for subsequent X-ray frame pairs to enable tracking of the coronary centerline throughout the angiographic series.

Both epipolar constraint-based methods and deformable models have been shown to faithfully reconstruct the 3D geometry of coronary arteries from X-ray angiography images. The main advantage of the epipolar constraint methods is their relatively low computational cost and speed, although these methods sometimes require manual interaction to match vessel branches, endpoints, or bifurcation points in multiple images. On the other hand, deformable models require less manual interaction, but these methods are more time-consuming due to their computational complexity. Regardless of the methodology, the trend in recent years has been toward the complete automation of coronary 3D reconstruction from X-ray angiography images.

5.7   SUMMARY AND DISCUSSION

X-ray coronary imaging is the gold standard to assess the presence of coronary artery disease; however, underestimation and overestimation of the severity of lesions (i.e., extent of stenosis) can occur due to human errors in assessment through visual estimation and imaging challenges such as inadequate radiographic view, noise, and artifacts. Automated techniques can help clinicians reduce the likelihood of diagnosing errors if they can overcome the imaging challenges. Many efforts have been made in recent years to deal with the challenges and improve the performance of analytical methods.

The methods presented here have taken a variety of approaches toward automatic vessel segmentation using machine learning and traditional image processing (non-learning method) approaches. Many of the non-learning methods are based on filtering techniques and more particularly, Frangi's method which highlights tubular-like objects (e.g., vessels) in images. After emphasizing these objects, other techniques such as region growing, thresholding and active contour could be used for segmenting vessels. Due to the challenges in analyzing X-ray images of coronary arteries, developing traditional methods that overcome the challenges is difficult. In general, the deep learning and hybrid methods were more accurate than the non-learning methods, perhaps because these were better equipped to handle the many imaging artifacts present in XCA images (i.e., the catheter, ribs, background noise, low contrast regions). The deep learning methods that performed the best used large datasets and data augmentation, as well as deep architectures with a combination of layers from multiple commonly used segmentation networks.

Thanks to deep neural networks, it is easier to create an analytical model that can consider variations in imaging data and better capture relationships between imaging data and desired outcomes. Therefore, in most of the research work, machine learning and, more specifically, deep learning methods are the main components of design systems. However, one of the challenges facing these methods is the lack of enough samples (in terms of size, quality, and diversity) and their corresponding labels which affect their robustness. For being successfully deployed into clinical practice, robust clinical evaluation besides technical evaluation is required.

REFERENCES

Anon. (2017). Cardiovascular Diseases (CVDs). Retrieved January 26, 2021, from https://www.who.int/news-room/fact-sheets/detail/cardiovascular-diseases-(cvds).

Antczak, K., & Liberadzki, Ł. (2018). Stenosis Detection with Deep Convolutional Neural Networks. *MATEC Web of Conferences*. https://doi.org/10.1051/matecconf/201821004001.

Au, B., Shaham, U., Dhruva, S., Bouras, G., Cristea, E., Lansky, A., Coppi, A., Warner, F., Li, S.-X., & Krumholz, H. (2018). Automated Characterization of Stenosis in Invasive Coronary Angiography Images with Convolutional Neural Networks. arXiv:1807.10597.

Badrinarayanan, V., Kendall, A., & Cipolla, R. (2017). SegNet: A Deep Convolutional Encoder-Decoder Architecture for Image Segmentation. *IEEE Transactions on Pattern Analysis and Machine Intelligence*, 39, 2481–2495. https://doi.org/10.1109/TPAMI.2016.2644615

Banerjee, A., Galassi, F., Zacur, E., De Maria, G. L., Choudhury, R. P., & Grau, V. (2020). Point-Cloud Method for Automated 3D Coronary Tree Reconstruction From Multiple Non-Simultaneous Angiographic Projections. *IEEE Transactions on Medical Imaging*, 39(4), 1278–1290. https://doi.org/10.1109/TMI.2019.2944092.

Blondel, C., Malandain, G., Vaillant, R., & Ayache, N. (2006). Reconstruction of Coronary Arteries from a Single Rotational X-Ray Projection Sequence. *IEEE Transactions on Medical Imaging*, *25*(5), 653–663. https://doi.org/10.1109/TMI.2006.873224.

Canero, C., Vilarino, F., Mauri, J., & Radeva, P. (2002). Predictive (Un)Distortion Model and 3-D Reconstruction by Biplane Snakes. *IEEE Transactions on Medical Imaging*, *21*(9), 1188–1201. https://doi.org/10.1109/TMI.2002.804421.

Cardenes, R., Novikov, A., Gunn, J., Hose, R., & Frangi, A. F. (2012). 3D Reconstruction of Coronary Arteries from Rotational X-Ray Angiography. *2012 9th IEEE International Symposium on Biomedical Imaging (ISBI)*. pp. 618–621.

Cervantes-Sanchez, F., Cruz-Aceves, I., & Hernández-Aguirre, A. (2016). Automatic Detection of Coronary Artery Stenosis in X-Ray Angiograms using Gaussian Filters and Genetic Algorithms. *AIP Conference Proceedings*. p. 020005. vol. 1747. AIP Publishing LLC.

Cervantes-Sanchez, F., Cruz-Aceves, I., Hernández-Aguirre, A., Hernandez-Gonzalez, M. A., & Solorio-Meza, S. E. (2019). Automatic Segmentation of Coronary Arteries in X-Ray Angiograms using Multiscale Analysis and Artificial Neural Networks. *Applied Sciences*, *9*(24), 5507. https://doi.org/10.3390/app9245507.

Chan, T. F., & Vese, L. A. (2001). Active Contours without Edges. *IEEE Transactions on Image Processing*, *10*(2), 266–277.

Chen, S. J., & Carroll, J. D. (2000). 3-D Reconstruction of Coronary Arterial Tree to Optimize Angiographic Visualization. *IEEE Transactions on Medical Imaging*, *19*(4), 318–336. https://doi.org/10.1109/42.848183.

Chen, Y., Zhang, Y., Yang, J., Cao, Q., Yang, G., Chen, J., Shu, H., Luo, L., Coatrieux, J.-L., & Feng, Q. (2015). Curve-like Structure Extraction Using Minimal Path Propagation with Backtracking. *IEEE Transactions on Image Processing*, *25*(2), 988–1003.

Chen, L.-C., Zhu, Y., Papandreou, G., Schroff, F., & Adam, H. (2018). Encoder-Decoder with Atrous Separable Convolution for Semantic Image Segmentation. *ECCV 2018*. arXiv:1802.02611.

Çimen, S., Gooya, A., Grass, M., & Frangi, A. F. (2016). Reconstruction of Coronary Arteries from X-Ray Angiography: A Review. *Medical Image Analysis*, *32*, 46–68. https://doi.org/10.1016/j.media.2016.02.007.

Cong, C., Kato, Y., Vasconcellos, H. D., Lima, J., & Venkatesh, B. (2019). Automated Stenosis Detection and Classification in X-Ray Angiography Using Deep Neural Network. *2019 IEEE International Conference on Bioinformatics and Biomedicine (BIBM)*. pp. 1301–1308.

Cong, W., Yang, J., Ai, D., Chen, Y., Liu, Y., & Wang, Y. (2015). Quantitative Analysis of Deformable Model-Based 3-D Reconstruction of Coronary Artery From Multiple Angiograms. *IEEE Transactions on Biomedical Engineering*, *62*(8), 2079–2090. https://doi.org/10.1109/TBME.2015.2408633.

Cruz-Aceves, I., Oloumi, F., Rangayyan, R. M., Aviña-Cervantes, J. G., & Hernández-Aguirre, A. (2016). Automatic Segmentation of Coronary Arteries Using Gabor Filters and Thresholding Based on Multiobjective Optimization. *Biomedical Signal Processing and Control*, *25*, 76–85. https://doi.org/10.1016/j.bspc.2015.11.001.

Fan, J., Yang, J., Wang, Y., Yang, S., Ai, D., Huang, Y., Song, H., Hao, A., & Wang, Y. (2018). Multichannel Fully Convolutional Network for Coronary Artery Segmentation in X-Ray Angiograms. *IEEE Access 6*, 44635–44643. https://doi.org/10.1109/ACCESS.2018.2864592.

Felfelian, B., Fazlali, H. R., Karimi, N., Soroushmehr, S. M. R., Samavi, S., Nallamothu, B., & Najarían, K. (2016). Vessel Segmentation in Low Contrast X-Ray Angiogram Images. *2016 IEEE International Conference on Image Processing (ICIP)*. pp. 375–379.

Galassi, F., Alkhalil, M., Lee, R., Martindale, P., Kharbanda, R. K., Channon, K. M., Grau, V., & Choudhury, R. P. (2018). 3D Reconstruction of Coronary Arteries from 2D Angiographic Projections Using Non-Uniform Rational Basis Splines (NURBS) for Accurate Modelling of Coronary Stenoses. *PLOS ONE, 13*(1), e0190650. https://doi.org/10.1371/journal.pone.0190650.

Grossman, W. (1986). *Cardiac Catheterization and Angiography*. Lea & Febiger.

Habijan, M., Babin, D., Galic, I., Leventic, H., Velicki, L., & Cankovic, M. (2020). Centerline Tracking of the Single Coronary Artery from X-Ray Angiograms. *2020 International Symposium ELMAR*. pp. 117–121. https://doi.org/10.1109/ELMAR4995 6.2020.9219025

Hao, D., Liu, Y., & Qin, B. (2019). Learning Saliently Temporal-Spatial Features for X-ray Coronary Angiography Sequence Segmentation. *Eleventh International Conference on Digital Image Processing (ICDIP 2019), 11179*, 1117930.

Hideo-Kajita, A., Wopperer, S., Beyene, S. S., Meirovich, Y. F., Melaku, G. D., Kuku, K. O., Brathwaite, E. J., Ozaki, Y., Dan, K., Torguson, R., Waksman, R., & Garcia-Garcia, H. M. (2019). Impact of Two Formulas to Calculate Percentage Diameter Stenosis of Coronary Lesions: From Stenosis Models (Phantom Lesion Model) to Actual Clinical Lesions. *The International Journal of Cardiovascular Imaging, 35*(12), 2139–2146. https://doi.org/10.1007/s10554-019-01672-z.

Hoffmann, K. R., Sen, A., Lan, L., Chua, K.-G., Esthappan, J., & Mazzucco, M. (2000). A System for Determination of 3D Vessel Tree Centerlines from Biplane Images. *The International Journal of Cardiac Imaging, 16*(5), 315–330. https://doi.org/10.1023/A:1 026528209003.

Iyer, K., Najarian, C. P., Fattah, A. A., Arthurs, C. J., Soroushmehr, S. M. R., Subban, V., Sankardas, M. A., Nadakuditi, R. R., Nallamothu, B. K., & Alberto Figueroa, C. (2021). AngioNet: A Convolutional Neural Network for Vessel Segmentation in X-Ray Angiography. *MedRxiv, 11*(1). https://doi.org/10.1101/2021.01.25.21250488.

Jandt, U., Schäfer, D., Grass, M., & Rasche, V. (2009). Automatic Generation of 3D Coronary Artery Centerlines Using Rotational X-Ray Angiography. *Medical Image Analysis, 13*(6), 846–858. https://doi.org/10.1016/j.media.2009.07.010.

Jo, K., Kweon, J., Kim, Y., & Choi, J. (2019). Segmentation of the Main Vessel of the Left Anterior Descending Artery Using Selective Feature Mapping in Coronary Angiography. *IEEE Access, 7*, 919–930. https://doi.org/10.1109/ACCESS.2018.2886009

Kerkeni, A., Benabdallah, A., Manzanera, A., & Bedoui, M. H. (2016). A Coronary Artery Segmentation Method Based on Multiscale Analysis and Region Growing. *Computerized Medical Imaging and Graphics, 48*, 49–61. https://doi.org/10.1016/ j.compmedimag.2015.12.004.

Krig, S. (2016). Image Pre-Processing. In S. Krig (Eds.), *Computer Vision Metrics: Textbook Edition* (pp. 35–74). Springer International Publishing.

Liao, R., Luc, D., Sun, Y., & Kirchberg, K. (2010). 3-D Reconstruction of the Coronary Artery Tree from Multiple Views of a Rotational X-Ray Angiography. *The International Journal of Cardiovascular Imaging, 26*(7), 733–749. https://doi.org/10. 1007/s10554-009-9528-0.

Libby, P., & Theroux, P. (2005). Pathophysiology of Coronary Artery Disease. *Circulation, 111*(25), 3481–3488. https://doi.org/10.1161/CIRCULATIONAHA.105.537878

Lv, T., Yang, G., Zhang, Y., Yang, J., Chen, Y., Shu, H., & Luo, L. (2019). Vessel Segmentation Using Centerline Constrained Level Set Method. *Multimedia Tools and Applications, 78*(12), 17051–17075. https://doi.org/10.1007/s11042-018-7087-x.

Ma, H., Hoogendoorn, A., Regar, E., Niessen, W. J., & van Walsum, T. (2017). Automatic Online Layer Separation for Vessel Enhancement in X-Ray Angiograms for Percutaneous Coronary Interventions. *Medical Image Analysis, 39*, 145–161. https:// doi.org/10.1016/j.media.2017.04.011.

Ma, G., Yang, J., & Zhao, H. (2020). A Coronary Artery Segmentation Method Based on Region Growing With Variable Sector Search Area. *Technology and Health Care*, *28*(S1), 463–472.

Maninis, K.-K., Pont-Tuset, J., Arbeláez, P., & Gool, L. V. (2016). Deep Retinal Image Understanding. *ArXiv:1609.01103 [Cs]*, *9901*, 140–148. https://doi.org/10.1007/978-3-319-46723-8_17.

Moccia, S., Momi, E. D., Hadji, S. E., & Mattos, L. S. (2018). Blood Vessel Segmentation Algorithms—Review of Methods, Datasets and Evaluation Metrics. *Computer Methods and Programs in Biomedicine*, *158*, 71–91. https://doi.org/10.1016/j.cmpb.2018.02.001.

Montalescot, G., Sechtem, U., Achenbach, S., Andreotti, F., Arden, C., Budaj, A., Bugiardini, R., Crea, F., Cuisset, T., & Di Mario, C. (2013). 2013 ESC Guidelines on the Management of Stable Coronary Artery Disease: The Task Force on the Management of Stable Coronary Artery Disease of the European Society of Cardiology. *European Heart Journal*, *34*(38), 2949–3003.

Moon, J. H., Cha, W. C., Chung, M. J., Lee, K.-S., Cho, B. H., & Choi, J. H. (2021). Automatic Stenosis Recognition from Coronary Angiography Using Convolutional Neural Networks. *Computer Methods and Programs in Biomedicine*, *198*, 105819. https://doi.org/10.1016/j.cmpb.2020.105819.

Movassaghi, B., Rasche, V., Grass, M., Viergever, M. A., & Niessen, W. J. (2004). A Quantitative Analysis of 3-D Coronary Modeling from Two or More Projection Images. *IEEE Transactions on Medical Imaging*, *23*(12), 1517–1531. https://doi.org/10.1109/TMI.2004.837340.

Nasr-Esfahani, E., Karimi, N., Jafari, M. H., Soroushmehr, S. M. R., Samavi, S., Nallamothu, B. K., & Najarian, K. (2018). Segmentation of Vessels in Angiograms Using Convolutional Neural Networks. *Biomedical Signal Processing and Control*, *40*, 240–251. https://doi.org/10.1016/j.bspc.2017.09.012.

Nowbar, A. N., Gitto, M., Howard, J. P., Francis, D. P., & Al-Lamee, R. (2019). Mortality From Ischemic Heart Disease. *Circulation. Cardiovascular Quality and Outcomes*, *12*(6), e005375–e005375. https://doi.org/10.1161/CIRCOUTCOMES.118.005375.

Ovalle-Magallanes, E., Avina-Cervantes, J. G., Cruz-Aceves, I., & Ruiz-Pinales, J. (2020). Transfer Learning for Stenosis Detection in X-Ray Coronary Angiography. *Mathematics*, *8*(9), 1510. https://doi.org/10.3390/math8091510.

Piegl, L., & Tiller, W. (1996). *The NURBS Book*. Springer Science & Business Media.

Qin, B., Jin, M., Hao, D., Lv, Y., Liu, Q., Zhu, Y., Ding, S., Zhao, J., & Fei, B. (2019). Accurate Vessel Extraction via Tensor Completion of Background Layer in X-Ray Coronary Angiograms. *Pattern Recognition*, *87*, 38–54. https://doi.org/10.1016/j.patcog.2018.09.015.

Ridler, T. W., & Calvard, S. (1978). Picture Thresholding Using an Iterative Selection Method. *IEEE Trans Syst Man Cybern*, *8*(8), 630–632. https://doi.org/10.1109/TSMC.1978.

Ronneberger, O., Fischer, P., & Brox, T. (2015). U-Net: Convolutional Networks for Biomedical Image Segmentation. In Navab, N., Hornegger, J., Wells, W. M., & Frangi, A. F. (Eds.), *Medical Image Computing and Computer-Assisted Intervention – MICCAI 2015* (Vol. 9351, pp. 234–241). Springer International Publishing.

Sameh, S., Azim, M. A., & AbdelRaouf, A. (2017). Narrowed Coronary Artery Detection and Classification Using Angiographic Scans.*2017 12th International Conference on Computer Engineering and Systems (ICCES)*. pp. 73–79. https://doi.org/10.1109/ICCES.2017.8275280

Samuel, P. M., & Veeramalai, T. (2021). VSSC Net: Vessel Specific Skip Chain Convolutional Network for Blood Vessel Segmentation. *Computer Methods and Programs in Biomedicine*, *198*. https://doi.org/10.1016/j.cmpb.2020.105769.

Shi, X., Du, T., Chen, S., Zhang, H., Guan, C., & Xu, B. (2020). *UENet: A Novel Generative Adversarial Network for Angiography Image Segmentation.* 42nd Annual International Conference of the IEEE Engineering in Medicine & Biology Society (EMBC). https://doi.org/10.1109/EMBC44109.2020.9175334.

Shin, S. Y., Lee, S., Yun, I. D., & Lee, K. M. (2019). Deep Vessel Segmentation by Learning Graphical Connectivity. *Medical Image Analysis*, *58*. https://doi.org/10.1016/j.media.2019.101556.

Shorten, C., & Khoshgoftaar, T. M. (2019). A Survey on Image Data Augmentation for Deep Learning. *Journal of Big Data*, *6*(1), 60. https://doi.org/10.1186/s40537-019-0197-0.

Simonyan, K., & Zisserman, A. (2015). Very Deep Convolutional Networks for Large-Scale Image Recognition. *ArXiv:1409.1556 [Cs].*

Tache, I.A., & Glotsos, D. (January, 2020). Vessel Segmentation and Stenosis Quantification from Coronary X-Ray Angiograms. *In International Conference on Medical Imaging and Computer-Aided Diagnosis,* (pp. 27–34). Springer, Singapore.

Thygesen, K., Alpert, J. S., White, H. D., Jaffe, A. S., Apple, F. S., Galvani, M., Katus, H. A., Newby, L. K., Ravkilde, J., & Chaitman, B. (2007). Universal Definition of Myocardial Infarction: Kristian Thygesen, Joseph S. Alpert and Harvey D. White on Behalf of the Joint ESC/ACCF/AHA/WHF Task Force for the Redefinition of Myocardial Infarction. *European Heart Journal*, *28*(20), 2525–2538. https://doi.org/10.1161/CIRCULATIONAHA.107.187397.

Vlontzos, A., & Mikolajczyk, K. (2018). Deep Segmentation and Registration in X-ray Angiography Video. *arXiv preprint*, arXiv:1805.06406.

Vukicevic, A. M., Çimen, S., Jagic, N., Jovicic, G., Frangi, A. F., & Filipovic, N. (2018). Three-Dimensional Reconstruction and NURBS-Based Structured Meshing of Coronary Arteries from the Conventional X-Ray Angiography Projection Images. *Scientific Reports*, *8*(1), 1711. https://doi.org/10.1038/s41598-018-19440-9.

Wan, T., Shang, X., Yang, W., Chen, J., Li, D., & Qin, Z. (2018). Automated Coronary Artery Tree Segmentation in X-Ray Angiography Using Improved Hessian Based Enhancement and Statistical Region Merging. *Computer Methods and Programs in Biomedicine*, *157*, 179–190. https://doi.org/10.1016/j.cmpb.2018.01.002.

Wang, L., Liang, D., Yin, X., Qiu, J., Yang, Z., Xing, J., Dong, J., & Ma, Z. (2020). Coronary Artery Segmentation in Angiographic Videos Utilizing Spatial-Temporal Information. *BMC Medical Imaging*, *20*(1), 110. https://doi.org/10.1186/s12880-020-00509-9.

Wu, W., Zhang, J., Xie, H., Zhao, Y., Zhang, S., & Gu, L. (2020). Automatic Detection of Coronary Artery Stenosis by Convolutional Neural Network with Temporal Constraint. *Computers in Biology and Medicine*, *118*, 103657. https://doi.org/10.1016/j.compbiomed.2020.103657.

Xia, S., Zhu, H., Liu, X., Gong, M., Huang, X., Xu, L., Zhang, H., & Guo, J. (2020). Vessel Segmentation of X-Ray Coronary Angiographic Image Sequence. *IEEE Transactions on Biomedical Engineering*, *67*(5), 1338–1348. https://doi.org/10.1109/TBME.2019.2936460.

Xu, C., & Prince, J. L. (1998). Generalized Gradient Vector Flow External Forces for Active Contours. *Signal Processing*, *71*(2), 131–139. https://doi.org/10.1016/S0165-1684(98)00140-6.

Yang, J., Cong, W., Chen, Y., Fan, J., Liu, Y., & Wang, Y. (2014). External Force Back-Projective Composition and Globally Deformable Optimization for 3-D Coronary Artery Reconstruction. *Physics in Medicine and Biology*, *59*(4), 975–1003. https://doi.org/10.1088/0031-9155/59/4/975.

Yang, S., Kweon, J., Roh, J.-H., Lee, J.-H., Kang, H., Park, L.-J., Kim, D. J., Yang, H., Hur, J., Kang, D.-Y., Lee, P. H., Ahn, J.-M., Kang, S.-J., Park, D.-W., Lee, S.-W., Kim, Y.-H., Lee, C. W., Park, S.-W., & Park, S.-J. (2019). Deep Learning Segmentation of Major Vessels in X-Ray Coronary Angiography. *Scientific Reports*, *9*(1), 1–11. https://doi.org/10.1038/s41598-019-53254-7.

Zhang, J., Wang, G., Xie, H., Zhang, S., HuangN., Zhang, S., & Gu, L. (2020). Weakly Supervised Vessel Segmentation in X-Ray Angiograms by Self-Paced Learning from Noisy Labels with Suggestive Annotation. *Neurocomputing, 417*, 114–127. https://doi.org/10.1016/j.neucom.2020.06.122.

Zhao, H., Shi, J., Qi, X., Wang, X., & Jia, J. (2017). Pyramid Scene Parsing Network. *ArXiv:1612.01105 [Cs]*.

Zheng, S., Meiying, T., & Jian, S. (2010). Sequential Reconstruction of Vessel Skeletons from X-Ray Coronary Angiographic Sequences. *Computerized Medical Imaging and Graphics, 34*(5), 333–345. https://doi.org/10.1016/j.compmedimag.2009.12.004.

Zhu, X., Cheng, Z., Wang, S., Chen, X., & Lu, G. (2020). Coronary Angiography Image Segmentation Based on PSPNet. *Computer Methods and Programs in Biomedicine, 105897*. https://doi.org/10.1016/j.cmpb.2020.105897.

6 Super-Resolution of 3D Magnetic Resonance Images of the Brain

Enrique Domínguez, Domingo López-Rodríguez,
Ezequiel López-Rubio, Rosa Maza-Quiroga,
Miguel A. Molina-Cabello, and
Karl Thurnhofer-Hemsi

CONTENTS

6.1   INTRODUCTION

Magnetic Resonance (MR) imaging is commonly used in medical procedures and diagnoses since it is a non-invasive technique that provides excellent soft-tissue contrast images. It does not require the use of ionizing radiation, together with full three-dimensional capabilities. Moreover, MR imaging allows performing functional, diffusion, and perfusion imaging.

Magnetic resonance images (MRI) with high-resolution (HR) can provide rich anatomical details, crucial for reliable computer-aided radiological diagnosis and image post-processing. Notwithstanding continuous improvements in the acquisition

DOI: 10.1201/9781003120902-6

technology, it is common to detect artifacts in the obtained image. The most common of these are blurring and the appearance of noise, limiting the quality of the produced images.

Besides the practical and operational limits to the acquisition time, the specific MR technology and the tissues' magnetic properties cause the image resolution to fall in the order of millimeters.

Hardware limitations, high signal-to-noise ratios (SNR), and patient motion (even the occasioned by heartbeat or by the patient's breathing) also limit image resolution. Note that MRI data is usually obtained with different voxel sizes in the typical clinical setting depending on the acquired modality. In particular, the in-plane resolution is usually higher than the out-plane one (i.e., in the slice direction), producing non-isotropic voxel sizes (i.e., rectangular voxels).

There are some recent studies on critical problems due to the use of low-resolution (LR) images. In Mulder et al., (2019), the authors investigated voxel geometry's influence by imaging simulated elliptical structures with voxels varying in shape and size. For each reconstructed structure, the authors calculated and analyzed the differences in volume and similarity between the labeled volume and the ellipsoid's predefined dimensions. As a result, the authors find that larger voxels typical of coarser-resolution images, and increasing anisotropy, end in more significant deviations of both volume and shape measures, clearly demonstrating the anatomical inaccuracies introduced in LR images.

Small but clinically important lesions in the brain may be challenging to visualize or characterize correctly when inspecting low-resolution MRIs.

Besides, the increasing use of functional imaging to examine intratumoral heterogeneity has led to a clinical need for improved spatial resolution for these inherently LR sequences.

Therefore, in some cases, the acquired images need an upsampling to decrease the voxel size and obtain higher resolution images, which will be post-processed or analyzed.

6.2 SUPER-RESOLUTION: DEFINITION

The formation of the LR image from an HR image follows a degradation model $I_{LR} = D(I_{HR}; \omega)$, where the degradation operator D acts on the HR image and whose parameters are represented by ω.

This degradation function D is usually defined as a sequential composition of blurs, downsampling, and the addition of noise. A complete model considers all these choices. The result is the following model (Zhang et al., 2018):

$$I_{LR} = (b * I_{HR}) \downarrow_s + \eta_\sigma \qquad (6.1)$$

where $b * I$ represents the convolution between a blur kernel b and a latent HR image I, $\downarrow_s$ is a subsequent downsampling operation with scale factor s, and η usually is additive white Gaussian noise with standard deviation (noise level) σ.

Image super-resolution aims at reconstructing the corresponding HR images from the LR images. We present a formal definition of the problem below.

Let us consider a three-dimensional (low-resolution) image I_{LR} where, without loss of generality, we will assume that voxel coordinates are linearly spaced in a cubic grid with $(0, 0, 0)$ as one vertex and the opposite vertex is given by (D_x, D_y, D_z). Voxel spacing is characterized by a vector (h_x, h_y, h_z), where h_s indicates the distance along the s-axis between the centers of two adjacent voxels. This means that voxel coordinates can be expressed in the form $(n_x h_x, n_y h_y, n_z h_z)$ for some $n_x, n_y, n_z \in N$.

The problem of super-resolution consists of determining an image I_{HR} with spacing $(\alpha h_x, \alpha h_y, \alpha h_z)$ for some $\alpha \in (0, 1)$ and such that $I_{HR} = I_{LR}$ in all the voxels where their corresponding grids coincide. The lower the value of α, the finer the resolution of I_{HR} is. The inverse of α, $\frac{1}{\alpha}$, is usually referred to as zoom factor and generally takes integer values. Using the "times" notation, commonly used zoom factors are 2x, 3x, 4x...

Note that this problem is ill-posed (López-Rubio, 2016) since, as it is defined, the solution is not unique. Multiple high-resolution images I_{HR} can be degraded to the same I_{LR}.

6.3   PREVIOUS WORKS

This section aims to give a comprehensive and brief overview of previous works in image super-resolution (SR). The process of generating high-resolution (HR) images from low-resolution (LR) images can be performed using a single image or multiple images. Most of the existing literature surveys are mainly focused on single image super-resolution, which has been extensively studied and reviewed in the following subsections. Additionally, in the next section, we provide an overview of recent advances in SR for higher-dimensional images (such as 3D scans).

SR methods can be divided into two main categories: traditional and deep learning methods. The former methods have been studied for decades, but now they are outperformed by the deep learning-based approaches, which have shown promising results in other fields in artificial intelligence, such as object classification and detection, natural language processing, audio signal processing, etcetera.

6.3.1   TRADITIONAL METHODS

The classical way of obtaining an SR image is by polynomial interpolation, which represents an arbitrary continuous function underlying the discrete samples that make up the LR image. In general, traditional SR methods are broadly classified as techniques based on the frequency domain, interpolation, or regularization, where the last two approaches are based on the spatial domain.

The methods based on the frequency domain are an intuitive way to enhance the details of the images by extrapolating the high-frequency information of LR images. These methods consist of transforming the input LR image(s) to the frequency domain, estimating the HR image, and then transforming back the HR image to the spatial domain. In general, we can divide these methods into two groups according to the transformation employed: algorithms based on Fourier or wavelet transformation. Detailed explanations of these techniques and a list of different provided approaches in the literature are described in Nasrollahi & Moeslund, (2014).

The approaches based on the interpolation construct an HR image by projecting all the information available from LR images. Usually, this reconstruction process consists of the following steps: image registration, multi-channel restoration, image fusion, and image interpolation. These techniques and the numerous algorithms available in the literature for image interpolation and SR are addressed in (Abd El-Samie et al., 2012), where several chapters are devoted to each stage, including simulation experiments along with the MATLAB® code.

Regularization-based methods consist of incorporating the prior knowledge of the unknown HR image by using a regularization strategy. This information can be extracted from the LR images, which is contained in the probability distribution of the unknown signal (HR image). From the Bayesian point of view, the HR image can be estimated by applying Bayesian inference to exploit the information provided by the LR images and the prior knowledge of the HR image. More details about the different regularization strategies provided in the literature and the related references are described in (Tian & Ma, 2011).

In the literature, a great variety of classical SR methods have been proposed, where most of them are surveying at several works (Abd El-Samie et al., 2012; Nasrollahi and Moeslund, 2014; van Ouwerkerk, 2006; Shah and Gupta, 2012; Tian & Ma, 2011). These surveys also provide different taxonomies covering all the types of traditional SR techniques and comparative discussions of the different methods.

6.3.2   Deep Neural Networks

SR models based on deep learning (DL) have been actively explored in recent years due to the rapid development of DL techniques, which have outperformed the state-of-the-art algorithms on various SR benchmarks. A great variety of DL methods have been proposed in the literature, which can be classified according to the most distinctive features in their model designs (e.g., network architecture, loss function, learning principles and strategies, etcetera.). Additionally, several taxonomies have been recently proposed (Anwar et al., 2020; Wang et al., 2020), covering the recent advances of SR techniques based on DL systematically and comprehensively.

Most of the existing SR works based on DL are focused on supervised learning, i.e., these models are trained with both LR images and the corresponding HR images. A taxonomy for these models grouped into nine categories is proposed in (Anwar et al., 2020), where a comparison between these models in terms of network complexity, memory footprint, model input and output, learning details, type of loss functions, and other architectural differences are also presented.

The Super-Resolution Convolutional Neural Network (SRCNN) was the pioneering work (Dong et al., 2014) using DL techniques that inspired several later works. Basically, it consists of convolutional layers where each layer is stacked together linearly with Rectified Linear Unit (ReLU). The functionality of these layers is different, and it varies from the feature extraction of the first layer to the aggregation of the features maps to the final HR image of the last layer. This model is trained by minimizing the difference between the reconstructed HR image and the ground truth HR image using Mean Squared Error (MSE). This is the earliest and simplest network design, which can be categorized as linear networks due to the linear network architecture. In this direction, several works can be found in the literature: a fast version (FSRCNN) which improves the speed and quality achieving a real-time rate (24 fps) of computation (Dong et al., 2016), and another fast SR approach using an efficient sub-pixel convolution capable of processing 1080p videos in real-time (Shi et al., 2016).

In addition to the above models, other further improvements have been proposed, which essentially differ in some components such as model frameworks, up-sampling methods, network design, or learning strategies. These works are addressed in (Anwar et al., 2020), which can be divided into several categories, including linear networks, residual networks, recursive networks, progressive reconstruction designs, densely connected networks, multi-branch designs, attention-based networks, multiple-degradation handling networks, and generative adversarial networks (GAN).

In general, most of the state-of-the-art SR models can be attributed to a combination of multiple strategies like the channel attention mechanism, sub-pixel up-sampling, residual learning, etcetera. These models are mostly focused on supervised learning, i.e., learning with LR-HR image pairs. However, in some real-world scenarios, it is not easy to collect images with different resolutions. Thus, datasets are constructed by performing some predefined degradations on HR images (and obtaining the paired LR image). The main drawback of these datasets is that the trained SR models actually learn a reverse process of the predefined degradation instead of the real-world LR-HR mapping. To avoid this behavior, unsupervised SR is needed, where only unpaired LR-HR images are provided for training. A summarized table including some representative models with their key strategies is provided in Wang et al. (2020), where unsupervised SR models are also discussed.

Apart from the above general SR works, there exist some other popular domain-specific applications where SR can serve to advance in those fields. Medical imaging is one of these fields which is rapidly evolving in increased resolution devices, demonstrating the potential of SR research into practical medical applications. Besides the existing SR works in medical imaging (Greenspan, 2009), recent promising advances to apply SR techniques in medical imaging applications are reviewed in the next section.

6.4 IMPROVED DEEP LEARNING METHODS

The enhancement of the results of deep learning super-resolution methods for 3D MRIs can be accomplished in two distinct ways. We consider these alternatives next.

First, it is possible to modify the architecture of the deep network in order to improve its super-resolution performance. There are several possibilities to do this, including the insertion or deletion of neural layers, the modification of the number of channels in the neural layers, a change in the stride parameter of the convolutional layers, etcetera. A novel strategy to obtain a better performing architecture consists of changing the loss function, as proposed in Thurnhofer-Hemsi et al. (2019) and Thurnhofer-Hemsi et al. (2020a). The standard loss function for super-resolution deep learning networks is the squared Euclidean norm of the difference between the high-resolution version of the original image (the ground truth) and its reconstruction that the network produces as output (the estimation). While the squared Euclidean norm has demonstrated its suitability for many purposes, some improvements can be obtained by employing other alternatives.

A possible alternative is the family of Lp norms. The members of this family are characterized by a parameter p that specifies to which power the absolute value of the error is raised. The value $p = 2$ corresponds to the usual squared Euclidean norm. On the one hand, higher values ($p > 2$) give more importance to extreme values of the absolute error. This is not convenient since there can be a small number of voxels in the MRI with gross errors or irrelevant features, which could have an outsized influence on the learning of the deep network parameters. On the other hand, smaller values ($p < 2$) give less importance to outliers than the standard squared Euclidean norm. Therefore, they are more promising to yield appropriate loss functions for super-resolution since they provide robustness with respect to outliers. Consequently, in Thurnhofer-Hemsi et al. (2019), the usage of Lp norms with $p < 2$ for 3D MRI super-resolution is proposed. It must be remembered that the mathematical definition of norm includes three conditions that must be fulfilled: non-negativity, linearity with respect to a multiplying factor, and the triangle inequality. The last condition is only held for $p \geq 1$. Given these considerations, Lp norms with values of p in the range $1 \leq p \leq 2$ are studied.

The derivation of a learning rule for an Lp norm is based on the definition of a loss function as the sum of the Lp norms of the absolute errors for all samples. Then the partial derivatives of the loss function are calculated with respect to the synaptic weights. Finally, the stochastic gradient descent method is employed in order to obtain a procedure that adaptively modifies the synaptic weights to minimize the loss function.

The usage of an Lp norm with $p < 2$ as the loss function can increase the resilience of the deep network to outliers. Nevertheless, there is the risk that the learning procedure does not pay enough attention to significant errors. In order to reconcile both requirements, it is proposed in Thurnhofer-Hemsi et al. (2020a) that a multiobjective optimization is carried out to learn the synaptic weights of the deep network. This is accomplished by setting two goals to be optimized. The first one is the minimization of the standard squared Euclidean norm, while the second one is the minimization of a suitable Lp norm with $p < 2$. In order to implement this method in practice, the scalarization strategy is advocated to obtain a single loss function out of the two norms. Two variants of scalarization can be applied to this purpose. The first one is weighted sum scalarization (Gass & Saaty, 1955). This is a straightforward approach where the multiobjective loss function is given as the weighted average of the squared Euclidean norm and the Lp norm. Here the weighting factors for the norms must be adjusted from the data. The second scalarization variant is weighted Chebyshev scalarization (Gong, 2011). This is a more elaborate approach that requires the estimation of an ideal point, which is an ordered pair formed by the minimum of the first norm and the minimum of the second norm, both computed over the full domain of both norms. The multi-objective loss function is then defined as the weighted Chebyshev distance to that point, where again we have two weighting factors that must be adjusted. For the purposes of combining the squared Euclidean norm and an Lp norm, the (0,0) point can be employed as the ideal point since neither of them can be negative. After the multiobjective loss function has been selected, the methodology proceeds as before. In other words, the partial derivatives of the loss function with respect to the synaptic weights are calculated, followed by its minimization by the stochastic gradient descent method.

The second family of methods to enhance the performance of deep convolutional neural network-based super-resolution of 3D MRIs is based on the combination of the output images obtained for shifted versions of the original input image. The rationale behind this strategy is that the behavior of the convolutional network is slightly different for those versions of the input, so that merging the associated outputs may remove the defects which are present in a minority of such outputs. In other words, a consensus image is built from the individual output images. Since the output images are shifted by the same displacement vector as its associated shifted input, it is necessary to undo the shift on each output image prior to the construction of the consensus image.

Two approximations can be distinguished for this shifting and consensus strategy. The first one is presented in Thurnhofer-Hemsi et al. (2018), and it is called random shifting. Here the displacement vector is modeled as a random variable that is uniformly distributed over a cube of possible integer displacements. Then the mathematical expectation of the unshifted output image is estimated over the distribution of the displacement vector, which is to be taken as the consensus final image. This estimation is carried out by averaging the unshifted images for a sample of randomly drawn displacements. As the number of samples tends to infinity, by the law of large numbers, it is known that such estimation converges to the true value of the mathematical expectation. Therefore, the more samples that are considered, the more accurate the consensus image is expected to be. This is because the mathematical expectation of the unshifted output image has no dependency on a specific displacement vector, while the unshifted images depend on the displacement vectors that have been used to produce them. Under this approach, two parameters must be optimized, namely the size of the cube of possible integer displacements and the number of samples.

The second approximation to shifting is given in Thurnhofer-Hemsi et al., (2020b). In this case, an underlying function is assumed to exist that takes the displacement vector as an argument and outputs the high-resolution 3D MRI. This underlying function is defined on the set of possible displacement vectors. In order to approximate the function, a regular lattice of displacement vectors is considered. This means that the procedure has two tunable parameters, namely the pixel stride that defines the distance between two consecutive points in the

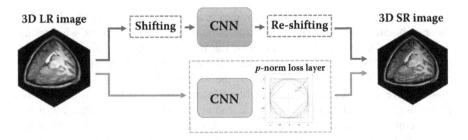

FIGURE 6.1   General overview of the latest super-resolution proposals.

lattice and the side of the cube of points in the lattice that will be considered. After these parameters are determined, the final consensus image is built by averaging the output images corresponding to the displacements that are associated with the points of the lattice within the chosen cube. This way, a zeroth-order approximation of the underlying function is implemented.

As shown above, the Lp norm and the multiobjective optimization approaches involve the proposal of new deep learning schemes, as shown in Figure 6.1. On the contrary, the random shifting and the regular shifting approaches are metamethods that can be applied to any 3D super-resolution deep learning architecture, which implies that they can enhance the results of new architectures as they are proposed.

6.5   EXAMPLES AND COMPARISONS

This Subsection describes several experimental issues such as the methods and datasets that can be employed in the experiments, the measures used to yield the performance of a selected method, or how to assess a fair analysis about the comparison between the competitor methods. Additionally, experiments have been carried out in order to evaluate the performance of each method.

6.5.1   METHODS

As previously mentioned, many algorithms that face the field of super-resolution of 3D Magnetic Resonance images can be found in the literature. Here some of them are presented:

- Spline. Bicubic spline interpolation as implemented in MATLAB (Mathworks Inc.).
- NLMU (non-local means upsampling) (Manjón et al., 2010). A sub-sampling coherence constraint combined with a data-adaptive patch-based reconstruction is used to recover some high-frequency information. It has been written in MATLAB.
- LRTV (low-rank total variation) (Shi et al., 2015). This method employs low-rank regularization and total variation techniques to integrate both local and global information for image reconstruction. It has been developed in MATLAB.

- SRCNN3D (Pham et al., 2017). A three-dimensional convolutional neural deep network has been trained with patches of HR brain images. It predicts a mapping from the LR space to the missing high-frequency components. This method is implemented in Python by using the Caffe package (Jia et al., 2014). A pre-trained network is available on its website. This model has been trained with ten images (in particular, images 33–42) from the Kirby dataset (Landman et al., 2011) over 470,000 iterations. Other parameters of that training process were a learning rate of 0.0001, a batch size of 256, and a momentum of 0.9, while the chosen model optimization was the stochastic gradient descent.
- SRCNN3D+RegSS (Thurnhofer-Hemsi et al., 2020b). A convolutional neural network is combined with a regularly spaced shifting mechanism over the input image. It has been implemented in Python.
- SRCNN3D+RndS (Thurnhofer-Hemsi et al., 2018). This method is a previous version of the SRCNN3D+RegSS approach based on the use of a random shifting technique.
- SRCNN3D-i50K-WCS, SRCNN3D-i50K-WSS (Thurnhofer-Hemsi et al., 2020a). They use a combination of Lp-norms in the loss layer where two multiobjective optimization techniques are employed to combine two cost functions: Weighted sum scalarization (WSS) and Weighted Chebyshev scalarization (WCS). The chosen convolutional neural network is the SRCNN3D method. In this case, that was trained over 50,000 iterations.
- SRCNN3D-i50K-p1.9, SRCNN3D-i50K-p2 (Thurnhofer-Hemsi et al., 2019). They are previous versions of SRCNN3D-i50K-WCS and SRCNN3D-i50K-WSS methods. They are based on the use of a p-norm loss layer to improve the learning process. Two versions with $p = 1.9$ and $p = 2.0$ are used in the experiments. Again, the chosen network is SRCNN3D, and it was trained over 50,000 iterations.

6.5.2 DATASETS

The experiments should be performed on well-known datasets, public if possible. Some datasets which can be found in the literature are:

- Kirby 21 (Landman et al., 2011). In particular, two T1-weighted MRI images of this dataset (10 and 11) were used in the experiments. These data were acquired on a 3-T MR scanner with a $1.0 \times 1.0 \times 1.2$ mm^3 voxel resolution over a field-of-view (FOV) of $240 \times 204 \times 256$ mm acquired in the sagittal plane.
- OASIS (Marcus et al., 2007). Specifically, experiments use two T1 images of this dataset (images 1 and 2 of the cross-sectional data). These data were acquired on a 1.5-T Vision scanner with a $1.0 \times 1.0 \times 1.25$ mm^3 voxel resolution over a FOV of 256×256 mm.
- IBSR (Worth, 2010). In the subsequent experiments, an image from this dataset is used. The features of this image are a size of $256 \times 256 \times 128$, with $1.5 \times 1.0 \times 1.0$ mm^3 voxel resolution.

- CIMES. A T1-weighted image was acquired at the Medical Research Center of the University of Malaga (CIMES) using a 3-T MR scanner with a $0.93 \times 0.93 \times 1.0$ mm^3 voxel resolution over a FOV of 256×256 mm.

Images used in the experimental test step must be different from those used to train the methods to establish a fair comparison.

All these datasets are formed by HR images. Due to the aim of described methods is to generate an HR image from an LR image, the HR images from the datasets are downsampled to obtain an LR version of them. The procedure to achieve these LR datasets is as follows. First, in order to avoid fractional values, HR images are cropped regarding the zoom factor to be applied. Then, a three-dimensional Gaussian filter with a standard deviation equal to 1 is applied. Finally, the LR image is generated by carrying out a downsampling method. This method can be an interpolation function such as the nearest neighbor approach, bicubic or bilinear interpolations, among others (Molina-Cabello et al., 2020). Some of these mentioned methods can be used through the imresize3 function of MATLAB. In particular, a cubic interpolation has been used in the experiments shown subsequently.

6.5.3 MEASURES

In order to compare the goodness of several competitor methods between them, several quality measures may be used to evaluate the performance of each method.

From a quantitative point of view, some quantitative measures may be considered. In particular, in the experiments carried out in this work, three well-known measures have been selected:

- Peak Signal-to-Noise Ratio (PSNR). It is measured in decibels (dB). Higher is better.

$$PSNR = 10\left(\frac{peak^2}{\|Y - \hat{Y}\|^2}\right) \qquad (6.2)$$

- Structural Similarity Index (SSIM) (Wang et al., 2004). It measures the structural similarities between images. Higher is better. It is defined as follows:

$$SSIM\,(x, y) = \frac{(2\mu_x\mu_y)(2\sigma_{xy} + c_2)}{(\mu_x^2 + \mu_y^2 + c_1)(\sigma_x^2 + \sigma_y^2 + c_2)} \qquad (6.3)$$

- Bhattacharyya coefficient (BC) (Bhattacharyya, 1946). It is focused on the closeness of the two discrete pixel probability distributions P and $\hat{P}$ corresponding to the ground truth (GT) and restored images with values in the range [0,255]:

$$BC = \sum_{j=0}^{255} P(j)\hat{P}(j) \qquad (6.4)$$

Other measures may be evaluated, such as CPU time.

Regarding a visual or qualitative point of view, besides the restored image produced by each competitor method, the residual image is also employed in the comparison. This residual image r is computed as the difference in absolute value between the original HR image h and the restored one s:

$$r = |h - s| \qquad (6.5)$$

Due to that difference might be close to zero, a darker residual image implies better performance. In the shown experiments, color maps were adjusted to obtain better visualization and discrimination between competitors.

6.5.4 RESULTS

Next, some experimental comparisons between the previously mentioned state-of-art methods are summarized.

First of all, we measure the performance from a quantitative point of view. Thus, employing the MPRAGE, IBSR, and CIMES images, we computed the mean and standard deviation of the PSNR, SSIM, and BC, and the results are presented in Figures 6.2 and 6.3. Within the deep networks, we should differentiate two groups of methods: SRCNN3D and SRCNN3D-i50K. The first has been trained for a larger number of iterations, 470,000, while the second for only 50,000. That's because the authors used a pre-trained model, which required a long time to be completed. Therefore, the results of the SRCNN3D-i50K based methods may not be good enough as the others, so a specific analysis is going to be done.

The first thing the reader can observe is the considerable difference between the deep neural network-based methods (SRCNN3D) and traditional methods (Spline, LRTV, NLMU). In Figure 6.2, where the results for zoom factor 2 are depicted, the general accuracy of the latter is clearly behind the newest ones in the three metrics. The shifting techniques proposed by Thurnhofer-Hemsi et al. (2018, 2020b) reported the best results for PSNR and SSIM and the second best with the BC. Nevertheless, the WCS and WSS-based methods (Thurnhofer-Hemsi et al., 2020a) also reached high PSNR values with a significantly low number of iterations, which means that the training procedure can be improved a lot. Although the SSIM and BC are not the best, they are also good. The best traditional method seems to be NLMU, although LRTV also yielded similar results.

In Figure 6.3, performances are reported for zoom factor 3. Here, the performance of the multiobjective optimization methods is remarkable, especially in terms of PSNR reaching 30dB, and BC, very close to 1. The SSIM of the traditional methods has improved, which means that they reconstruct adequately more complex problems (larger-scale factors), although they are still below in the rest of the measures.

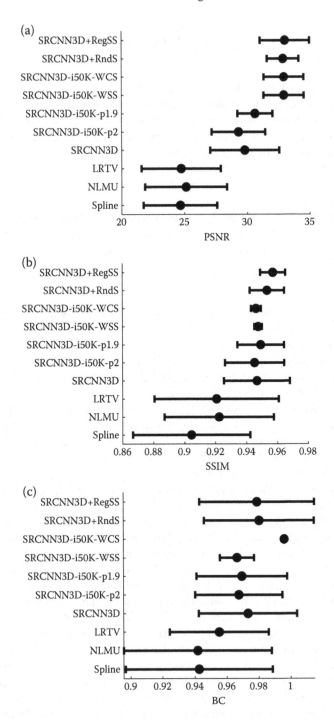

FIGURE 6.2 Comparison of the PSNR, SSIM, BC (higher is better) for the ten methods. Mean and standard deviation of the results for all the test images using zoom factor 2.

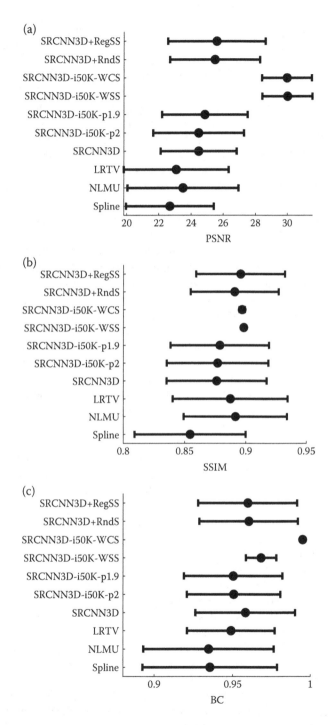

FIGURE 6.3 Comparison of the PSNR, SSIM, BC (higher is better) for the ten methods. Mean and standard deviation of the results for all the test images using zoom factor 3.

If we carry out a detailed inspection of the types of methods, the comparisons of the Lp norm conclude that $p = 1.9$ yielded better results and the traditional Euclidean norm for all cases and reconstruction scales. On the other hand, the regular shifting technique (SRCNN3D+RegSS) is slightly better than the random version, which might be stretchily related to the brain structures. The combination of the shifting method with the Lp norm may produce better results.

From a qualitative point of view, the outcomes of the compared methods are presented in Figures 6.4, 6.5, and 6.6. The outputs of the SRCNN3D-i50K-WSS and SRCNN3D-i50K-WCS methods are not available since both methods were trained with these images but not tested. First, Figure 6.4 depicts a sectional 3D view of the KKI2009-11-MPRAGE T1-weighted image. Restored and residual images are presented, as well as the original HR image and the downsampled one. Spline and LRTV methods generate blurred super-resoluted images, which can be clearly observed in Figures 6.4 (c) and (d). In order to differentiate better the rest of the outcomes, the residual images are more appropriate. With them, the NLMU method can also be discarded, as well as the $L2$ norm-based version (SRCNN3D and SRCNN3D-i50K-p2). The best restorations are obtained by the image shifting methods.

Second, Figures 6.5 and 6.6 depict the restored and residual images, respectively, for a coronal section of the CIMES image using a zoom factor equal to 3. The main difficulty of this case is that the number of voxels to be reconstructed is higher, so the differences between the methods are very tiny. In this example, the regular shifting does not provide the best outcome, which is yielded by the Lp norm-based methods. Concretely, $p = 1.9$ seems to be the most accurate. Nevertheless, the random shifting also worked well, but not the traditional techniques.

6.6   CONCLUSIONS

Obtaining HR by LR using single images has been widely considered. On the other hand, approaches in SR for multidimensional images show more promising results. At present, traditional methods of SR are outperformed by the deep learning-based method. SR based on DL shows multiple encouraging taxonomies in last year. DL's pioneering work utilized convolutional networks (SRCNN), and the recent future shows improvements with works that vary in some components like GAN's works.

Many 2D models were developed in the last years, but 3D models are not plenty studied. This chapter reports four recent 3D methodologies that upgrade the current state of the art. Globally, there exist two ways to enhance a model: modify the model architecture and vary the input images.

Two robust three-dimensional super-resolution methods for MRIs were presented based on the use of a p-norm loss layer ($1 \le p \le 2$), at first, and combinations of Lp norm cost functions ($p < 2$) using the weighted sum *and* the weighted Chebyshev scalarizations, in the second, alternately of the conventional Euclidean formulation ($p = 2$). Qualitatively, restored images look more refined and with less structural degradation. Future research with deeper neural networks might increase the efficiency with lower values of p. Besides, adding more p-norms might improve quality images. Nevertheless, complex optimization problems must be solved, which consists of the arduous task of choosing p-norm.

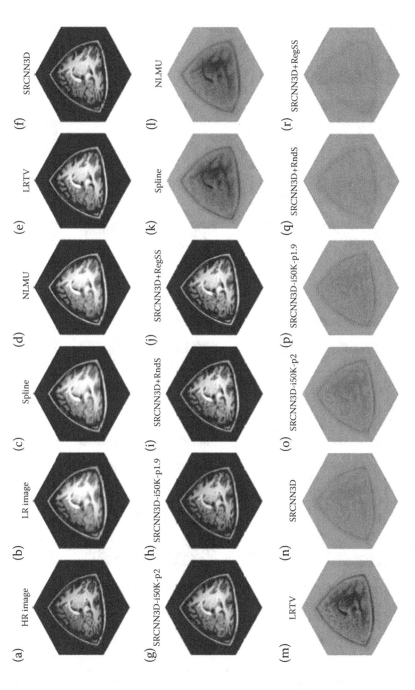

FIGURE 6.4 Qualitative results for KKI2009-11-MPRAGE image for eight methods, applied with zoom factor 2. Three-dimensional images are shown, where the XY plane corresponds to a slice of the axial view, XZ to a slice of the sagittal view, and YZ to a slice of the coronal view.

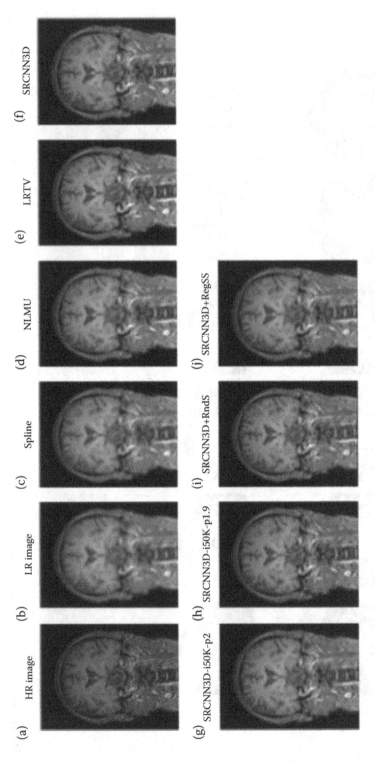

FIGURE 6.5 Restored T1-weighted images from CIMES for eight methods, applied with zoom factor 3. A coronal view is shown. The first row displays the original HR image and the input LR image.

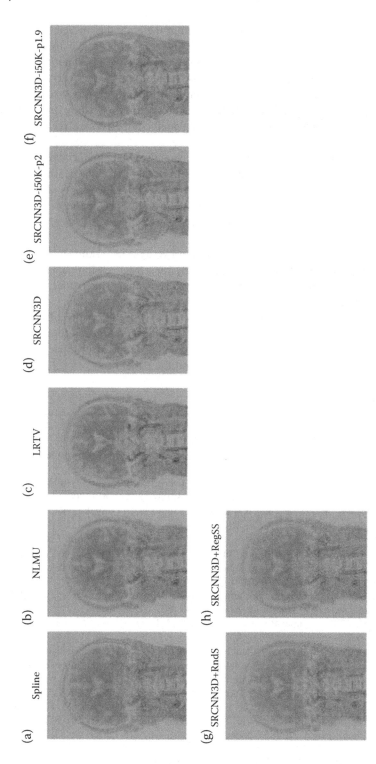

FIGURE 6.6 Residual T1-weighted images from CIMES for eight methods, applied with zoom factor 3. The coronal view is shown.

On the other hand, two MRI super-resolution methods that merge two methodologies have been presented: restoring a low-resolution image by convolutional neural network processing and quality increasing by random, in the first, and regular, in the second, shifting to the input images. The over-smoothing was weathered by using different zoom factors. Blooming results were reached that overcome other state-of-art methods. Future lines of study incorporate the development of additional tuned filtering methods performed on the three-dimensional shift space.

Lp norm and multiobjective optimization generate new DL architectures, and the random/regular shifting approach improves DL results. Machine learning could draw on the applicability of this method. Other neural networks could include these methodologies to improve the quality of the outputs. The line most directly is applying other neural networks to improve the image's quality in widespread tasks like noise removal or segmentation.

The problems of acquiring quality MRI images are widely known, like health budget for a specific machine, with high-quality long MRI sessions fewer patients are intervened, patients' claustrophobic fear of spending a long time in the resonance machine, and some more.

The finding of a methodology that allows improving the quality of an image will reduce healthcare costs and interventions duration and, therefore, increase the number of patients review daily. It supposes an immediate impact both for healthcare, including costs and physicians, and the patient.

REFERENCES

Abd El-Samie, F.E., Hadhoud, M., & El-Khamy, S. (2012). *Image Super-Resolution and Applications.* CRC Press.

Anwar, S., Khan, S., & Barnes, N. (2020). A Deep Journey into Super-resolution. *ACM Computing Surveys (CSUR), 53*, 1–34.

Bhattacharyya, A. (1946). On a Measure of Divergence Between Two Multinomial Populations. *Sankhyā: The Indian Journal of Statistics, 7*(4), 401–406.

Dong, C., Loy, C.C., He, K., & Tang, X. (2014). Learning a Deep Convolutional Network for Image Super-Resolution. *ECCV, 8692*, 184–199.

Dong, C., Loy, C.C., & Tang, X. (2016). Accelerating the Super-Resolution Convolutional Neural Network. *ECCV*, 391–407. https://doi.org/10.1007/978-3-319-46475-6_25.

Gass, S., & Saaty, T. (1955). The Computational Algorithm for the Parametric Objective Function. *Naval Research Logistics Quarterly, 2*, 39–45.

Gong, X.H. (2011). Chebyshev Scalarization of Solutions to the Vector Equilibrium Problems. *Journal of Global Optimization, 49*, 607–622.

Greenspan, H. (2009). Super-Resolution in Medical Imaging. *The Computer Journal, 52*, 43–63.

Jia, Y., Shelhamer, E., Donahue, J., Karayev, S., Long, J., Girshick, R.B., Guadarrama, S., & Darrell, T. (2014). Caffe: Convolutional Architecture for Fast Feature Embedding. *Proceedings of the 22nd ACM international conference on Multimedia*, 675–678.

Landman, B., Huang, A., Gifford, A., Vikram, D.S., Lim, I., Farrell, J., Bogovic, J., Hua, J., Chen, M., Jarso, S., Smith, S., Joel, S., Mori, S., Pekar, J., Barker, P., Prince, J., & Zijl, P. (2011). Multi-parametric Neuroimaging Reproducibility: A 3-T Resource Study. *NeuroImage, 54*, 2854–2866.

López-Rubio, E. (2016). Superresolution from a Single Noisy Image by the Median Filter Transform. *SIAM Journal on Imaging Sciences*, *9*, 82–115.

Manjón, J., Coupé, P., Buades, A., Fonov, V., Collins, D., & Robles, M. (2010). Non-local MRI Upsampling. *Medical Image Analysis*, *14*(6), 784–792.

Marcus, D., Wang, T.H., Parker, J., Csernansky, J., Morris, J., & Buckner, R. (2007). Open Access Series of Imaging Studies (OASIS): Cross-sectional MRI Data in Young, Middle Aged, Nondemented, and Demented Older Adults. *Journal of Cognitive Neuroscience*, *19*, 1498–1507.

Molina-Cabello, M.A., García-González, J., Luque-Baena, R.M., & López-Rubio, E. (2020). The Effect of Downsampling–Upsampling Strategy on Foreground Detection Algorithms. *Artificial Intelligence Review*, *53*, 4935–4965.

Mulder, M., Keuken, M., Bazin, P., Alkemade, A., & Forstmann, B. (2019). Size and Shape Matter: The Impact of Voxel Geometry on the Identification of Small Nuclei. *PLOS ONE*, *14*(4).

Nasrollahi, K., & Moeslund, T. (2014). Super-Resolution: A Comprehensive Survey. *Machine Vision and Applications*, *25*, 1423–1468.

Ouwerkerk, J. (2006). Image Super-resolution Survey. *Image and Vision Computing*, *24*, 1039–1052.

Pham, C., Ducournau, A., Fablet, R., & Rousseau, F. (2017). Brain MRI Super-resolution Using Deep 3D Convolutional Networks. *2017 IEEE 14th International Symposium on Biomedical Imaging (ISBI 2017)*, 197–200.

Shah, A.J., & Gupta, S. (2012). Image Super Resolution-A Survey. *2012 1st International Conference on Emerging Technology Trends in Electronics, Communication & Networking*, 1–6.

Shi, F., Cheng, J., Wang, L., Yap, P., & Shen, D. (2015). LRTV: MR Image Super-Resolution with Low-Rank and Total Variation Regularizations. *IEEE Transactions on Medical Imaging*, *34*, 2459–2466.

Shi, W., Caballero, J., Huszár, F., Totz, J., Aitken, A., Bishop, R., Rueckert, D., & Wang, Z. (2016). Real-Time Single Image and Video Super-Resolution Using an Efficient Sub-Pixel Convolutional Neural Network. *2016 IEEE Conference on Computer Vision and Pattern Recognition (CVPR)*, 1874–1883.

Thurnhofer-Hemsi, K., López-Rubio, E., Roé-Vellvé, N., Domínguez, E., & Molina-Cabello, M.A. (2018). Super-Resolution of 3D Magnetic Resonance Images by Random Shifting and Convolutional Neural Networks. *2018 International Joint Conference on Neural Networks (IJCNN)*, 4008–4015.

Thurnhofer-Hemsi, K., López-Rubio, E., Roé-Vellvé, N., & Molina-Cabello, M.A. (2019). Deep Learning Networks with p-norm Loss Layers for Spatial Resolution Enhancement of 3D Medical Images. *IWINAC*, 287–296. https://doi.org/10.1007/978-3-030-19651-6_28.

Thurnhofer-Hemsi, K., López-Rubio, E., Roé-Vellvé, N., & Molina-Cabello, M.A. (2020a). Multiobjective Optimization of Deep Neural Networks with Combinations of Lp-norm Cost Functions for 3D Medical Image Super-resolution. *Integrated Computer-Aided Engineering*, *27*, 233–251.

Thurnhofer-Hemsi, K., López-Rubio, E., Domínguez, E., Luque-Baena, R.M., & Roé-Vellvé, N. (2020b). Deep Learning-based Super-resolution of 3d Magnetic Resonance Images by Regularly Spaced Shifting. *Neurocomputing*, *398*, 314–327.

Tian, J., & Ma, K. (2011). A Survey on Super-resolution Imaging. *Signal, Image and Video Processing*, *5*, 329–342.

Wang, Z., Bovik, A., Sheikh, H.R., & Simoncelli, E.P. (2004). Image Quality Assessment: From Error Visibility to Structural Similarity. *IEEE Transactions on Image Processing*, *13*, 600–612.

Wang, Z., Chen, J., & Hoi, S. (2020). Deep Learning for Image Super-resolution: A Survey. *IEEE Transactions on Pattern Analysis and Machine Intelligence*, *43*(10). https://doi.org/10.1109/TPAMI.2020.2982166.

Worth, A.J. (2010).MGH CMA Internet Brain Segmentation Repository (IBSR). http://www.cma.mgh.harvard.edu/ibsr/.

Zhang, K., Zuo, W., & Zhang, L. (2018). Learning a Single Convolutional Super-Resolution Network for Multiple Degradations. *2018 IEEE/CVF Conference on Computer Vision and Pattern Recognition*, 3262–3271.

7 Head CT Analysis for Intracranial Hemorrhage Segmentation

*Heming Yao, Negar Farzaneh, Jonathan Gryak,
Craig Williamson, Kayvan Najarian, and
Reza Soroushmehr*

CONTENTS

7.1 INTRODUCTION

Traumatic brain injury (TBI) is caused by a number of principal mechanisms, in-cluding motor-vehicle crashes, falls, and assaults (Taylor et al., 2017), and is a major cause of injury-related deaths worldwide. Incidence, primarily driven by falls and road injuries, is increasing due to an aging population and increased use of motor vehicles, motorcycles, bicycles, and population density. In addition to falls and vehicle-related collisions, violence, sports injuries, and combat injuries in-cluding explosive blasts are common events causing TBI (Van Voorhees et al., 2018). TBI is recognized as a global health priority that requires complex and expensive medical care (GBD, 2016; Traumatic Brain Injury and Spinal Cord Injury Collaborators, 2019).

DOI: 10.1201/9781003120902-7

A number of complications such as neurologic complications (e.g., seizures, dementia, Alzheimer's disease, and cranial nerve injuries), psychiatric complications (e.g., depression, posttraumatic stress disorder, generalized anxiety disorder, and obsessive-compulsive disorder) and neurosurgical complications (e.g., hematoma) might occur following TBI that could significantly increase the comorbidity of the patients (Ahmed et al., 2017). Computed tomography (CT) scanning is the gold standard technique to identify and assess the severity of TBI in a timely and accurate fashion. Each head CT scan is comprised of a sequence of two-dimensional radiographic images that collectively represent the intracranial region in a three-dimensional fashion.

Hemorrhage within the intracranial compartment has five subtypes depending on its anatomical location: epidural hemorrhage (EDH), subdural hematoma (SDH), subarachnoid hemorrhage (SAH), intraparenchymal hemorrhage (IPH), and intraventricular hemorrhage (IVH). IVH is defined as bleeding inside or around the ventricles in the brain, while IPH is defined as non-traumatic bleeding into the brain parenchyma (Naidech, 2011). Moreover, SAH refers to bleeding into the space between the pia and the arachnoid membranes (Naidech, 2011) while SDH refers to the accumulation of blood under the skull and outside of the brain arachnoid mater. As blood accumulates in the subdural region, it compresses brain tissue which can lead to unconsciousness or death; thus, its early detection and management are of utmost importance. The texture and so the attenuation of blood on a CT scan varies based on the age of blood. Thus, depending on the chronicity of SDH, this type of hematoma can be categorized into three main categories, including chronic, subacute, and acute, as well as a mixture of the three, each of which is represented with different attenuation of blood on a scan. Acute EDHs are identified on a CT scan as a hyperdense collection in the epidural space, located between the inner table of the skull and the dura mater (Kulesza et al., 2020).

TBI patients are categorized according to a score called the Glasgow Coma Scale (GCS). GCS is the summation of three scores; motor performance, verbal responses, and eye-opening and has a range between 3 and 15 where 3–8 indicates severe TBI, 9–12 defines moderate TBI and 13–15 shows mild TBI (Yamamoto et al., 2018).

Early detection and severity assessment of TBI are critical for optimal triage and management of trauma patients. Current trauma protocols utilize CT assessment of injuries in a subjective and qualitative (v.s. quantitative) fashion, shortcomings that could both be addressed by automated image processing systems capable of generating real-time, reproducible, and quantitative information.

As visual assessment of CT images and quantitative measurement of injuries (e.g., volume of SDH) are time-consuming and prone to human error, automated image processing, computational and learning methods could help clinicians assess CT images and provide quantitative measurements.

Current image processing methods can be categorized into three main categories: (1) classical image processing techniques that are mainly driven by clinical domain knowledge about the underlying disease. (2) Deep learning techniques that are end-to-end black box models. Deep convolutional neural networks have recently been successfully applied to medical image segmentation tasks and have been proven to outperform classical models in terms of average performance. However, they require relatively large training sets to effectively learn distinctive patterns. Otherwise, these approaches may fail to reflect the unseen spectrum of the real-world data distribution. (3) Hybrid of deep learning and classical image processing techniques where classical image processing approaches are integrated with deep models to overcome the limitations of each method.

This chapter outlines multiple techniques for CT image pre-processing and segmentation. In addition, the commonly used performance evaluation metrics are explained in detail. Finally, the current challenges in hematoma segmentation tasks are discussed.

7.2 HEMATOMA SEGMENTATION

7.2.1 PRE-PROCESSING

Common pre-processing techniques include skull removal, intensity adjustment, noise reduction, etc. Data pre-processing is important to standardize the image data from different sources. The image contrast will be adjusted, and the region-of-interest will be extracted to boost the performance of image segmentation techniques.

7.2.1.1 Skull Removal

The skull removal step is proposed to extract the soft brain tissue and narrow down the region of interest. As the skull region usually has a higher intensity, a simple skull removal technique can be applied using thresholding (Shahangian & Pourghassem, 2016; Tu et al., 2019). In Farzaneh et al. (2017) and Yao et al. (2019), the seed for soft brain tissue was initialized by thresholding and then the soft tissue was segmented using the distance regularized level set evolution (DRLSE) method.

7.2.1.2 Intensity Adjustment

The gray value of CT scans can be converted to Hounsfield units by parameters in digital imaging and communication in medicine (DICOM) format. Hounsfield units describe radiodensity. Different body tissues are within a specific HU range (e.g., the HU value of the distilled water under standard pressure is defined as 0). A commonly used HU window to visualize soft brain tissue is from 40 to 80. By applying a pre-defined HU window to CT images, the contrast of the target tissue will be enhanced. In Guo et al. (2020) and Kwon et al. (2019), multiple HU windows were applied to extract richer information for segmentation.

7.2.1.3 Noise Reduction

Noises exist in CT scans from image acquisition, coding, or transmission. A median filter was applied to reduce noise while maintaining the shape edges (Shahangian & Pourghassem, 2016). Besides a median filter, denoising was performed by minimizing the total variation norms in Soroushmehr et al. (2015).

7.2.1.4 Image Standardization

Image standardization is performed to reduce the variations of images from different cohorts. The soft brain tissue will be cropped and resampled to ensure the same voxel resolution. In Sharrock et al. (2020), the 3D volume registration was performed to register the target brain volumes to a template with voxel resolution 1.5mm × 1.5mm × 1.5mm.

7.2.2 HEMATOMA SEGMENTATION TECHNIQUES

7.2.2.1 Classical Image Processing Techniques

Conventional image processing techniques have been applied for hematoma segmentation (Table 7.1). In Soroushmehr et al. (2015), the expectation-maximization was applied on a Gaussian Mixture Model to segment four components, including hematoma regions, normal tissues, white-matter regions, and catheters. After that, post-processing was performed by removing bright objects and outlier edges. In Ray et al. (2018), the thresholding technique is used to find the brain tissue and hematoma region clusters. The intensity distribution of pixels is analyzed to segment the hematoma regions. After that, morphological operations were performed as post-processing to get rid of outliers. In Roy et al. (2015), a two-class dictionary for normal tissue and hematoma regions was built using patches from "atlas" CT scans and corresponding manual hematoma segmentation. For a given new CT scan, patches were modeled as a combination of the "atlas" patches in the built dictionary to generate hematoma segmentation. The proposed algorithm was evaluated on CT scans from 25 patients with TBI, and the algorithm has a median Dice score of 0.85.

A two-stage fully-automated segmentation method has been applied in previous literature for hematoma segmentation on 2D CT slices, including initialization and contour evolving. For the initialization stage, Tu et al. (2019) applied a nonlocal regularized spatial fuzzy C-means clustering to segment a coarse hematoma contour. After that, an active contour without edges method was used to refine the contour. CT scans from 30 subjects with different hematoma sizes, shapes, and locations were used to evaluate the proposed method, and a Dice score of 0.92 was produced. Similarly, in Singh et al. (2018), the output from fuzzy C-means clustering was used as the initialization of the modified version of DRLSE. In Kumar et al. (2020), fuzzy c-means clustering and entropy-based thresholding with morphological operations were used to initialize the DRLSE model. The algorithm was developed and tested on CT scans from 35 patients with ICH and achieved an average Dice score of 0.93.

Hematoma segmentation maps can be generated by voxel-wise classification. Scherer Moritz et al. (2016) extracted first- and second-order statistics, texture, and threshold features to describe each voxel and the contextual information. After that, a random forest classification algorithm was developed. The 30 CT scans from patients with spontaneous intracerebral hemorrhage were used to train and test the algorithm, respectively. The calculated volumes of segmented spontaneous intracerebral hemorrhage were compared with volumes from manual segmentation

TABLE 7.1

Hematoma Segmentation Using Image Processing Techniques

Ref	Hematoma Type	Dataset (Training/Test)	Inputs of the Method	Pre-Processing	Segmentation Technique
(Kumar et al., 2020)	ICH	CT scans from 35 patients, provided by SGRR Institute of Medical & Health Sciences and SMI Hospital, Dehradun, Uttarakhand, India.	2D CT slice	Skull removal	Fuzzy c-mean clustering, entropy-based thresholding, and edge-abased DRLSE
(Yao et al., 2019)	Acute ICH	35 CT scans from ProTECT dataset and 27 CT scans collected from the University of Michigan Health System	Super-pixels over-segmented from 2D CT slices	Intensity normalization, skull removal, superpixel generation	Hand-craft feature extraction and SVM model with active learning strategy. Boundary refined by region-based active contour model.
(Tu et al., 2019)	Intracerebral hemorrhage	CT scans from 30 subjects with ICH acquired in hospital emergency rooms	2D CT slice	Skull removal	Nonlocal regularized spatial fuzzy C-means and active contour without edges
(Singh et al., 2018)	ICH	20 CT scans were collected from patients with brain hematomas.	2D CT slice	N/A	Fuzzy c mean clustering and modified version of DRLSE
(Ray et al., 2018)	ICH	590 CT images from 22 patients, provided by the radiology department of Postgraduate Institute of Medical Education and Research, Chandigarh, India.	3D CT scan	Skull and background removal	Fusion of knowledge of brain anatomy with intensity distribution information of CT brain image

(Continued)

TABLE 7.1 (Continued)
Hematoma Segmentation Using Image Processing Techniques

Ref	Hematoma Type	Dataset (Training/Test)	Inputs of the Method	Pre-Processing	Segmentation Technique
(Farzaneh et al., 2017)	SDH	866 axial CT images from 42 patients with SDH admitted at University of Michigan Hospital	Voxels from 3D CT scan	Intensity normalization, skull removal	Hand-crafted feature extractions with tree bagger classifier
(Scherer Moritz et al., 2016)	Spontaneous intracerebral hemorrhage	60 CT scans from Heidelberg University HospitalTrain: 30 casesTest: 30 cases	Voxels from 3D CT scan	N/A	Voxel-wise random forest classification. Features include first-and second-order statistics and texture features
(Shahangian & Pourghassem, 2016)	EDH SDH Intracerebral hemorrhage	627 CT images were collected in Kashani hospital spiral CT-scan imaging center	2D CT slice	Skull and brain ventricle removal, noise reduction and intensity normalization, soft-tissue edema removal	Modified DRLSE
(Sun et al., 2015)	ICH	20 cases from Department of Neurosurgery, CPLA No. 98 Hospital, Huzhou, China	3D CT scan	Skull removal	Supervoxel segmentation and graph cuts
(Soroushmehr et al., 2015)	ICH	CT images from 11 patients with TBI at Virginia Commonwealth Health system	2D CT slice	Intensity normalization, noise reduction	Gaussian Mixture Model and Expectation maximization
(Roy et al., 2015)	IPH	25 cases with mild, moderate, and severe TBI.	3D patch from CT scans	Skull removal	A two-class dictionary was built by patches from the atlas and its manual segmentation

and had a correlation coefficient of 0.95, which is superior to the volumes estimated using ABC/2 (the correlation coefficient is 0.77). Similarly, Farzaneh et al. (2017) performed voxel-wise classification for SDH segmentation. Statistical, textural, and geometrical features of voxels from 3D CT scans were extracted, followed by a tree bagger model. The geometrical features provide location information, which is important as SDHs are typically located on the surface of the brain. Based on the voxel classification, small segmented regions that are less likely to be SDHs were removed iteratively. The segmentation method was trained and validated on 866 CT images from 42 patients with SDHs. The algorithm produced a sensitivity of 0.85 and a specificity of 0.74.

Some methods over-segment superpixels and supervoxels from CT scans and regard them as hematoma region candidates. In Yao et al. (2019), 2D CT slices were first over-segmented into super-pixels by the simple linear iterative clustering (SLIC) algorithm (Achanta et al., 2012). A number of statistical and textural features for each superpixel sample were extracted, and an SVM classifier was applied to classify whether each of the superpixels belongs to hematoma regions. An active learning strategy was proposed to overcome the limited size of annotated CT images. After superpixel classification, a region-based active contour model was used to refine the hematoma boundary. 35 CT scans from the PROTECT dataset and 27 CT scans collected from the University of Michigan Health System were used to validate the proposed method. Those CT scans are with acute hematoma of different sizes, types, and shapes. At the beginning of the superpixel classification model training process, only one CT slice from the training set was annotated. From experimental results, the proposed active learning strategy achieved comparable segmentation performance with five times fewer labeled training samples. In Sun et al. (2015), the 3D volume was reconstructed from a stack of 2D slices and then was over-segmented into super-voxels using an extended SLIC algorithm. After that, the graph cuts algorithm with a novel energy function was developed for hematoma segmentation.

7.2.2.2 Deep Learning Techniques

Deep learning techniques have been widely applied in medical image analysis. A number of deep learning architectures such as fully convolutional neural networks (FCN) have been developed to integrate 2D and 3D contextual information for hematoma segmentation (Table 7.2).

A multiscale dilated 2D Convolutional Neural Network (CNN) was proposed in Yao et al. (2020) to extract hematoma regions of varied volumes. A robust loss function was proposed to improve the generalization of the model to CT scans collected from different medical centers and protocols. The segmentation system was trained and validated on 120 CT scans from patients with TBI enrolled in multiple medical centers. This study focuses on challenging acute TBI cases, where the shape and size of hematoma may vary from tiny scatter dots to large volumes. From the experimental validation, the proposed algorithm achieved an average Dice score of 0.70, which is better than the conventional UNet architecture. The volumes of segmented hematoma were further used as features and significantly improved the 6-month mortality prediction on a dataset containing clinical data of 828 patients.

TABLE 7.2
Hematoma Segmentation Using Deep Learning Techniques

Ref	Hematoma Type	Dataset (Training/Test)	Inputs of the Method	Pre-Processing	Segmentation Technique
(Kellogg et al., 2021)	SDH	108 CT scans collected from patients with subdural hematomasTrain: 102 casesTest: 26 cases	3D CT scan	N/A	3D CNN with a modified U-Net architecture
(Yao et al., 2020)	ICH	100 CT scans from ProTECT dataset, collected from patients with mild-to-severe Traumatic Brain Injury and 20 CT scans from University of Michigan Health SystemTrain: 100 casesTest: 20 cases	2D CT slice	Intensity normalization	Multi-view CNN architecture with robust training.
(Monteiro et al., 2020)	IPH, extra-axial hemorrhage, IVH	One dataset consists of 98 CT scans from 27 patients in the same medical center. The other dataset includes 839 CT scans from 512 patients and 38 centers across Europe.Train and validation: 282 casesTest: 655 cases	Image patch of 110mm * 110mm * 110 mm from 2D image slice	Resampled the image to 1 mm * 1 mm * 1 mm volume;	DeepMedic network
(Guo et al., 2020)	ICH	CT scans from 1176 patients and three hospitalsTrain: 706 casesValidation: 235 casesTest: 235 cases	2D CT slice	CT scans were truncated by three windows and then stacked as a three-channel image.	Multi-task FCN which jointly performs ICH detection, subtype classification, and ICH segmentation

Reference	Application	Dataset	Input	Preprocessing	Method
(Sharrock et al., 2020)	Intracerebral hemorrhage	Data are from phase II and phase III MISTIE multi-center randomized controlled clinical trials.Train: 100 casesValidation: 12 casesTest: 500 cases	3D CT scan	Skull removal, 3D volume registration	VNet with dropout
(Barros et al., 2020)	SDH	775 CT slices collected from patients with aneurysmal subdural hematomas Training: 302 slides test: 473 slices	2D CT image patch	Intracranial segmentation	CNN
(Wang et al., 2020)	ICH	Labeled: 318 CT slices with detected hemorrhage from 36 patients diagnosed with ICH. Unlabeled data: 100000 CT slices from patients diagnosed with ICH in the Radiological Society of North America (RSNA) ICH dataset	2D CT slice	Down-sample the image to 256 * 256	Semi-supervised Multi-task Attention-based UNet
(Ironside Natasha et al., 2019)	Spontaneous intracerebral hemorrhage	CT scans from 300 patients enrolled in the Intracerebral Hemorrhage Outcomes ProjectTrain: 260 casesTest: 40 cases	2D CT slice	Intensity normalization	UNet
(Patel et al., 2019)	Spontaneous intracerebral hemorrhage	The PATCH dataset consisting of CT scans from 120 patients and the Radboudumc dataset consisting of CT scans from 51 patients with intracerebral hemorrhage training: 50 cases from PATCH and 21 cases from RadboudumcTest: 50 cases from Radboudumc	3D patch from CT scans	Intensity normalization, cranial cavity removal	3D CNN with UNet architecture

(Continued)

TABLE 7.2 (Continued)
Hematoma Segmentation Using Deep Learning Techniques

Ref	Hematoma Type	Dataset (Training/Test)	Inputs of the Method	Pre-Processing	Segmentation Technique
		PATCH and 25 cases from Radboudumc			
(Cho, Park, et al., 2019)	ICH	CT scans from 5702 patients, of which 2647 cases are bleeding. Five-fold cross-validation was applied.	2D CT slice	Two different approaches on CT window setting were conducted for building cascade deep learning models	Cascaded CNN for positive hematoma classification and Dual FCN for hematoma segmentation
(Remedios et al., 2019)	ICH	45 CT scans collected from patients with acute TBI from multiple sites training: 27 casesTest: 18 cases	2D CT slice	N/A	CNN with Inception Net-based architecture with multi-site learning
(Cho, Choi, et al., 2019)	The majority is SAH, with a few of ICH and IVH	CT scans were acquired from Kyungpook National HospitalTrain: 6,000 patches from 60 CT scan validation: 19 imagesTest: 13 images	64 * 64 image patch from 2D CT slice	Skull removal, multi-intensity windowing method with 3 levels	Learning the adjacent pixel connectivity (affinity graph) from a U-Net and then generating the segmentation mask
(Kwon et al., 2019)	ICH	275 cases from Kyungpook National University Hospital.Train: 250 casesTest: 25 cases	64*64 image patch from 2D CT slice	Skull removal, brain volume registration	Siamese UNet

(Chang et al., 2018)	EDH/SDH, IPH, SAH	The training cohort consists of CT scans from patients enrolled at the study institution between Jan 1, 2018Train: 10159 casesTest: 682 sets	A stack of CT slices	N/A	Mask R-CNN with a custom hybrid 3D/2D variant of the feature pyramid network
(Yao et al., 2018)	Acute ICH	35 CT scans from ProTECT dataset and 27 CT scans collected from University of Michigan Health SystemTrain: 48 casesTest:14 cases	2D CT slice	Intensity normalization	CNN with UNet architecture and dilated convolution
(Grewal et al., 2018)	ICH	Data from two local hospitals train: 185Validation: 67Test: 77	A sequence of 2D CT slices	Images were resampled to 256*256	Recurrent Attention DenseNet

Deep learning techniques usually require a large annotated training dataset to ensure the model's generalization. To overcome the challenge of the limited size of the labeled dataset and make use of the large-scale unlabeled dataset, Wang et al. (2020) proposed a semi-supervised multi-task attention-based UNet model. The segmentation task was performed by a UNet architecture and trained using the labeled dataset. The encoder part of the UNet was shared for the unsupervised model, which was trained using the unlabeled dataset to reconstruct the foreground and the background. The reconstruction loss was calculated using soft predictions from the segmentation network. The semi-supervised network can leverage the information from the large-scale unlabeled dataset and regularize the segmentation network. In this study, the proposed network was trained by 256 CT slices with hemorrhage and 100,000 unlabeled CT slices from patients with diagnosed ICH. From the experimental results, with a small set of the labeled dataset, the model achieved an average Dice score of 0.67, which is 6% higher than training the segmentation model only with the labeled dataset.

Kellogg et al. (2021) applied a 3D CNN to segment the chronic type of SDH in pre- and post-operative CT scans. This study uses an internal dataset of 128 non-contrast CT scans, all from patients with chronic SDH. A 3D UNet-based model was trained using the weighted Dice loss function to overcome the class imbalance. CT scans from 128 patients were used for training and validation, while ten CT scans were held out for validation. This model achieved a Dice score of 0.84 on the test dataset.

Patel et al. (2019) compared the deep learning segmentation performance with human inter-rater and intra-rater variability. A 3D UNet was applied for spontaneous intracerebral hemorrhage segmentation, and patches from 3D CT scans were used as input to the network. The algorithm achieves a median Dice of 0.91 on the Radboudumc test dataset consisting of CT scans from 25 patients, and a median Dice of 0.90 on the PATCH dataset consisting of CT scans from 50 patients. Five CT scans from the test set were annotated by two trained observers, and one of the observers annotated the same set of images for a second time after 2 weeks to evaluate the inter-observer and intra-observer variability. From the experimental results, the proposed algorithm achieved a comparable level to the observer variability.

Several studies investigated the performance of deep learning models on large-scale datasets. Monteiro et al. (2020) applied DeepMedi network (Kamnitsas et al. 2017), an efficient 3D CNN incorporating multi-scale information. The network was validated in a large-scale dataset consisting of 655 CT scans from 38 medical centers across Europe. Estimated hemorrhage volumes were compared with the volumes from manual segmentation. The mean volume differences for IPH, extra-axial hemorrhage, and IVH are 0.86 ml, 1.83 ml, and 2.09 ml, respectively. The hemorrhage detection was further validated using an external dataset CQ500 dataset containing 491 CT scans (Chilamkurthy et al., 2018). The experimental results show that the CNN network can accurately detect and segment brain hemorrhage. In Chang et al. (2018), a mask R-CNN architecture was applied with a custom feature extractor combined with 3D and 2D contextual information. The detected bounding boxes for hematoma regions were further segmented by a segmentation branch.

The model was trained by 10,159 CT scans and evaluated on 862 CT scans. The model produced a Dice score of 0.931, 0.863, and 0.772 for intraparenchymal hemorrhage, EDH/SDH, SAH, respectively.

7.2.2.3 Hybrid of Deep Learning and Classical Image Processing Techniques

A hybrid of deep learning techniques and image processing techniques is appealing to take advantage of both techniques (Table 7.3).

Farzaneh et al. (2020), which is the continuation of work proposed in Farzaneh et al. (2017), investigates the use of integration of CNN, machine learning, and

TABLE 7.3

Hematoma Segmentation Using Hybrid Techniques

Ref	Hematoma Type	Dataset (Training/ Test)	Inputs of the Method	Pre-processing	Segmentation Technique
(Farzaneh et al., 2020)	SDH	110 CT scans were collected from patients with subdural hematomas. 10-fold cross-validation was performed.	Superpixels from 2D CT slices	Intensity normalization, skull removal, intracranial segmentation, super-pixel generation	Hand-crafted feature extraction + features from CNN + random forest model
(Nag et al., 2019)	ICH	48 CT scans were collected from EKO diagnostics, medical college campus, Kolkata, India. Train: 24 casesTest: 24 cases	2D CT slice to CNN; 3D CT scan to Chan Vese model	Skull removal; Contrast enhancement	Autoencoder and Chan Vese model
(Islam et al., 2019)	Intracerebral hemorrhage	89 CT scans with ICH were collected from the Singapore General Hospital. 5-fold cross-validation was performed	2D CT slice to CNN; 3D CT scan to conditional random field	Intensity normalization, resample to 1 mm * 1 mm * 1 mm volume; Skull removal	CNN with PixelNet architecture and 3D conditional random field

classical image processing techniques for segmenting all different types of SDH, including chronic, subacute, acute, and the mixture of three. This study used an internal dataset from the University of Michigan Health System, including 110 non-contrast CT scans, 98 from patients with SDH patients, and the remaining 12 from patients without any indication of SDH on their CT scans. For the segmentation purpose, for each superpixel, a combination of domain knowledge-driven handcrafted features and data-driven deep features were extracted. Deep features were generated using the U-net architecture. Next, a random forest model was trained to classify each superpixel into SDD or non-SDH superpixel using extracted features. Next, during the post-processing step, the segmented mask was refined using morphological operations and Gaussian kernel smoothing to ensure the 3D spatial coherence. This model achieved an average Dice score of 0.81 on the validation set and an average Dice score of 0.84 on the test dataset. Ten-fold cross-validation was employed to train and evaluate the model throughout the whole process. The results showed that the integration of classical image processing techniques and deep models could achieve greater average performance and robustness compared to each of these techniques alone. This model achieved 0.75, 0.79, and 0.76 for the Dice score, recall, and precision, respectively.

In Islam et al. (2019), ICH-Net, a pixel-wise classification model, was proposed for hematoma segmentation. For each iteration, a number of pixels from CT slices were randomly sampled, and the feature descriptor of each pixel consists of activations from multiple convolutional layers. After a coarse hematoma prediction map was generated, the 3D conditional random field method was applied to refine the final segmentation. Five-fold cross-validation was performed on the collated CT scans from 89 patients with ICH. The algorithm produced an average Dice score of 0.88.

7.3 SEGMENTATION PERFORMANCE EVALUATION

Let's denote V_{TP} as the volume of true positive pixels, V_{TN} as the volume of true negative pixels, V_{FP} as the volume of false positive pixels, V_{FN} as the volume of false negative pixels.

For pixel-wise prediction evaluation, the sensitivity, specificity, precision, and accuracy scores can be calculated as:

$$Sensitivity = V_{TP}/(V_{TP} + V_{FN})$$

$$Specificity = V_{TN}/(V_{TN} + V_{FP})$$

$$Precision = V_{TP}/(V_{TP} + V_{FP})$$

$$Accuracy = (V_{TP} + V_{TN})/(V_{TP} + V_{TN} + V_{FP} + V_{FN})$$

For image-wise or volume-wise segmentation evaluation, the Dice score, Jaccard score can be calculated as:

$$Dice\ score\ =\ 2V_{TP}/(2V_{TP}\ +\ V_{FP}\ +\ V_{FN})$$

$$Jaccard\ index = V_{TP}/(V_{TP}\ +\ V_{FP}\ +\ V_{FN})$$

The segmented hematoma contour can be further evaluated using distance-based metrics. Average perpendicular distance (APD) measures the average distance from the predicted contour to the annotated contour. Hausdorff distance is the longest distance from a point in the predicted contour to the closest point in the annotated contour.

Hematoma segmentation can be used to quantitatively estimate the hematoma volumes, which is an important predictor for the outcome. To evaluate the accuracy of the volume estimation, the correlation graph and Bland–Altman plot can be used to visualize the relationship between the estimated hematoma volume and the volumes from manual annotations. An intraclass correlation coefficient can be calculated to quantitatively evaluate the volume correlation.

7.4 DISCUSSION

In previous literature, multiple image processing techniques and deep learning architectures have been applied for hematoma segmentation for 2D CT slices or 3D CT scans. Recently, deep learning techniques have been widely applied to hematoma segmentation and achieved a good performance on data from cohorts. However, it is challenging to directly compare the segmentation performance of methods proposed in different works because the majority of them were trained and validated on internal datasets. The Dice scores of algorithms are highly related to the dataset size and the heterogeneity in the dataset, as well as the hematoma types, shapes, and volumes.

A large-scale public dataset with good-quality annotated hematoma contours would be a key to hematoma segmentation algorithm development, with which a more comprehensive algorithm evaluation and stratified comparison can be performed. For example, it is possible that some algorithms work better for small hematoma volumes while some other algorithms work better for a specific hematoma type. The CQ500 dataset (Chilamkurthy et al., 2018) is publicly available, which consists of 491 CT scans with 193,317 slices. However, it only provides an image-wise label for hematoma existence rather than pixel-wise segmentation.

As mentioned, the Dice scores are related to hematoma volumes and types. It is hard to define whether the segmentation performance is satisfactory or reliable for downstream usages solely on metrics values. For some challenging cases, human reviewers may have a high inter-rater or intra-rater variability. In Patel et al. (2019) and Farzaneh et al. (2020), annotations from multiple reviewers or the same review from different time points were collected. From the results shown in

Farzaneh et al. (2020), the average inter-rater Dice score on 2D SDH cases is 0.74, while the stratified inter-rater Dice score on SDH is smaller than 25 cc (cubic centimeter) and larger than 25 cc are 0.50, 0.78, respectively. The high variability indicates that a better-quality hematoma region annotation may be derived by finding consensus among annotations from multiple reviewers. It also shows that the inter-rater and intra-rater variability are helpful references for algorithm performance evaluation.

REFERENCES

Achanta, R., Shaji, A., Smith, K., Lucchi, A., Fua, P., & Süsstrunk, S. (2012). SLIC Superpixels Compared to State-of-the-Art Superpixel Methods. *IEEE Transactions on Pattern Analysis and Machine Intelligence*, *34*(11), 2274–2282.

Ahmed, S., Venigalla, H., Mekala, H. M., Dar, S., Hassan, M., & Ayub, S. (2017). Traumatic Brain Injury and Neuropsychiatric Complications. *Indian Journal of Psychological Medicine*, *39*(2), 114–121.

Barros, R. S., van der Steen, W. E., Boers, A. M. M., Zijlstra, Ij., van den Berg, R., El Youssoufi, W., Urwald, A., Verbaan, D., Vandertop, P., Majoie, C., Olabarriaga, S. D., & Marquering, H. A. (2020). Automated Segmentation of Subarachnoid Hemorrhages With Convolutional Neural Networks. *Informatics in Medicine Unlocked*, *19*, 100321. https://doi.org/10.1016/j.imu.2020.100321.

Chang, P. D., Kuoy, E., Grinband, J., Weinberg, B. D., Thompson, M., Homo, R., Chen, J., Abcede, H., Shafie, M., Sugrue, L., Filippi, C. G., Su, M.-Y., Yu, W., Hess, C., & Chow, D. (2018). Hybrid 3D/2D Convolutional Neural Network for Hemorrhage Evaluation on Head CT. *American Journal of Neuroradiology*, *39*(9), 1609–1616. https://doi.org/10.3174/ajnr.A5742.

Chilamkurthy, S., Ghosh, R., Tanamala, S., Biviji, M., Campeau, N. G., Venugopal, V. K., Mahajan, V., Rao, P., & Warier, P. (2018). Development and Validation of Deep Learning Algorithms for Detection of Critical Findings in Head CT Scans. *ArXiv:1803.05854 [Cs]*. http://arxiv.org/abs/1803.05854.

Cho, J., Choi, I., Kim, J., Jeong, S., Lee, Y.-S., Park, J., Kim, J., & Lee, M. (2019). Affinity Graph Based End-to-End Deep Convolutional Networks for CT Hemorrhage Segmentation. In T. Gedeon, K. W. Wong, & M. Lee (Eds.), *Neural Information Processing* (pp. 546–555). Springer International Publishing. https://doi.org/10.1007/978-3-030-36708-4_45.

Cho, J., Park, K.-S., Karki, M., Lee, E., Ko, S., Kim, J. K., Lee, D., Choe, J., Son, J., Kim, M., Lee, S., Lee, J., Yoon, C., & Park, S. (2019). Improving Sensitivity on Identification and Delineation of Intracranial Hemorrhage Lesion Using Cascaded Deep Learning Models. *Journal of Digital Imaging*, *32*(3), 450–461. https://doi.org/1 0.1007/s10278-018-00172-1.

Farzaneh, N., Soroushmehr, S. M. R., Williamson, C. A., Jiang, C., Srinivasan, A., Bapuraj, J. R., Ward, K. R., Korley, F. K., & Najarian, K. (2017). Automated Subdural Hematoma Segmentation for Traumatic Brain Injured (TBI) patients. 2017 39th Annual International Conference of the IEEE Engineering in Medicine and Biology Society (EMBC), 3069–3072. https://doi.org/10.1109/EMBC.2017.8037505.

Farzaneh, N., Williamson, C. A., Jiang, C., Srinivasan, A., Bapuraj, J. R., Gryak, J., Najarian, K., & Soroushmehr, S. M. R. (2020). Automated Segmentation and Severity Analysis of Subdural Hematoma for Patients with Traumatic Brain Injuries. *Diagnostics*, *10*(10), 773. 1. https://doi.org/10.3390/diagnostics10100773.

GBD 2016 Traumatic Brain Injury and Spinal Cord Injury Collaborators. (2019). Global, Regional, and National Burden of Traumatic Brain Injury and Spinal Cord Injury,

1990–2016: A Systematic Analysis for the Global Burden of Disease Study 2016. *The Lancet. Neurology*, *18*(1), 56–87. https://doi.org/10.1016/S1474-4422(18)30415-0.

Grewal, M., Srivastava, M. M., Kumar, P., & Varadarajan, S. (2018). RADnet: Radiologist Level Accuracy Using Deep Learning for Hemorrhage Detection in CT Scans. 2018 IEEE 15th International Symposium on Biomedical Imaging (ISBI 2018), 281–284. https://doi.org/10.1109/ISBI.2018.8363574.

Guo, D., Wei, H., Zhao, P., Pan, Y., Yang, H., Wang, X., Bai, J., Cao, K., Song, Q., Xia, J., Gao, F., & Yin, Y. (2020). Simultaneous Classification and Segmentation of Intracranial Hemorrhage Using a Fully Convolutional Neural Network. 2020 IEEE 17th International Symposium on Biomedical Imaging (ISBI), 118–121. https://doi.org/10.1109/ISBI45749.2020.9098596.

Islam, M., Sanghani, P., See, A. A. Q., James, M. L., King, N. K. K., & Ren, H. (2019). ICHNet: Intracerebral Hemorrhage (ICH) Segmentation Using Deep Learning. In A. Crimi, S. Bakas, H. Kuijf, F. Keyvan, M. Reyes, & T. van Walsum (Eds.), *Brainlesion: Glioma, Multiple Sclerosis, Stroke and Traumatic Brain Injuries* (pp. 456–463). Springer International Publishing. https://doi.org/10.1007/978-3-030-11723-8_46.

Kamnitsas, K., Ledig, C., Newcombe, V. F. J., Simpson, J. P., Kane, A. D., Menon, D. K., Rueckert, D., & Glocker, B. (2017). Efficient Multi-scale 3D Cnn With Fully Connected CRF for Accurate Brain Lesion Segmentation. *Medical Image Analysis*, *36*, 61–78. https://doi.org/10.1016/j.media.2016.10.004.

Kellogg, R. T., Vargas, J., Barros, G., Sen, R., Bass, D., Mason, J. R., & Levitt, M. (2021). Segmentation of Chronic Subdural Hematomas Using 3D Convolutional Neural Networks. *World Neurosurgery*, *148*, e58–e65. https://doi.org/10.1016/j.wneu.2020.12.014.

Kulesza, B., Mazurek, M., Rams, Ł., & Nogalski, A. (2020). Acute Epidural and Subdural Hematomas After Head Injury: Clinical Distinguishing Features. *Indian Journal of Surgery*, 83(S1), 96–104. https://doi.org/10.1007/s12262-020-02304-w.

Kumar, I., Bhatt, C., & Singh, K. U. (2020). Entropy Based Automatic Unsupervised Brain Intracranial Hemorrhage Segmentation Using CT Images. *Journal of King Saud University - Computer and Information Sciences*. https://doi.org/10.1016/j.jksuci.2020.01.003.

Kwon, D., Ahn, J., Kim, J., Choi, I., Jeong, S., Lee, Y.-S., Park, J., & Lee, M. (2019). Siamese U-Net with Healthy Template for Accurate Segmentation of Intracranial Hemorrhage. In D. Shen, T. Liu, T. M. Peters, L. H. Staib, C. Essert, S. Zhou, P.-T. Yap, & A. Khan (Eds.), *Medical Image Computing and Computer Assisted Intervention – MICCAI 2019* (pp. 848–855). Springer International Publishing. https://doi.org/10.1007/978-3-030-32248-9_94.

Nag, M. K., Chatterjee, S., Sadhu, A. K., Chatterjee, J., & Ghosh, N. (2019). Computer-Assisted Delineation of Hematoma from CT Volume Using Autoencoder and Chan Vese Model. *International Journal of Computer Assisted Radiology and Surgery*, *14*(2), 259–269. https://doi.org/10.1007/s11548-018-1873-9.

Naidech, A. M. (2011). Intracranial Hemorrhage. *American Journal of Respiratory and Critical Care Medicine*, *184*(9), 998–1006.

Natasha, I., Ching-Jen, C., Simukayi, M., Sim, J. L., Saurabh, M., David, R., Dale, D., Mayer, S. A., Angela, L., & Sander, C. E. (2019). Fully Automated Segmentation Algorithm for Hematoma Volumetric Analysis in Spontaneous Intracerebral Hemorrhage. *Stroke*, *50*(12), 3416–3423. https://doi.org/10.1161/STROKEAHA.119.026561.

Monteiro, M., Newcombe, V. F. J., Mathieu, F., Adatia, K., Kamnitsas, K., Ferrante, E., Das, T., Whitehouse, D., Rueckert, D., Menon, D. K., & Glocker, B. (2020). Multiclass Semantic Segmentation and Quantification of Traumatic Brain Injury Lesions on Head CT Using Deep Learning: An Algorithm Development and Multicentre Validation

Study. *The Lancet Digital Health*, 2(6), e314–e322. https://doi.org/10.1016/S2589-75 00(20)30085-6.

Moritz, S., Jonas, C., Alexander, Y., Yasemin-Aylin, S., Michael, G., Markus, M., Christian, S., Julian, B., Andreas, U., Klaus, M.-H., & Berk, O. (2016). Development and Validation of an Automatic Segmentation Algorithm for Quantification of Intracerebral Hemorrhage. *Stroke*, 47(11), 2776–2782. https://doi.org/10.1161/STROKEAHA.116.013779.

Patel, A., Schreuder, F. H. B. M., Klijn, C. J. M., Prokop, M., Ginneken, B. van, Marquering, H. A., Roos, Y. B. W. E. M., Baharoglu, M. I., Meijer, F. J. A., & Manniesing, R. (2019). Intracerebral Haemorrhage Segmentation in Non-Contrast CT. *Scientific Reports*, 9(1), 17858. https://doi.org/10.1038/s41598-019-54491-6.

Ray, S., Kumar, V., Ahuja, C., & Khandelwal, N. (2018). Intensity Population Based Unsupervised Hemorrhage Segmentation from Brain CT images. *Expert Systems with Applications*, 97, 325–335. https://doi.org/10.1016/j.eswa.2017.12.032.

Remedios, S., Roy, S., Blaber, J., Bermudez, C., Nath, V., Patel, M. B., Butman, J. A., Landman, B. A., & Pham, D. L. (2019). Distributed Deep Learning for Robust Multi-site Segmentation of CT Imaging After Traumatic Brain Injury. *Medical Imaging 2019: Image Processing*, 10949, 109490A. https://doi.org/10.1117/12.2511997.

Roy, S., Wilkes, S., Diaz-Arrastia, R., Butman, J. A., & Pham, D. L. (2015). Intraparenchymal Hemorrhage Segmentation From Clinical Head CT of Patients With Traumatic Brain Injury. *Medical Imaging 2015: Image Processing*, 9413, 94130I. https://doi.org/10.1117/12.2082199.

Shahangian, B., & Pourghassem, H. (2016). Automatic Brain Hemorrhage Segmentation and Classification Algorithm Based on Weighted Grayscale Histogram Feature in a Hierarchical Classification Structure. *Biocybernetics and Biomedical Engineering*, 36(1), 217–232. https://doi.org/10.1016/j.bbe.2015.12.001.

Sharrock, M. F., Mould, W. A., Ali, H., Hildreth, M., Awad, I. A., Hanley, D. F., & Muschelli, J. (2020). 3D Deep Neural Network Segmentation of Intracerebral Hemorrhage: Development and Validation for Clinical Trials. *MedRxiv*, 19(3), 403–415. https://doi.org/10.1101/2020.03.05.20031823.

Singh, P., Khanna, V., & Kamal, M. (2018). Hemorrhage Segmentation by Fuzzy C-mean with Modified Level Set on CT imaging. 2018 5th International Conference on Signal Processing and Integrated Networks (SPIN), 550–555. https://doi.org/10.1109/SPIN.2 018.8474166.

Soroushmehr, S. M. R., Bafna, A., Schlosser, S., Ward, K., Derksen, H., & Najarian, K. (2015). CT Image Segmentation in Traumatic Brain Injury. 2015 37th Annual International Conference of the IEEE Engineering in Medicine and Biology Society (EMBC), 2973–2976. 1. https://doi.org/10.1109/EMBC.2015.7319016.

Sun, M., Hu, R., Yu, H., Zhao, B., & Ren, H. (2015). Intracranial Hemorrhage Detection by 3D Voxel Segmentation on Brain CT Images. 2015 International Conference on Wireless Communications Signal Processing (WCSP), 1–5. https://doi.org/10.1109/ WCSP.2015.7341238.

Taylor, C. A., Bell, J. M., Breiding, M. J., & Xu, L. (2017). Traumatic Brain Injury–related Emergency Department Visits, Hospitalizations, and Deaths—United States, 2007 and 2013. *MMWR Surveillance Summaries*, 66(9), 1.

Tu, W., Kong, L., Karunamuni, R., Butcher, K., Zheng, L., & McCourt, R. (2019). Nonlocal Spatial Clustering in Automated Brain Hematoma and Edema Segmentation. *Applied Stochastic Models in Business and Industry*, 35(2), 321–329. https://doi.org/10.1002/ asmb.2431.

Van Voorhees, E. E., Moore, D. A., Kimbrel, N. A., Dedert, E. A., Dillon, K. H., Elbogen, E. B., & Calhoun, P. S. (2018). Association of Posttraumatic Stress Disorder and Traumatic Brain Injury With Aggressive Driving in Iraq and Afghanistan Combat Veterans. *Rehabilitation Psychology*, 63(1), 160–166. https://doi.org/10.1037/rep0000178.

Wang, J. L., Farooq, H., Zhuang, H., & Ibrahim, A. K. (2020). Segmentation of Intracranial Hemorrhage Using Semi-Supervised Multi-Task Attention-Based U-Net. *Applied Sciences*, *10*(9), 3297. https://doi.org/10.3390/app10093297.

Yamamoto, S., Levin, H. S., & Prough, D. S. (2018). Mild, Moderate and Severe: Terminology Implications for Clinical and Experimental Traumatic Brain Injury. *Current Opinion in Neurology*, *31*(6), 672–680. Scopus. https://doi.org/10.1097/WCO.0000000000000624.

Yao, H., Williamson, C., Soroushmehr, R., Gryak, J., & Najarian, K. (2018). Hematoma Segmentation Using Dilated Convolutional Neural Network. 2018 40th Annual International Conference of the IEEE Engineering in Medicine and Biology Society (EMBC), 5902–5905.

Yao, H., Williamson, C., Gryak, J., & Najarian, K. (2019). Brain Hematoma Segmentation Using Active Learning and an Active Contour Model. In I. Rojas, O. Valenzuela, F. Rojas, & F. Ortuño (Eds.), *Bioinformatics and Biomedical Engineering* (pp. 385–396). Springer International Publishing. https://doi.org/10.1007/978-3-030-17935-9_35.

Yao, H., Williamson, C., Gryak, J., & Najarian, K. (2020). Automated Hematoma Segmentation and Outcome Prediction for Patients with Traumatic Brain Injury. *Artificial Intelligence in Medicine*, *107*, 101910. https://doi.org/10.1016/j.artmed.2020.101910.

8 Wound Tissue Classification with Convolutional Neural Networks

Rafael M. Luque-Baena, Francisco Ortega-Zamorano, Guillermo López-García, and Francisco J. Veredas

CONTENTS

8.1 INTRODUCTION

8.1.1 WOUND DIAGNOSIS

The healing of a chronic or acute wound is a complex process that constitutes one of the main problems in the field of healthcare. The main types of wounds include chronic wounds, such as pressure ulcers, diabetic foot ulcers, and arterial and venous ulcers, as well as acute wounds, such as surgical wounds, burns, etc. It has been estimated that around 451 million people in the world are affected by different types of diabetes (Cho et al. 2018) and could suffer from diabetic foot ulcers, which is one of the main diseases associated with diabetes. The global prevalence of diabetic foot ulcers has been estimated at 6.3% (Zhang et al. 2017), while it is estimated that around 15% of diabetics will develop a diabetic foot ulcer in their lifetime (Reiber 1996). On the other hand, it is estimated that 1% (Ruckley 1997; Bergqvist et al. 1999) of the population suffering from chronic venous insufficiency – which has an overall prevalence of 25–40% and 10–20% in women and men, respectively (Al Shammeri et al. 2014) – is affected by venous ulcers. In the specific case of pressure ulcers, they are responsible for a high mortality rate, close to 30%, among the older population (Landi et al. 2007). Acute and chronic wound care has a high impact on the budgets of health systems. In a recent study carried out in the United States, it has been estimated that the annual cost of wound care in that country rises in a range of between 28.1 billion dollars and 96.8 billion dollars, affecting more than eight millions of people who have suffered from these wounds (Sen 2019).

Accurate wound evaluation is a fundamental task carried out by healthcare specialists who are responsible for establishing suitable diagnoses and treatments. Traditionally, wound evaluation has been accomplished by means of the visual inspection of the wounds and the use of standardized classification scales (Beeckman et al. 2007), which has been shown as an inaccurate way of carrying out wound diagnosis (Edsberg 2007). On the other hand, due to the usual limitations in the availability of primary-care personnel specialized in wound evaluation – especially in rural areas and small towns – many patients do not have access to an appropriate diagnosis and care for the wounds they suffer from, which is especially dramatic in the case of chronic wounds. For all these reasons, the development of telemedicine and automatic diagnostic tools that utilize image processing techniques (Bowling et al. 2011; Chan and Lo 2020) makes it possible to provide these patients with diagnostic services that they would otherwise be deprived of. The growing use of portable devices (such as smart mobile phones, tablets, etc.) and the increase in artificial intelligence applications allow the development of intelligent diagnosis and prognosis systems based on digital imaging for wound care. The use of automatic assessment systems based on image processing makes it possible to improve the precision of the diagnoses, adjust the treatments to the characteristics of each wound and each patient, reduce the costs derived from wound care and, in general, increase the quality of care provided to patients (E. Pasero & Castagneri 2017).

However, one of the main problems we face with natural wound imaging in general and pressure ulcer images in particular is the very nature of these wounds, which results in a high variability in the images used. Thus, wounds tend to have very heterogeneous colorations, related to the skin color of the patients and other abnormalities present in the wound – such as erythema or striae of the skin – as well as very imprecise and irregular separation edges between different regions or tissues present in the wound (Veredas et al. 2015), which greatly hinder the image segmentation process necessary for the separation of the different regions of interest in it, as well as the identification and classification of tissues necessary for automatic diagnosis (Veredas et al. 2010).

8.1.2 WORKFLOW FOR WOUND IMAGE EVALUATION

The usual workflow for diagnosing a wound on the basis of a 2D or 3D image begins with a phase in which a set of features is extracted, which are then used as input to a wound classification system (e.g., classification of the type of chronic wound, or the presence or absence of infection in the wound) or for the classification of the different tissues present in the wound (granulation tissues, sloughs, necrotic tissue, etc.). In the specific literature, a great diversity of techniques has been used for the extraction of characteristics from wound images, among which several rule-based techniques can be mentioned (Papazoglou et al. 2010; Filko et al. 2010; Rao et al. 2013; Ahmad Fauzi et al. 2015; Poon and Friesen 2015; E. Pasero and Castagneri 2017; Kumar and Sudharsan 2018), systems based on machine learning (Kolesnik and Fexa 2004, 2005, 2006; Wantanajittikul et al. 2012; Song and Sacan 2012; Hani et al. 2012; M. K. Yadav et al. 2013; L. Wang et al. 2017) and, lately, deep learning algorithms (C. Wang et al. 2015; M. Goyal et al. 2017; Liu et al. 2017; F. Li et al. 2018). Wound classification, for its part, consists of the classification of the image as a whole (distinguishing between venous, diabetic, pressure ulcers, etc.) or the differentiation of certain wound conditions, such as the presence of ischemia or infection. Most of the approaches used for wound classification employ techniques from the field of machine learning (Begoña Acha et al. 2005; Begoña Acha et al. 2013; Serrano et al. 2005, 2015; C. Wang et al. 2015; D. P. Yadav et al. 2019; Abubakar et al. 2020) and, more currently, deep learning (Manu Goyal et al. 2018; Aguirre Nilsson and Velic 2018; Shenoy et al. 2018; Despo et al. n.d.; Alzubaidi et al. 2020).

Finally, wound tissue classification constitutes the case of this current study we are presenting herein and consists of the analysis of the characteristics of each pixel in the image – or of a group of pixels (called super-pixel) – so that different pixels are assigned to different types of tissues (granulation, healing tissue, slough, necrotic, etc.).

In this work, the application of deep learning convolutional networks for wound tissue classification and wound image segmentation is proposed. The main contributions of this work are:

- The use of data augmentation strategies for increasing the size of the dataset used for model training. Given the reduced number of patterns or segmented images available in the original wound image dataset (with only several dozens of patterns), we proceeded with a first stage consisting in the extraction of thousands of regions of interest (ROIs), with size 128×128 pixels, for which a class or tissue type is assigned to each ROI. The labeled samples generated this way are used to train convolutional networks for wound tissue classification.
- The application of supervised classification convolutional networks, of known architectures, to classify ROIs from wound images into different classes or tissue types. In the second stage, the previously trained classifiers are applied to overlapping 128×128 regions extracted from each image.

- The reconstruction of the final segmented image from the classification labels of the overlapped regions. Having that each pixel is associated with several overlapping regions as a result of the previous stage, a consensus strategy is required to decide the label (i.e., tissue type) assigned to that pixel.
- The comparative analysis of the performance of the proposed convolutional networks when they are applied to tissue classification on a wound image dataset of reference, for which classical machine learning techniques were originally used and reported in the literature (Veredas et al. 2015).

The rest of the work is structured as follows. Section 2 describes the related works on segmentation of wound images. Section 3 shows the proposed methodology, emphasizing the different phases that comprise it. Section 4 defines the experiments performed and outlines the results obtained. Finally, Section 5 presents the conclusions of the proposal.

8.2 RELATED WORKS

8.2.1 MACHINE LEARNING FOR WOUND IMAGE EVALUATION

Traditionally, various machine learning techniques have been used to perform tissue classification on wound images into a series of pre-established categories. In one of the seminal papers that used machine learning to classify tissues from wound images (Wannous et al. 2007) used an SVM classifier to classify wound regions of interest into three different tissue types – granulation, slough, and necrotic – using a set of 20 color and texture descriptors extracted from the images and selected using a forward selection method (SFS). In later work, these same authors followed an image segmentation-driven classification approach that turned out to be more appropriate than pixel-level classification (Wannous et al. 2008). For this, they again used SVM classifiers on characteristics of textures and color extracted from the image. The work of this research team from the *PRISME Institute* (France) focused later on telemedicine and continued with the use of C-SVM classifiers to obtain higher robustness and performance rates (Wannous et al. 2010) under uncontrolled lighting conditions, different image framing, and variable camera settings. Later, in (Wannous et al. 2011) these same authors followed a multi-view strategy that combined dimensional measurements obtained from a 3D model with tissue classification using SVM classifiers, from color and texture descriptors computed after non-supervised image segmentation. The tissue labels obtained that way are finally mapped onto the 3D model, so that the classification results are merged together. In this way, the authors are able to compensate for the perspective distortions introduced in the different images, as well as the uncontrolled lighting conditions and the different framings used to acquire the images. For their part, in their initial work at the University of Malaga (Spain) (Veredas et al. 2010) proposed a hybrid approach, based on binary classifiers that combined multilayer neural networks and Bayesian classifiers, to perform a cascade classification, tissue by tissue, in order to differentiate regions of tissues in the images – namely, healthy skin, sloughs, granulation, scar and necrotic tissue – which were previously segmented using a segmentation algorithm based on mean-shift smoothing and region growth algorithms. For the training of the models, they used a data set consisting of 113 images segmented and labeled by a group of clinical experts, following consensus techniques in order to reduce inter- and intra-observer variability. Along this same line of work, Veredas et al. (2015) explored various machine learning strategies in order to make a comparison of the performance of different tissue classification models, namely, neural networks, SVM, and random-forest decision trees. The authors reported higher performance rates for decision trees and SVM classifiers, with lower performance rates for neural networks. On the other hand, in (Mukherjee et al. 2014) Bayesian and SVM classifiers were used to classify three types of tissues – granulation, slough, and necrotic – in six different types of chronic wounds – burns, diabetic ulcers, pressure ulcers, vascular ulcers, malignant ulcers, and pyoderma gangrenosum – using color and texture characteristics extracted from the images, which were previously segmented by fuzzy divergence-based thresholding by minimizing edge ambiguity. The best classification results were achieved using SVM classifiers with third-order polynomial kernels.

8.2.2 DEEP LEARNING MODELS FOR WOUND IMAGE EVALUATION

More recently, deep learning strategies have been employed for classifying tissues in wound imaging. Thus, in Zahia et al. (2018), a method is proposed for the classification of tissues in pressure ulcer images that uses convolutional neural networks, tested on a data set composed of 22 images that were previously labeled by clinical specialists. The pipeline proposed by these authors includes a preprocessing phase in which the regions of interest in the images are identified, the reflections produced by the lighting of the scene are filtered and patches of size 5×5 pixels are extracted from each region of interest. These patches, together with the ground truth labels, are given as input to a nine-layer neural network – that includes three convolutional layers – in order to classify three different types of tissue: granulation, necrotic, and slough. The worst performance rates in the classification, as reported in the article, are obtained for the latter type of tissue. For their part, in Rajathi et al. (2019), the authors also propose the use of convolutional neural networks for tissue classification, in this case from vascular ulcer images. To do this, they use a data set consisting of 1,250 images collected at an Indian medical school and labeled by experts. In this study, an approach similar to that presented in Zahia et al. (2018) is followed, through a pipeline that contains the first phase of image preprocessing, a segmentation phase – using the technique of active contours – aimed at the separation of the wound area from the healthy skin that surrounds it, and a final phase based on classification by using convolutional neural networks. In the pre-processing phase, they use thresholding strategies to eliminate reflections produced by the flash during image capture. In the classification phase, they use a four-layer convolutional network to classify tissues into four different types: granulation, slough, necrotic, and epithelial tissue. The first study in which deep learning strategies are used to classify a number of tissues greater than five is

Nejati et al. (2018), in which a classifier capable of distinguishing between seven different types of tissues is trained. To do this, the authors use a data set consisting of 350 images of chronic wounds to fine-tune a pre-training AlexNet architecture for the extraction of image characteristics and their subsequent classification using SVM classifiers that work at the image-area level. For their part, in Eros Pasero & Castagneri (2017) these same authors extended their previous work on ulcer image segmentation presented in E. Pasero & Castagneri (2017), proposing a self-organized map for tissue classification, trained on a set of images that were larger than the one they used in their previous study. More recently, in Blanco et al. (2020) a method has been proposed for the analysis of images of dermatological ulcers, called QTDU, which combines directed segmentation models at the super-pixel level with deep neural networks for the classification of tissues present in a dataset composed of 217 images of arterial and venous ulcers. The authors propose a three-stage system, with the first stage of data preparation in which the images are segmented, by building and increasing super-pixels, to go to a second stage in which convolutional neural networks are trained – specifically, pre-trained Inceptionv3 and ResNet architectures to which six new layers are added at the end – from more than 44K super-pixel patterns obtained as a result of the previous stage, thus obtaining a tissue classification model that distinguishes between four different tissues: fibrin, granulation, necrotic, and non-wound. The authors conclude that pre-trained networks on the ImageNet dataset manage to be trained in a faster and more efficient way than the corresponding architectures trained from scratch (with random weight initialization), obtaining better results with the ResNet architecture, in comparison with the Inceptionv3 model, and surpassing these results to those obtained by these same authors using traditional machine learning approaches. Finally, 3D convolutional networks have also been used to classify tissues present in wound images. Specifically, in García-Zapirain et al. (2018), this type of deep learning architecture is used to classify granulation, necrotic, and slough tissues in images of pressure ulcers. To do this, the authors start from regions of interest identified in the images to extract characteristics from the original images – as well as from HSI images convolved with 3D Gaussian kernels – that are combined with first-order models of visual appearance. The framework developed by these authors is trained and tested on a set of 193 ulcer images, obtaining results, in terms of AUC (area under the ROC curve), DSC (Dice similarity coefficient) and PAD (percentage area distance), that the authors of this study themselves consider very promising.

8.3 METHODOLOGY

Having a sufficient amount of data is essential to train any computational learning model. In the last decade, the proliferation of deep learning models has been exponential and, as it is known, these models require a large amount of training data for the inference of the trained model to be effective. However, as discussed in section 8.2, most wound-image datasets have very few patterns (i.e., images), several hundred at most, which limits model training and makes it difficult for the model to generalize during the inference phase. There are data augmentation techniques that partially alleviate the problem, but sometimes they are not enough

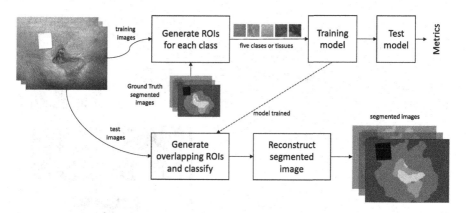

FIGURE 8.1 Framework of the wound image segmentation approach. In the upper part, the training of the neural learning model is carried out while the lower part details the segmentation process using the model trained from the input image.

for the development of effective models. In this proposal, the key idea is to further augment the input data by generating thousands of regions of interest (ROIs) that are used to train deep neural network models. This is to improve the robustness, generalization, and efficiency of wound-image segmentation models.

Therefore, this section describes our proposed framework for wound image segmentation, which is scheduled in Figure 8.1. As mentioned above, the segmentation process is carried out in two sequential phases. In the first phase, an ROI generation process is accomplished for each one of the N classes (i.e., tissues) present in the images (see subsection 8.3.1). Subsequently in this same first phase, a Convolutional Neural Network (CNN) model for classification is trained on the samples consisting of the previously generated ROIs. For comparative analysis purposes, we train and compare four different CNN models (see subsection 8.3.2). In the second phase, each model previously trained is applied to the individual pixels in each input image, as a convolution operation. Finally, and given the overlap between regions containing nearby pixels in the image, the final segmentation is reconstructed to obtain a robust output image that is free, to a greater extent, of isolated classification errors (see subsection 8.3.3).

8.3.1 ROIs GENERATION

The first stage consists of the generation of regions of interest (ROIs) from the training images. In this case, a small dataset with few available pressure ulcer images is used (113 labeled images) and a segmentation process is performed based on five preexisting categories, one for each relevant tissue that appears in the images. Namely, *skin*, *healing*, *granulation*, *slough*, and *necrotic*. The *skin* class corresponds to healthy tissue, while the remaining tissues appear depending on the severity of the wound analyzed. In order to obtain a large volume of data from each training image, hundreds of ROIs are generated using the procedure shown in Figure 8.2.

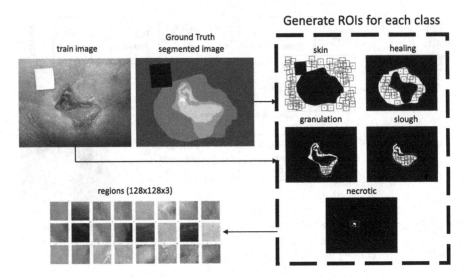

FIGURE 8.2 Scheme of the ROI generation process. Dozens of regions are obtained for each image distributed in five tissue types.

For each input image, of size $1632 \times 1224 \times 3$, there exists an ideal segmentation and labeling (ground truth, or GT) provided by a group of clinical specialists. This segmentation is carried out by following a consensus process for the experts to assign one of the five given classes to each differentiated area in the image. For each given class, tens or hundreds of ROIs of size $128 \times 128 \times 3$ are randomly generated from the areas manually segmented by the experts, where each ROI is labeled by the class its central pixel belongs to. This process is graphically illustrated on the right side of Figure 8.2. The ROI's size of 128×128 has been chosen to take advantage of the power and characteristics of convolutional networks that are subsequently applied. There is a reduced overlap between regions to avoid redundancy, at least for the more frequent tissue types found in the images. Due to the characteristics of the images, there are classes that consist of a higher number of samples than others since there are more pixels in the images labeled with that class label than with the others. For example, there are usually more skin-tissue pixels than necrotic-tissue ones. This implies that the distribution of tissue classes is unbalanced, thus we have a greater number of regions for *skin* or *healing* than for *slough* and *necrotic*. However, this imbalance is counteracted as much as possible by increasing the number of ROIs obtained from low-frequent classes and reducing it for high-frequent classes.

Finally, this process has allowed the generation of ~100 ROIs from each wound image in the dataset, thus giving a final dataset consisting of 11,435 samples that are used to train the learning model. The distribution of these training patterns among the five classes is shown in Figure 8.3. As it can be observed, the dataset is unbalanced and the classification model should cope with this contingency during the training phase.

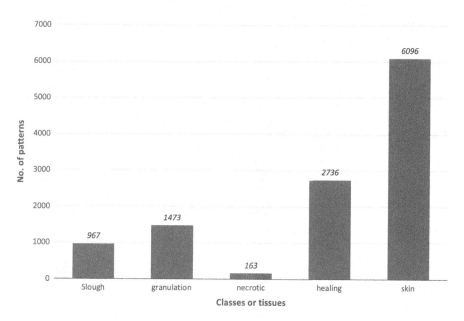

FIGURE 8.3   Distribution of the number of regions (ROIs) for each class for the data set.

8.3.2   ROIs Classification

The appearance of deep neural networks has meant a profound change in the way in which any task related to image analysis is approached as an end-to-end process. The main advantage of these models is that it is no longer necessary to define or extract a set of ad-hoc features from the image, but the model itself manages to extract them along the hierarchical transformations performed by the network as part of the classification task it has been trained for. For the problem addressed herein, it means that an additional preprocessing stage for image regionalization – i.e., the previous differentiation of regions in the image belonging to different tissue types, as it is addressed in Veredas et al. (2015) – and engineering and extraction of features from those predetermined regions is not needed before proceeding with the classification task performed by the network.

In this section, the application of various types of convolutional networks for the classification of the ROIs extracted from the wound images is presented. On the one hand, three pre-trained networks are used and analyzed, whose training has been previously performed on the public dataset called COCO. On the other hand, a network designed from scratch, which we call *custom*, is also presented. For the latter, some recommendations for the design of deep convolutional neural networks have been followed (Krizhevsky et al. 2012), such as the suggestion of using ReLU activation units or setting a pooling function after each convolutional layer to reduce the size of the learned feature map. Thus, the objective after each block in the convolutional network is to get a feature map from the previous block with increasing depth and decreasing size (width and height). Finally, two fully connected layers are included after the convolutional blocks, where the last one has as many neurons as there are classes (i.e., tissues) to be classified. Figure 8.4 shows the architecture of the proposed *custom* model.

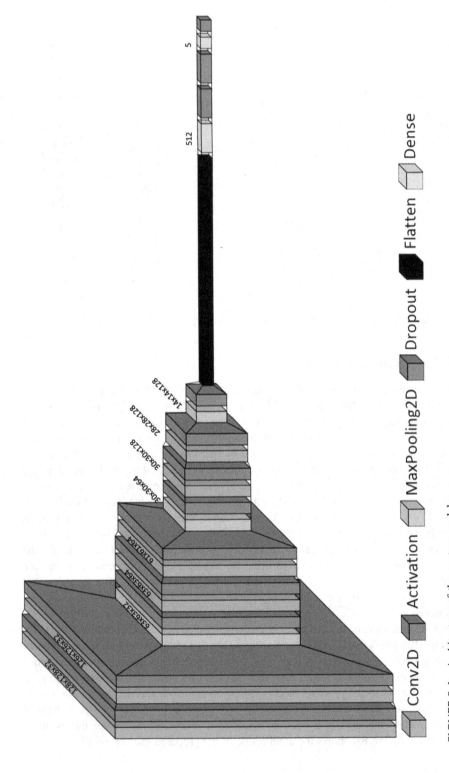

FIGURE 8.4 Architecture of the *custom* model.

Among the pre-trained networks, *vgg16* (Simonyan and Zisserman 2015) is chosen, whose main characteristic is that it is made up of 16 convolutional layers, with a modular and uniform architecture. Only 3×3 convolutions are applied so that the training of the model is less abrupt and not too much information is lost in the first layers, a fact that occurs when the size of the filters in the convolution is too large. Another of the models studied is *ResNet* (He et al. 2016), which is mainly characterized by the way it learns by computing the residual or the difference between the input and the output of the function to be estimated. That is why these types of networks are called residual networks. Furthermore, the use of fully connected layers in the final layers is avoided, due to the high number of parameters that their inclusion would generate. In this case, the 50-layer model is used, which is usually called ResNet50. Finally, within the pre-trained networks, we analyze *exception* (Chollet 2017), consisting of a modified version of the well-known Inception network (Szegedy et al. 2015), which incorporates a modified depthwise separable convolution, where pointwise convolution followed by the depthwise convolution to each channel is applied.

All the pre-trained convolutional models have focused on image (or ROI) classification, although they have also been used as feature extractors or as internal modules for proposals applied to other areas. Since we perform the fine-tuning of each model, the pre-trained weights of each network serve as the initial values for the subsequent training process carried out with the wound-image dataset.

8.3.3 SEGMENTATION PROCESS

Once we have a trained model, the process ends up with the segmentation of the complete image. Since the model works on ROIs of size $128 \times 128 \times 3$ as input, overlapping regions of that size are obtained with a stride of N pixels, that is, overlapping regions are selected with an interval of N pixels between them. To make the analysis approachable in a reduced time, the value of N has been assigned to 5. In total and for each wound image of 1632×1224 pixels, 6,486 regions are generated that are tested by the model. A class or label is obtained for each ROI.

Given the above process, each pixel of the image is included within several evaluated regions. Thus, for each pixel, there are as many labels or classes as regions that include that pixel, which requires a consensus strategy to decide the label (or tissue) assigned to that pixel. Thus for each pixel, there are at most 64 regions (with corresponding labels), so that the strategy followed consists of weighting the output of each region by the proximity of the center of that region to the analyzed pixel. The closer the center of the region is to the pixel, the higher the weight it is assigned. Finally, we keep the most frequent class (after the weighting process) as the output class for that pixel. This strategy provides a more reliable and robust output for each pixel, avoiding isolated outlayers and misclassifications.

8.4 EXPERIMENTAL RESULTS

This section presents the results obtained after applying our approach to the wound image dataset. As it has been already mentioned in previous sections, this dataset consists of 113 pressure ulcer images that were manually segmented by clinical experts into regions of five different classes: *skin, healing, granulation, slough,* and *necrotic*.

The proposed methodology has been illustrated in Figure 8.1 and has been presented in the previous section. To avoid classification bias, a stratified K-fold cross-validation process has been carried out with the number of folds K = 5, taking the 20% of the dataset apart for model testing. Furthermore, this process has been repeated 5 times to increase the robustness of the results. For each fold, there are 67 training images, 23 validation images (in total 80% of the dataset), and 23 test images (20% of the dataset). After the resampling to get the partition of the images in the dataset corresponding to each fold, the process described in section X.3.1 is applied, generating on average 6,652 ROIs for training, 2,362 for validation, and 2,305 for testing.

The validation set is used for early stop and for monitoring the performance of the model during the training process. Since four different convolutional networks have been analyzed (*custom, vgg16, ResNet50 and xception*), 25 training processes have been carried out for each particular convolutional model (5 repetitions of 5 folds). The processes have been launched on a computer with an NVIDIA Titan Xp GPU card with 12Gb of memory, and the training of each model took between 2 and 3 hours to finish.

Since the data set is not balanced (see Figure 8.3), two types of strategies are considered to improve the classification process: cost-sensitive learning and data augmentation, which are applied during the training process of the learning model.

The cost-sensitive learning strategy (Johnson and Khoshgoftaar 2019; Zhou and Liu 2006) consists of weighting the minority classes with respect to the classes with a greater number of patterns, including these weights in the energy function to be minimized. With this, it is possible to improve the performance of the classifier for the classes with the least number of patterns (mainly necrotic tissue regions), at the cost of slightly reducing the optimization over the classes with the highest population (skin regions).

In addition, in order to improve the accuracy of the classifiers and avoid overfitting, a data augmentation process has been applied only during training. The transformation operations applied to the original ROIs extracted from the images to augment the training dataset with new transformed patterns are rotation, zoom in – of up to 20% magnification – and horizontal and vertical flipping. No shift operator is included because each region is identified with the class of the center pixel, thus shifting its position would not be appropriate.

Several hyperparameters have been considered for all the deep learning classifiers tested. The batch size has been fixed to 32 because of the hardware limitations, whereas the number of epochs for early-stop is 30. Both the optimizer and the learning rate (*lr*) are the two hyperparameters adjusted for each classifier. The possible values for the optimizer are *SGD, Adam, RMSprop,* and *AdaDelta*, while for the learning rate: 0.01, 0.001, 0.0001. The Hyperband algorithm (L. Li et al. 2018) is used for the selection of the best combination of these hyperparameters, which does not require a complete training phase for each combination to be tested, but rather it gradually carries out a progressive training while maintaining

TABLE 8.1
Study of the Most Frequently Selected Hyperparameters for each Method.
The Value in each Cell Indicates the Number of Times that Combination
Was a Winner During Hyperparameter Optimization. For each Method,
25 Executions (5 × 5) Have Been Carried Out

	vgg16					*custom*			
		Learning Rate					Learning Rate		
		0.01	0.001	0,0001			0.01	0.001	0,0001
Optimizer	adam	0	0	1	Optimizer	adam	0	0	**17**
	rmsprop	0	0	0		rmsprop	0	3	5
	SGD	0	5	1		SGD	0	0	0
	adadelta	**18**	0	0		adadelta	0	0	0
	xception					*ResNet50*			
		Learning Rate					Learning Rate		
		0.01	0.001	0,0001			0.01	0.001	0,0001
Optimizer	adam	0	0	9	Optimizer	adam	0	0	4
	rmsprop	0	0	**16**		rmsprop	0	0	**12**
	SGD	0	0	0		SGD	5	0	0
	adadelta	0	0	0		adadelta	4	0	0

the most promising combinations. Table 8.1 shows the most frequently selected combinations from the 25 training sessions performed by each classifier. Thus, it is observed that the most suitable combination for the *xception* and *ResNet50* classifiers is (*RMSprop*, *lr* = 0.0001), while for *vgg16* and *custom* they are (*AdaDelta*, *lr* = 0.01) and (*Adam*, *lr* = 0.0001) respectively. In general, it can be stated that *AdaDelta* and *SGD* optimizers perform better with high learning rates (0.01) while *Adam* and *RMSprop* perform best with slower training processes (*lr* = 0.0001).

The performance results obtained from the application of the proposed convolutional networks to the classification of the ROIs extracted from the images in the test set of each fold are shown in Figures 8.5 and 8.6 as well as in Tables 8.2–8.6. For each tissue type the following set of performance measures is defined:

$$\text{Accuracy} = (\text{TP} + \text{TN})/(\text{TP} + \text{TN} + \text{FP} + \text{FN}) \tag{8.1}$$

$$\text{Prevalence} = (\text{TP} + \text{FN})/(\text{TP} + \text{TN} + \text{FP} + \text{FN}) \tag{8.2}$$

$$\text{Sensitivity} = \text{TP}/(\text{TP} + \text{FN}) \tag{8.3}$$

$$\text{Specificity} = \text{TN}/(\text{TN} + \text{FP}) \tag{8.4}$$

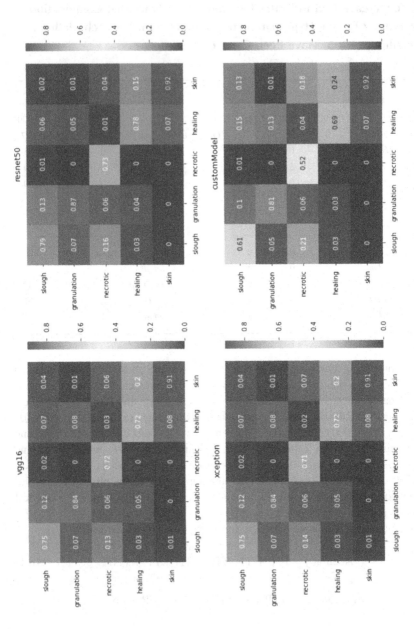

FIGURE 8.5 Confusion matrices for each method analyzed. Each cell shows the mean and standard deviation (in brackets) of the sensitivity measure (equation 8.3). The redder the cell on the diagonal of each matrix, the better the method used.

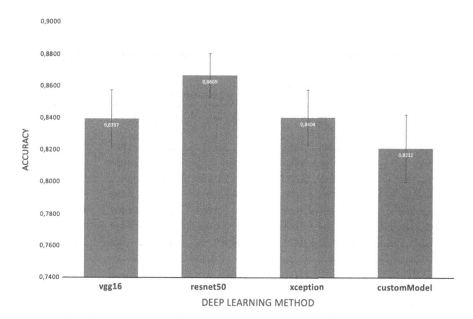

FIGURE 8.6 Balanced accuracy (equation 8.9) for each method over the entire ROIs dataset.

TABLE 8.2

Quantitative Results for each Method for the Classification of the Regions Associated with the *Slough* Tissue

Tissue (prevalence)	*Slough* (0.0837)			
DL method	*vgg16*	*ResNet50*	*xception*	*custom*
Sensitivity	**0.7854 ± 0.012**	0.7747 ± 0.011	0.7476 ± 0.03	0.6505 ± 0.055
Specificity	0.9757 ± 0.002	0.9777 ± 0.003	**0.9815 ± 0.003**	0.9721 ± 0.005
Pos Pred value	0.7476 ± 0.018	0.7607 ± 0.023	**0.7878 ± 0.022**	0.6823 ± 0.039
Neg Pred value	**0.9803 ± 0.001**	0.9794 ± 0.001	0.9771 ± 0.003	0.9682 ± 0.005
Detection rate	**0.0657 ± 0.001**	0.0648 ± 0.001	0.0625 ± 0.003	0.0544 ± 0.005
Detection prevalence	**0.0880 ± 0.003**	0.0853 ± 0.003	0.0795 ± 0.005	0.0800 ± 0.007
Balanced Accuracy	**0.8806 ± 0.005**	0.8762 ± 0.006	0.8645 ± 0.014	0.8113 ± 0.027

$$\text{Positive predictive value (PPV)} = (\text{sensitivity} \times \text{prevalence})$$
$$/((\text{sensitivity} \times \text{prevalence})$$
$$+ ((1 - \text{specificity}) \times (1 - \text{prevalence})))$$
$$(8.5)$$

TABLE 8.3

Quantitative Results for each Method for the Classification of the Regions Associated with the *Granulation* Tissue

Tissue (prevalence)	*Granulation* (0.1292)			
DL method	*vgg16*	*ResNet50*	*xception*	*custom*
Sensitivity	0.8443 ± 0.016	0.8581 ± 0.02	**0.8687 ± 0.015**	0.8181 ± 0.03
Specificity	**0.9761 ± 0.004**	0.9751 ± 0.002	0.9735 ± 0.002	0.9750 ± 0.007
Pos Pred value	**0.8406 ± 0.02**	0.8368 ± 0.009	0.8298 ± 0.009	0.8320 ± 0.037
Neg Pred value	0.9769 ± 0.002	0.9789 ± 0.003	**0.9804 ± 0.002**	0.9731 ± 0.004
Detection rate	0.1091 ± 0.002	0.1108 ± 0.003	**0.1122 ± 0.002**	0.1057 ± 0.004
Detection prevalence	0.1299 ± 0.005	0.1325 ± 0.004	**0.1353 ± 0.004**	0.1274 ± 0.01
Balanced Accuracy	0.9102 ± 0.007	0.9166 ± 0.009	**0.9211 ± 0.006**	0.8965 ± 0.012

TABLE 8.4

Quantitative Results for each Method for the Classification of the Regions Associated with the *Necrotic* Tissue

Tissue (prevalence)	*Necrotic* (0.0141)			
DL method	*vgg16*	*ResNet50*	*xception*	*custom*
Sensitivity	0.7800 ± 0.052	**0.8425 ± 0.033**	0.7975 ± 0.019	0.7900 ± 0.029
Specificity	0.9968 ± 0.001	0.9963 ± 0.0	**0.9970 ± 0.001**	0.9913 ± 0.002
Pos Pred value	0.7809 ± 0.04	0.7646 ± 0.015	**0.7972 ± 0.042**	0.5740 ± 0.07
Neg Pred value	0.9968 ± 0.001	**0.9977 ± 0.0**	0.9971 ± 0.0	0.9970 ± 0.0
Detection rate	0.0110 ± 0.001	**0.0119 ± 0.0**	0.0113 ± 0.0	0.0112 ± 0.0
Detection prevalence	0.0142 ± 0.002	0.0156 ± 0.001	0.0142 ± 0.001	**0.0197 ± 0.002**
Balanced Accuracy	0.8884 ± 0.026	**0.9194 ± 0.016**	0.8973 ± 0.009	0.8907 ± 0.015

$$\text{Negative predictive value (NPV)} = (\text{specificity} \times (1 - \text{prevalence}))$$
$$/(((1 - \text{sensitivity}) \times \text{prevalence})$$
$$+ (\text{specificity} \times (1 - \text{prevalence}))) \quad (8.6)$$

$$\text{Detection rate} = TP/(TP + TN + FP + FN) \quad (8.7)$$

$$\text{Detection prevalence} = (TP + FP)/(TP + TN + FP + FN) \quad (8.8)$$

$$\text{Balanced accuracy} = (\text{sensitivity} + \text{specificity})/2 \quad (8.9)$$

TABLE 8.5

Quantitative Results for each Method for the Classification of the Regions Associated with the *Healing* Tissue

Tissue (prevalence)	*Healing* (0.2382)			
DL method	*vgg16*	*ResNet50*	*xception*	*custom*
Sensitivity	0.7615 ± 0.005	**0.7770 ± 0.012**	0.7538 ± 0.012	0.6657 ± 0.058
Specificity	0.9268 ± 0.001	0.9247 ± 0.003	**0.9358 ± 0.004**	0.9173 ± 0.009
Pos Pred value	0.7649 ± 0.003	0.7636 ± 0.006	**0.7859 ± 0.012**	0.7160 ± 0.013
Neg Pred value	0.9255 ± 0.002	**0.9299 ± 0.003**	0.9240 ± 0.004	0.8981 ± 0.015
Detection rate	0.1814 ± 0.001	**0.1851 ± 0.003**	0.1795 ± 0.003	0.1585 ± 0.014
Detection prevalence	0.2371 ± 0.001	**0.2424 ± 0.005**	0.2285 ± 0.004	0.2216 ± 0.02
Balanced Accuracy	0.8441 ± 0.003	**0.8509 ± 0.005**	0.8448 ± 0.007	0.7915 ± 0.025

TABLE 8.6

Quantitative results for each method for the classification of the regions associated with the *skin* tissue

Tissue (prevalence)	*Skin* (0.5349)			
DL method	*vgg16*	*ResNet50*	*xception*	*custom*
Sensitivity	0.9123 ± 0.004	0.9066 ± 0.009	**0.9243 ± 0.007**	0.8957 ± 0.019
Specificity	0.9077 ± 0.008	**0.9155 ± 0.006**	0.8964 ± 0.008	0.8446 ± 0.036
Pos Pred value	0.9192 ± 0.007	**0.9251 ± 0.005**	0.9113 ± 0.006	0.8699 ± 0.025
Neg Pred value	0.9000 ± 0.003	0.8952 ± 0.008	**0.9116 ± 0.007**	0.8763 ± 0.017
Detection rate	0.4879 ± 0.002	0.4849 ± 0.005	**0.4944 ± 0.004**	0.4791 ± 0.01
Detection prevalence	0.5309 ± 0.005	0.5242 ± 0.008	0.5426 ± 0.007	**0.5514 ± 0.024**
Balanced Accuracy	0.9100 ± 0.003	**0.9111 ± 0.002**	0.9104 ± 0.003	0.8701 ± 0.014

where TP, FP, TN, and FN stand for true positive, false positive, true negative, and false negative for each tissue type, respectively.

Thus, Figure 8.5 shows the confusion matrix for each model and tissue type (mean and standard deviation), where the redder the diagonal of the matrix, the better the classification performed. The measure shown in the cells of the matrices is sensitivity (equation 8.3), since it is the one that better determines the differences between the models analyzed. The information shown in Figure 8.5 can be complemented with the average number of ROIs per class in the test set, or *support*: *skin* (1273), *healing* (505), *granulation* (327), *slough* (234) and *necrotic* (45).

Figure 8.6 displays the mean (vertical bar) and the standard deviation (vertical thin line associated with each bar) of the overall success or balanced accuracy (equation 8.9) for each convolutional model. Unlike accuracy (equation 8.1), this measure is more suitable for problems with imbalanced data. In Tables 8.2–8.6, the results of the performance measures are shown for each tissue type. In the first row of each table the prevalence (equation 8.2) of the corresponding tissue type is displayed.

In general, it is observed that the performance of the *ResNet50* model slightly improves the competitors (balanced accuracy 0.8945), surpassing the following method by almost 0.01. This result is supported by the numbers in the confusion matrices shown in Figure 8.5, where a significant difference is observed in the detection of *necrotic* tissue (sensitivity 0.84 vs 0.79 of the second-best method). In general, it is observed that the best-classified tissues are *skin* and *granulation*, the first probably due to the number of *skin* regions available for training (prevalence 0.5349), the second because of the intrinsic characteristics of granulation regions, which make them easily distinguishable from the others. On the other hand, it is observed that, despite the fact that *healing* tissue has a high prevalence (0.2382) and consequently a high number of available ROIs for training, the classification results are not very satisfactory (0.76, 0.75, and 0.67 for *vgg16*, *xception* and *custom*, respectively), only improving by *ResNet50* (0.78). Furthermore, it should be noted that the classifiers confuse *healing* mainly with *skin*, since most of the time (0.15 in *ResNet50*), the regions associated with *healing* are misclassified as *skin*. The main reason why this occurs might be the variability of colors of *healing* regions found in different images, which can make the classifiers less robust for the classification of this kind of region. The difficulty in classifying *necrotic* tissues is also noteworthy since, although it does not obtain the poorest performance results, this class has the highest value of variability for all the models analyzed (standard deviation 0.14 as the best result for *ResNet50*). In this case, the difficulty lies in the small number of samples available for this tissue (prevalence 0.0141) and the small size (number of pixels) of the ROIs of this tissue type, sometimes mixing up *necrotic*-tissue pixels with *slough*-tissue pixels in the same ROI, especially in *vgg16* and *xception* methods. This conclusion is supported also by the confusion matrices, where the *necrotic* regions are shown to be confused with *slough* 12–13% of times on average.

Tables 8.2–8.6 and Figure 8.6 show that, in general, the *ResNet50* model exceeds the rest of the proposals. However, and carrying out an analysis by classes or tissues, it is observed that *vgg16* is the best classifier to detect *slough*-tissue patterns (balanced accuracy 0.8806), *xception* improves the results of its competitors for the *granulation* tissue (balanced accuracy 0.9211), while *necrotic* and *healing* tissues are better discriminated by the *ResNet50* network (balanced accuracy 0.9194 and 0.8509, respectively). For the *skin* class, *vgg16*, *ResNet50,* and *xception* models obtain similar results. According to Figure 8.6, it is also remarkable that the results for the *vgg16* and *xception* models are practically identical, so that between them we would choose the *xception* model given the lower complexity and size of the final trained network. These results can be compared with any of the previous approaches existing in the literature that analyze the same wound-image dataset. Thus, it is notable that the performance results obtained in this approach are slightly better than those shown

TABLE 8.7

P-value Between each Pair of Methods on the Balanced Accuracy Measure. Values less than 0.05 Indicate that the Difference Between Methods Is Significant. The * Indicates that the Method Has Significant Differences from all Competitors

Balanced Accuracy	p-value	vgg16	ResNet50	xception	custom
0,8865	vgg16	–	–	–	–
0,8945	ResNet50*	0,009417425	–	–	–
0,8852	xception	0,903626797	0,003507060	–	–
0,8520	custom*	0,000045739	0,000013898	0,000036243	–

in Tables 4 and 6 of (Veredas et al. 2015) for this same dataset. Moreover, the approach proposed in that study is much more complex than ours as it requires several preprocessing steps for invalid-regions and wound-bed separation, followed by an ad-hoc regionalization process of the wound images, along with feature engineering, extraction, and selection derived from the extensive knowledge of the problem domain. This last drawback can be avoided with proposals based on deep learning, like ours, for which it is not yet needed to engineer and extract specific features from the images, since the network itself is capable of extracting high-level information on which it bases the classification process.

Table 8.7 has been added to reflect the p-value after applying the Mann–Whitney–Wilcoxon non-parametric statistical test on the balanced accuracy measure (equation 8.9) of each pairwise classifier. It can be seen that the differences between *xception* and *vgg16* are not significant (p-value 0.9036), while the method with the best performance (*ResNet50*) does show significant differences with respect to all competitors, given that the p-value with all of them is below 0.05.

Figures 8.7 and 8.8 show both qualitative and quantitative results of the complete image segmentation and region classification process given the trained convolutional models analyzed in this study. Both the ideal segmentation of the wound image and the output proposed by the best model (*ResNet50*) can be seen for three wound examples in Figure 8.7, corresponding to images #95, #5, and #112 from the dataset (which contains 113 wound images), respectively (rows from top to bottom). It is observed how the convolutional classifier followed by the consensus process between the regions associated with each classified pixel generate a valid and plausible wound segmentation, given the high complexity of the problem. However, the details and definition of each segmented region are limited, causing the appearance of jagged edges mainly due to the value of the overlap between regions (N = 5).

The comparison of the segmentation given by the models with the ideal segmentation image (GT) is performed at the pixel level, that is, the matching pixels are counted and divided by the total pixels in the image. This accuracy value for

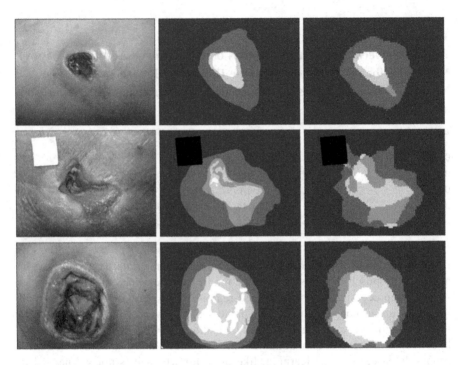

FIGURE 8.7 Some qualitative results by applying the *ResNet50* model. The first column shows the original image, the second the GT or ideal segmentation, and the third the result of the segmentation after applying our proposal. The rows, from top to bottom, correspond to the wound images, 95, 5, and 112, respectively.

each image is averaged and calculated for the training, validation and test sets, only for the best model in the ROIs classification task (*ResNet50*). Figure 8.8 shows that, in general, the results in training tend to be a little higher, although there is not too much overfitting given that the results in tests do not vary significantly from those obtained for training and validation. Note that depending on each fold the results may also vary slightly. Thus, it is possible to observe that the accuracy over the test set is estimated on average at 0.795.

8.5 CONCLUSIONS

In this work, the development of a framework for wound image segmentation has been proposed. Given the small size of the available dataset, with a reduced number of images labeled and segmented by a group of clinical experts, the methodology has been divided into two phases. The first phase consisted of generating regions of interest (ROIs) and training a deep learning model for their classification into five different tissue types. In the second phase, the segmentation of the complete image is achieved by using the overlapping labeled regions given by the classifiers in the previous phase as part of a consensus method designed for the classification of each individual pixel in the final segmented image.

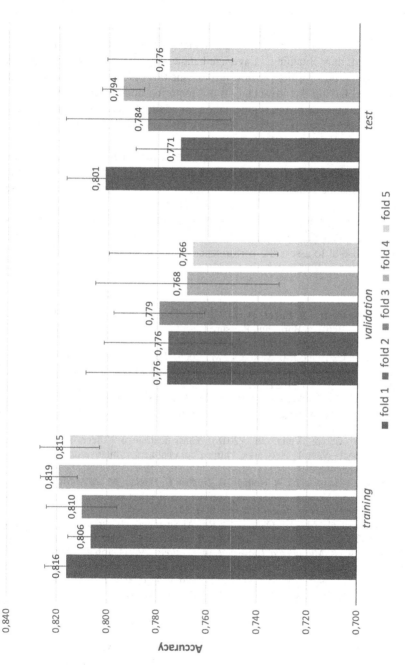

FIGURE 8.8 Quantitative results of accuracy in the segmentation of the wound-image dataset using *ResNet50* model. Each block of bars represents the mean of accuracy on the sets of training, validation and test images, divided into the different execution k-folds.

Four deep learning models have been evaluated, from three pre-trained convolutional networks, *ResNet50,* and *xception* – to one convolutional network implemented from scratch (*custom*). Given the results obtained on a wound-image dataset previously used in the literature and consisting of 113 images, it is observed that *ResNet50* is the network with the greatest classification performance among the models analyzed, thus obtaining an average balanced accuracy of 0.8945, which is comparable to the best results previously published in the literature, but with the *ResNet50* model having the advantage of not requiring domain knowledge or feature extraction stages for image segmentation. In addition, qualitative results show that the final output of our proposed framework for wound image segmentation and tissue classification is acceptable and consistent.

As future lines to work, it is proposed to improve the robustness of the classification of low-prevalence tissues, such as *necrotic,* either by applying test-time augmentation techniques or including additional domain information in the design of the architecture of the convolutional networks. On the other hand, it would be interesting to assess the possibility of smoothing or improving the edges of the final segmentation. Finally, we consider that making a comparison with image segmentation deep learning models would be very useful to assess the performance of our proposal on small datasets.

REFERENCES

Abubakar, A., Ugail, H., & Bukar, A. M. (2020). Can Machine Learning Be Used to Discriminate Between Burns and Pressure Ulcer? *Intelligent Systems and Applications* (pp. 870–880). Springer International Publishing.

Acha, B., Serrano, C., Acha, J. I., & Roa, L. M. (2005). Segmentation and Classification of Burn Images by Color and Texture Information. *Journal of Biomedical Optics, 10*(3). https://doi.org/101117/1.1921227.

Acha, B., Serrano, C., Fondón, I., & Gómez-Cía, T. (2013). Burn Depth Analysis Using Multidimensional Scaling Applied to Psychophysical Experiment Data. *IEEE Transactions on Medical Imaging, 32*(6), 1111–1120.

Aguirre Nilsson, C., & Velic, M. (2018). Classification of Ulcer Images Using Convolutional Neural Networks. https://odr.chalmers.se/bitstream/20.500.12380/255746/1/255746.pdf.

Al Shammeri, O., AlHamdan, N., Al-Hothaly, B., Midhet, F., Hussain, M., & Al-Mohaimeed, A. (2014). Chronic Venous Insufficiency: Prevalence and Effect of Compression Stockings. *International Journal of Health Sciences, 8*(3), 231–236.

Alzubaidi, L., Fadhel, M. A., Oleiwi, S. R., Al-Shamma, O., & Zhang, J. (2020). DFU_QUTNet: Diabetic Foot Ulcer Classification Using Novel Deep Convolutional Neural Network. *Multimedia Tools and Applications, 79*(21), 15655–15677.

Beeckman, D., Schoonhoven, L., Fletcher, J., Furtado, K., Gunningberg, L., Heyman, H., Lindholm, C., Paquay, L., Verdú, J., & Defloor, T. (2007). EPUAP Classification System for Pressure Ulcers: European Reliability Study. *Journal of Advanced Nursing, 60*(6), 682–691.

Bergqvist, D., Lindholm, C., & Nelzén, O. (1999). Chronic Leg Ulcers: The Impact of Venous Disease. *Journal of Vascular Surgery, 29*(4), 752–755.

Blanco, G., Traina, A. J. M., Traina, C. Jr., Azevedo-Marques, P. M., Jorge, A. E. S., de Oliveira, D., & Bedo, M. V. N. (2020). A Superpixel-Driven Deep Learning Approach for the Analysis of Dermatological Wounds. *Computer Methods and Programs in Biomedicine, 183* (January), 105079.

Bowling, F. L., King, L., Paterson, J. A., Hu, J., Lipsky, B. A., Matthews, D. R., & Boulton, A. J. M. (2011). Remote Assessment of Diabetic Foot Ulcers Using a Novel Wound Imaging System. *Wound Repair and Regeneration: Official Publication of the Wound Healing Society [and] the European Tissue Repair Society*, *19*(1), 25–30.

Chan, K. S., & Lo, Z. J. (2020). Wound Assessment, Imaging and Monitoring Systems in Diabetic Foot Ulcers: A Systematic Review. *International Wound Journal*, *17*(6). https://doi.org/10.1111/iwj.13481.

Cho, N. H., Shaw, J. E., Karuranga, S., Huang, Y., da Rocha Fernandes, J. D., Ohlrogge, A. W., & Malanda, B. (2018). IDF Diabetes Atlas: Global Estimates of Diabetes Prevalence for 2017 and Projections for 2045. *Diabetes Research and Clinical Practice*, *138*, 271–381. https://doi.org/10.1016/j.diabres.2018.02.023.

Chollet, F. (2017). Xception: Deep Learning With Depthwise Separable Convolutions (2017) *Proceedings – 30th IEEE Conference on Computer Vision and Pattern Recognition, CVPR, January*, pp. 1800–1807.

Despo, O., Yeung, S., Jopling, J., Pridgen, B., Sheckter, C., Silberstein, S., Fei-Fei, L., & Milstein, A. (n.d.) BURNED: Towards Efficient and Accurate Burn Prognosis Using Deep Learning. Accessed March 8 (2021). http://cs231n.stanford.edu/reports/2017/pdfs/507.pdf.

Edsberg, L. E. (2007). Pressure Ulcer Tissue Histology: An Appraisal of Current Knowledge. *Ostomy/wound Management*, *53*(10), 40–49.

Fauzi, A., Faizal, M., Khansa, I., Catignani, K., Gordillo, G., Sen, C. K., & Gurcan, M. N. (2015). Computerized Segmentation and Measurement of Chronic Wound Images. *Computers in Biology and Medicine*, *60*(May), 74–85.

Filko, D., Antonic, D., & Huljev, D. (2010). WITA — Application for Wound Analysis and Management. The 12th IEEE International Conference on E-Health Networking, Applications and Services. https://doi.org/10.1109/health.2010.5556533.

García-Zapirain, B., Elmogy, M., El-Baz, A., & Elmaghraby, A.S. (2018). Classification of Pressure Ulcer Tissues with 3D Convolutional Neural Network. *Medical & Biological Engineering & Computing*, *56*(12), 2245–2258.

Goyal, M., Reeves, N. D., Davison, A. K., Rajbhandari, S., Spragg, J., & YapM. H. (2018). Dfunet: Convolutional Neural Networks for Diabetic Foot Ulcer Classification. *IEEE Transactions on Emerging Topics in Computational Intelligence*, *4*(5), 728–739.

Goyal, M., Yap, M. H., Reeves, N. D., Rajbhandari, S., & Spragg, J. (2017). Fully Convolutional Networks for Diabetic Foot Ulcer Segmentation. 2017 IEEE International Conference on Systems, Man, and Cybernetics (SMC), 618–623.

Hani, A. F. M., Arshad, L., Malik, A. S., Jamil, A., & Bin, F. Y. B. (2012). Haemoglobin Distribution in Ulcers for Healing Assessment. 2012 4th International Conference on Intelligent and Advanced Systems (ICIAS2012). https://doi.org/10.1109/icias.2012.6306219.

He, K., Zhang, X., Ren, S., & Sun, J. (2016). Deep Residual Learning for Image Recognition. Proceedings of the IEEE Computer Society Conference on Computer Vision and Pattern Recognition, pp. 770–778.

Johnson, J. M., & Khoshgoftaar, T. M. (2019). Survey on Deep Learning With Class Imbalance. *J Big Data*, *6*, 27.

Kolesnik, M., & Fexa, A. (2004). Segmentation of Wounds in the Combined Color-Texture Feature Space. *Medical Imaging 2004: Image Processing* (Vol. 5370, pp. 549–556). International Society for Optics and Photonics.

Kolesnik, M., & Fexa, A. (2005). Multi-Dimensional Color Histograms for Segmentation of Wounds in Images. *Image Analysis and Recognition* (pp. 1014–1022). Springer Berlin Heidelberg.

Kolesnik, M., & Fexa, A. (2006). How Robust Is the SVM Wound Segmentation? Proceedings of the 7th Nordic Signal Processing Symposium - NORSIG 2006. https://doi.org/10.1109/norsig.2006.275274.

Krizhevsky, A., Sutskever, I., & Hinton, G.E. (2012). ImageNet Classification With Deep Convolutional Neural Networks. *Advances in Neural Information Processing Systems*, *2*, 1097–1105.

Kumar, U. S., & Sudharsan, N. M. (2018). Enhancement Techniques for Abnormality Detection Using Thermal Image. *The Journal of Engineering*, *5*, 279–283. https://doi.org/10.1049/joe.2017.0899.

Landi, F., Onder, G., Russo, A., & Bernabei, R. (2007). Pressure Ulcer and Mortality in Frail Elderly People Living in Community. *Archives of Gerontology and Geriatrics*, *44*(1), 217–223.

Li, L., Jamieson, K., DeSalvo, G., Rostamizadeh, A., & Talwalkar, A. (2018). Hyperband: A Novel Bandit-based Approach to Hyperparameter Optimization. *Journal of Machine Learning Research*, *18*, 1–52.

Li, F., Wang, C., Liu, X., Peng, Y., & Jin, S. (2018). A Composite Model of Wound Segmentation Based on Traditional Methods and Deep Neural Networks. *Computational Intelligence and Neuroscience*, *2018*(May), 4149103.

Liu, X., Wang, C., Li, F., Zhao, X., Zhu, E., & Peng, Y. (2017). A Framework of Wound Segmentation Based on Deep Convolutional Networks. 2017 10th International Congress on Image and Signal Processing, BioMedical Engineering and Informatics (CISP-BMEI), 1–7.

Mukherjee, R., Manohar, D. D., Das, D. K., Achar, A., Mitra, & Chakraborty, C. (2014). Automated Tissue Classification Framework for Reproducible Chronic Wound Assessment. *BioMed Research International*, *2014* (July), 851582.

Nejati, H., Ghazijahani, H. A., Abdollahzadeh, M., Malekzadeh, T., N. - Cheung, K. -. Lee, and L. -. Low. (2018). Fine-Grained Wound Tissue Analysis Using Deep Neural Network. 2018 IEEE International Conference on Acoustics, Speech and Signal Processing (ICASSP), 1010–1014.

Papazoglou, E. S., Zubkov, L., Mao, X., Neidrauer, M., Rannou, N., & Weingarten, M. S. (2010). Image Analysis of Chronic Wounds for Determining the Surface Area. *Wound Repair and Regeneration: Official Publication of the Wound Healing Society [and] the European Tissue Repair Society*, *18*(4), 349–358.

Pasero, E., & Castagneri, C. (2017). Application of an Automatic Ulcer Segmentation Algorithm. 2017 IEEE 3rd International Forum on Research and Technologies for Society and Industry (RTSI), 1–4.

Pasero, Eros., & Castagneri, C. (2017). Leg Ulcer Long Term Analysis. *Intelligent Computing Theories and Application* (pp. 35–44). Springer International Publishing.

Poon, T. W. K., & Friesen, M. R. (2015). Algorithms for Size and Color Detection of Smartphone Images of Chronic Wounds for Healthcare Applications. *IEEE Access*, *3*, 1799–1808. https://doi.org/10.1109/access.2015.2487859,

Rajathi, V., Bhavani, R. R., & Jiji, G. W. (2019). Varicose Ulcer(C6) Wound Image Tissue Classification Using Multidimensional Convolutional Neural Networks. *The Imaging Science Journal*, *67*(7), 374–384.

Rao, K. N., Srinivasa, R. P., Rao, A. A., & Sridhar, G. R. (2013). Sobel Edge Detection Method to Identify and Quantify the Risk Factors for Diabetic Foot Ulcers. *International Journal of Computer Science and Information Technology*, *5*(1). https://doi.org/10.5121/ijcsit.2013.5103.

Reiber, G. E. (1996). The Epidemiology of Diabetic Foot Problems. *Diabetic Medicine*, *59*(6). https://doi.org/10.1002/dme.1996.13.s1.6.

Ruckley, C. V. (1997). Socioeconomic Impact of Chronic Venous Insufficiency and Leg Ulcers. *Angiology*, *48*(1), 67–69.

Sen, C. K. (2019). Human Wounds and Its Burden: An Updated Compendium of Estimates. *Advances in Wound Care: The Journal for Prevention and Healing*, *8*(2), 39–48.

Serrano, C., Acha, B., Gómez-Cía, T., Acha, J. I., & Roa, L. M. (2005). A Computer Assisted Diagnosis Tool for the Classification of Burns by Depth of Injury. *Burns: Journal of the International Society for Burn Injuries, 31*(3), 275–281.

Serrano, C., Boloix-Tortosa, R., Gómez-Cía, T., & Acha, B. (2015). Features Identification for Automatic Burn Classification. *Burns: Journal of the International Society for Burn Injuries, 41*(8), 1883–1890.

Shenoy, V. N., Foster, E., Aalami, L., Majeed, B., & Aalami, O. (2018). Deepwound: Automated Postoperative Wound Assessment and Surgical Site Surveillance through Convolutional Neural Networks. *2018 IEEE International Conference on Bioinformatics and Biomedicine (BIBM)*, 1017–1021.

Simonyan, K., & Zisserman, A. (2015). Very Deep Convolutional Networks For Large-scale Image Recognition. *3rd International Conference on Learning Representations. ICLR 2015 – Conference Track Proceedings.* https://arxiv.org/abs/1409.1556

Song, B., & Sacan, A. (2012). Automated Wound Identification System Based on Image Segmentation and Artificial Neural Networks. *2012 IEEE International Conference on Bioinformatics and Biomedicine*, 1–4.

Szegedy, C., Liu, W., Jia, Y., Sermanet, P., Reed, S., Anguelov, D., Erhan, D., Vanhoucke, V., & Rabinovich, A. (2015). Going Deeper With Convolutions. *Proceedings of the IEEE Computer Society Conference on Computer Vision and Pattern Recognition*, 07-12-June-2015, art. no. 7298594, pp. 1–9.

Veredas, F. J., Luque-Baena, R. M., Martín-Santos, F. J., Morilla-Herrera, J. C., & Morente, L. (2015). Wound Image Evaluation with Machine Learning. *Neurocomputing, 164* (September), 112–122.

Veredas, F. J., Mesa, H., & Morente, L. (2010). Binary Tissue Classification on Wound Images with Neural Networks and Bayesian Classifiers. *IEEE Transactions on Medical Imaging, 29*(2), 410–427.

Wang, L., Pedersen, P. C., Agu, E., Strong, D. M., & Tulu, B. (2017). Area Determination of Diabetic Foot Ulcer Images Using a Cascaded Two-Stage SVM-Based Classification. *IEEE Transactions on Bio-Medical Engineering, 64*(9), 2098–2109.

Wang, C., Yan, X., Smith, M., Kochhar, K., Rubin, M., Warren, S. M., Wrobel, J., & Lee, H. (2015). A Unified Framework for Automatic Wound Segmentation and Analysis with Deep Convolutional Neural Networks. *Annual International Conference of the IEEE Engineering in Medicine and Biology Society. Conference* 2015. 2415–2418.

Wannous, H., Lucas, Y., & Treuillet, S. (2008). Efficient SVM Classifier Based on Color and Texture Region Features for Wound Tissue Images. *Medical Imaging 2008: Computer-Aided Diagnosis* (Vol. 6915, pp. 69152T). International Society for Optics and Photonics.

Wannous, H., Treuillet, S., & Lucas, Y. (2007). Supervised Tissue Classification from Color Images for a Complete Wound Assessment Tool. *Annual International Conference of the IEEE Engineering in Medicine and Biology Society.* 6032–6035.

Wannous, H., Treuillet, S., & Lucas, Y. (2010). Robust Tissue Classification for Reproducible Wound Assessment in Telemedicine Environments. *Journal of Electronic Imaging, 19*(2), 023002.

Wannous, H., Lucas, Y., & Treuillet, S. (2011). Enhanced Assessment of the Wound-Healing Process by Accurate Multiview Tissue Classification. *IEEE Transactions on Medical Imaging, 30*(2), 315–326.

Wantanajittikul, K., Auephanwiriyakul, S., Theera-Umpon, N., & Koanantakool, T. (2012). Automatic Segmentation and Degree Identification in Burn Color Images. *4th 2011 Biomedical Engineering International Conference*, 169–173.

Yadav, M. K., Manohar, D. D., Mukherjee, G., & Chakraborty, C. (2013). Segmentation of Chronic Wound Areas by Clustering Techniques Using Selected Color Space. *Journal of Medical Imaging and Health Informatics, 3*(1), 22–29.

Yadav, D. P., Sharma, A., Singh, M., & Goyal, A. (2019). Feature Extraction Based Machine Learning for Human Burn Diagnosis From Burn Images. *IEEE Journal of Translational Engineering in Health and Medicine*, 7(July), 1800507.

Zahia, S., Sierra-Sosa, D., Garcia-Zapirain, B., & Elmaghraby, A. (2018). Tissue Classification and Segmentation of Pressure Injuries Using Convolutional Neural Networks. *Computer Methods and Programs in Biomedicine*, 159(June), 51–58.

Zhang, P., Lu, J., Jing, Y., Tang, S., Zhu, D., & Bi, Y. (2017). Global Epidemiology of Diabetic Foot Ulceration: A Systematic Review and Meta-Analysis †. *Annals of Medicine*, 49(2), 106–116.

Zhou, Z.-H. & Liu, X.-Y. (2006). On Multi-Class Cost-sensitive Learning. Proceedings of the 21st National Conference on Artificial intelligence – Volume 1 (AAAI'06). AAAI Press, 567–572.

9 Artificial Intelligence Methodologies in Dentistry

Reza Soroushmehr, Winston Zhang, Jonathan Gryak, Kayvan Najarian, Najla Al Turkestani, Lucia Cevidanes, Romain Deleat-Besson, Celia Le, and Jonas Bianchi

CONTENTS

9.1 INTRODUCTION

Artificial Intelligence (AI) is a general term describing the use of a computer to model intelligent behavior with minimal human intervention (Ramesh et al., 2004). In healthcare systems, applications of AI include automated robotic arms, smart sensors, or by software-type algorithms which support clinical decision making (Shan et al., 2020). AI techniques can be categorized as either machine learning or non-learning methods. Machine learning (ML) is a subfield of AI with a focus on constructing computer systems that automatically improve through experience and identify fundamental statistical and computational information-theoretic laws that govern all learning systems, including computers, humans, and organizations (Jordan & Mitchell, 2015). After identifying patterns within data through training, ML models apply these patterns to new data to make predictions. In contrast, non-learning AI methods perform tasks using human-designed rules. Applications of

ML methods are rapidly increasing in dentistry due to the increasing data acqui-sition, recent advances in their ability to automatically extract patterns from data (compared to non-learning methods where many trials and errors might be needed to identify those patterns), and their benefit toward clinicians' decision making.

Here, we provide basic definitions and then review AI techniques developed for dental applications.

Machine learning models have different structures. ***Artificial neural network (ANN)*** is a subset of machine learning that mimics the human brain through a set of algorithms and has three node layers/depths including an input, a hidden layer (weights, a bias, or threshold), and an output. ***Deep neural networks*** are a form of machine learning with the backbone of ANN but have more than three-node layers. Consequently, they can learn non-linear patterns in structured and un-structured data. There are different types of deep learning models or deep neural networks designed for different applications. One of these networks that are widely used in image analysis is called Convolutional Neural Network (CNN), where a number of convolutional and subsampling layers are employed and that could be followed by fully connected layers. Due to their performance, deep neural networks are widely used in many applications. If a machine learning model is well designed, trained, and validated and some characteristics of data, such as adequacy and quality, are satisfied, developing rules and algorithms to capture relationships between inputs and outputs would not be necessary. Such learning models not only could reduce the development time, but could also improve the captured input-output relationships, especially for complex data. In the past 4 years, most of the AI-based methods with applications in Dentistry that have been published have employed machine learning and, in particular, deep learning techniques (Arifin et al., 2019; Balbin et al., 2019; Bianchi et al., 2020; Boiko et al., 2019; Bozkurt & Karagol, 2020; H. Chen et al., 2019; Y. Chen et al., 2020; Cui et al., 2021; Datta & Chaki, 2020; Dong et al., 2019; Guijarro-Rodríguez et al., 2020; Harrison et al., 2019; Hung et al., 2019; Janardanan & Logeswaran, 2019; Jaskari et al., 2020, 2020; Kwak et al., 2020; Lakshmi & Chitra, 2020; Lee et al., 2019; Li et al., 2020; Majanga & Viriri, 2020; Muresan et al., 2020; Tian et al., 2019; Zheng et al., 2020). In Dentistry, AI-based tech-niques using the CNN architecture have been developed with applications in diagnosis, treatment, and prognosis for Oral and Maxillofacial surgery, Cariology, Endodontics, Periodontics, Temporomandibular Joint disorders, Orthodontics, and Prosthodontics (Shan et al., 2020). As illustrated in Figure 9.1, these applications include tooth segmentation, tooth structure detection, biofilm classification, caries detection, tooth classification, sex classification, age esti-mation, dental pathology detection, cephalometric landmark detection, period-ontal inflammation detection, maxilla and mandible segmentation, image quality enhancement/assessment, atherosclerotic carotid plaques detection, root mor-phology classification, osteoporosis detection, and bone loss detection (Schwendicke et al., 2019).

AI Dentistry and Craniofacial Applications

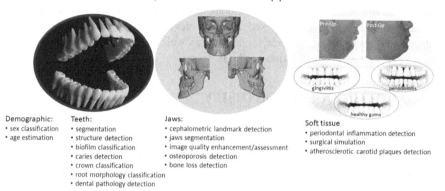

Demographic:
• sex classification
• age estimation

Teeth:
• segmentation
• structure detection
• biofilm classification
• caries detection
• crown classification
• root morphology classification
• dental pathology detection

Jaws:
• cephalometric landmark detection
• jaws segmentation
• image quality enhancement/assessment
• osteoporosis detection
• bone loss detection

Soft tissue
• periodontal inflammation detection
• surgical simulation
• atherosclerotic carotid plaques detection

FIGURE 9.1 Current applications of AI-based techniques have been developed in the diagnosis, treatment, and prognosis of dental (teeth), skeletal (jaw bones), and facial soft tissues.

To analyze, evaluate, and compare AI-based methods, different parameters, such as image type, number of images used in total or for training/testing, AI model structure, evaluation metrics, and number of human annotators assessing each image, need to be carefully considered, as shown in Figure 9.2. These parameters pose training challenges and an important concept is the generalizability of the findings from each CNN. With the ultimate goal of improving patients' oral health, patient-specific disease classification and prediction is challenging when datasets from different clinical centers or practices pose individual variability not previously included in the training models.

Previous studies (Prasaanth et al., 2021; Schwendicke et al., 2019; Shan et al., 2020) reviewed the application of AI-based methods and discussed the challenges

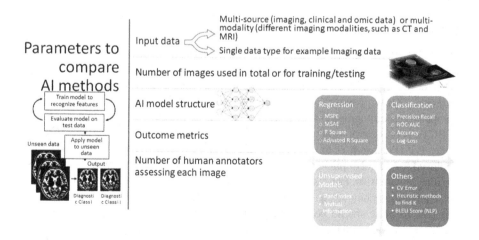

FIGURE 9.2 Parameters to compare AI methods.

facing AI techniques in Dentistry. In Shan et al. (2020), these applications are categorized into diagnosis, disease treatment, and disease prediction. In this chapter, we review state-of-the-art AI-based methods developed for dental applications, categorizing them based on their input data, as well as providing details on their evaluation metrics and validation methods.

9.2   AI TECHNIQUES IN DENTISTRY

In this section, we review AI techniques for dental applications and, based on their input data, categorize them into two groups: imaging data and integrative multi-source data. The first group has been widely described in the literature due to the existence of a large number of imaging exams in the dental and medical clinical and research field, such as magnetic resonance (MRI), conventional radiographic images, CBCT and CT. For the use of multi-source data, studies usually address complex diseases, and use multiple types of data such as clinical, imaging, and omic (e.g. genomics, proteomics, and metabolomics) from a patient to better diagnose his/her disease. As in many applications imaging data is a highly valuable resource for diagnostics, we dedicate one section for reviewing image processing techniques. These techniques either implement machine learning or traditional image processing methods and might need pre-processing (e.g. denoising, contrast enhancement) and/or post-processing (e.g. removing tiny objects) to refine/improve the results. As deep neural networks require a large volume of diverse data, data augmentation (e.g. translations, flipping, cropping, padding, Gaussian noise, Gaussian blur, rotation, and Gamma contrast (Muresan et al., 2020)) is performed before feeding the data to the model to increase its size and diversity. When deploying a deep learning model, the model might learn relationships between inputs and outputs and handle some tasks such as noise reduction and image enhancement automatically. Therefore, in some deep learning models, no pre-processing is performed besides data augmentation.

9.2.1   AI TECHNIQUES USING IMAGING DATA

Medical imaging data plays an important role in clinical research and clinical applications such as diagnosis, prognosis, surgery planning, and computer-aided therapy (D.L. et al., 2000) and various imaging modalities are used in Dentistry for these purposes. X-ray, panoramic radiographs/X-ray, cephalometric radiographs/X-ray, computerized tomography (CT), magnetic resonance imaging (MRI), cone-beam CT (CBCT), positron emission tomography (PET), optical coherence tomography (OCT), micro CT (Prasaanth et al., 2021) and ultrasound imaging are common modalities deployed for dental applications. Each of these modalities requires specific techniques for imaging analysis. For instance, speckle noise could be observed in ultrasound and OCT imaging techniques and, hence, noise reduction techniques capable of reducing this type of noise should be applied for their enhancement.

One of the main image processing techniques used in many medical applications is image segmentation that facilitates the delineation of regions of interest such as root canals, condyle, and other anatomical structures (Figure 9.3). Segmented

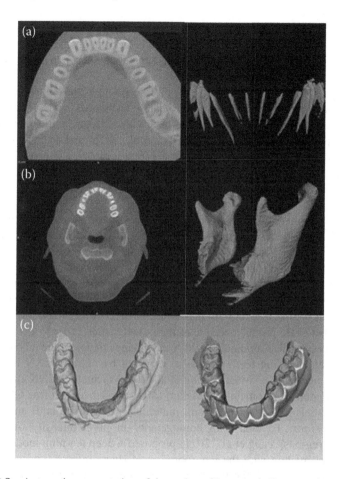

FIGURE 9.3 Automatic segmentation of the region of interest. **A.** Root canal segmentation in the small field of view cone-beam computed tomography images and 3D rendering of the segmentation; **B.** Mandibular condyle and rami in the large field of view of the full head cone-beam computed tomography images and 3D rendering of the segmentation; **C.** 3D mesh rendering and teeth-gingiva-gum labeling.

regions could be used for further analysis such as quantitative image analysis of the root canal cavities' shape and volume, predicting the progression of TMJ OA and so on (Bianchi et al., 2020). Vranckx et al. (2020) developed an automated method to segment the mandibular molars on panoramic radiographs and extracted the molar orientations in order to predict the third molars' eruption potential. Quantitative analysis based on medical image segmentation especially on a large volume of data would not be possible without automated techniques as manual delineation is costly, time-consuming, and prone to human errors. However, automated dental image segmentation is challenging due to factors such as variations of patient-to-patient anatomy, artifacts used for restorations and prostheses, noise, low contrast, homogeneity in teeth, and space due to missing teeth (Silva et al., 2018). Moreover, manual labels and ground-truth images are required for training a model and/or evaluating its performance. On the other hand, there might be variations among experts performing the ground truth image segmentation which makes computational learning more challenging.

In the past three decades, with the aforementioned challenges to delineate images, many automated and semi-automated techniques with applications in Dentistry have been developed. Silva et al. reviewed segmentation methods employed in dental imaging applications (Silva et al., 2018) and categorized them to region-based, threshold-based, cluster-based (e.g., Fuzzy C-means), boundary-based (e.g., edge detection, active contour), and watershed-based methodologies. They evaluated these methods using accuracy, specificity, precision, recall, and f-score metrics.

One of the dental applications of AI techniques that has been studied by many research groups is teeth detection/segmentation and classification (H. Chen et al., 2019; Y. Chen et al., 2020; Muresan et al., 2020; Tian et al., 2019; Wang et al., 2020). Teeth detection can facilitate 3D modeling, orthodontic treatment, and automatic feature extraction that could be used in automated disease progression analysis and forensic identification (H. Chen et al., 2019). Teeth segmentation methods could be divided into 3D mesh-based and 2D image-based methods.

In (H. Chen et al., 2019), a deep learning model, called faster R-CNN, followed by post-processing was applied on 800 X-ray images to determine a bounding box for each tooth. Missing teeth were also detected by another CNN in the post-processing stage. Dense U-Net (Zheng et al., 2020) is another CNN structure that integrates anatomical domain knowledge to the model to segment CBCT scans with multi labels (lesion, bone, teeth, materials, and background) and to detect periapical lesions. To deal with homogeneity in pixel intensity of teeth and background regions, Yang et al. (2021) proposed a deep learning model followed by a level set method with the incorporation of shape information including the size and position of each tooth to limit the segmentation curve to evolve around the shape prior information. They only used two CBCT scans for training a U-net to segment dental pulps for detecting the center of the tooth. Although the reported performance of this model was high, the model is not reliable and generalizable due to the lack of enough training samples. Nishitani et al. (2021) developed a deep learning model using U-net to segment teeth from panoramic X-ray images. As the tooth size might vary from one patient to another due to the image size and patient's imaging position, they extracted regions containing all teeth automatically using another deep learning model to normalize the teeth sizes. They used a loss function by combining cross-entropy values of both the entire image and teeth edges. TSegNet (Cui et al., 2021) is another CNN model with a two-stage neural network including a tooth centroid prediction subnetwork and a single tooth segmentation subnetwork. In this network, a loss function is defined based on a distance-aware voting scheme to accurately localize tooth objects. TSASNet (Zhao et al., 2020) is a deep learning model with a two-stage segmentation strategy to deal with low-contrast dental panoramic X-ray images. In this model, the location of dental regions is identified and then a fully convolutional network is used to search for the exact dental region in the second stage and segment teeth. Moreover, a hybrid loss function is considered that combines structural similarity as well as binary cross-entropy loss.

3D mesh segmentation is another approach for teeth segmentation (Xu et al., 2019; Zhang et al., 2020). In Xu et al. (2019), 600-dimension geometry features are extracted for each mesh face and packed into a 20 × 30 image. Then, two CNNs, one for teeth-gingiva labeling and the other one for inter-teeth labeling, were employed to deal with data imbalance and improve boundary accuracy. A boundary-aware mesh simplification algorithm and a correspondence-free mapping algorithm were applied to pre-process and post-process the dental meshes. Finally, a fuzzy clustering boundary refinement algorithm was applied to smooth the boundary of segmentation. A very similar approach to Xu et al. (2019) was proposed in Zhang et al. (2020) where a 3D tooth model was first mapped isomorphically to a 2D harmonic parameter space and then converted into an image. After that, a U-net model was applied to segment the 2D images and finally, the segmentation boundary became smoother by employing a post-processing method.

9.2.2 AI TECHNIQUES USING INTEGRATIVE MULTI-SOURCE DATA MODELS

Integrating a multitude of heterogeneous patients' data in a computational framework/model is an approach to mimic how a clinician makes a more accurate clinical judgment. Such frameworks can help personalized medicine and achieve a more informative analysis of disease compared to using a single data source. The integrative models can utilize machine learning methods to derive novel biomarker signatures for different health states (Schwendicke et al., 2020; Shomorony et al., 2020). These models might need to process heterogeneous data (usually with a large volume) that require advanced platforms and computational techniques to consider the challenges such as data variations in terms of type, size, complexity, and noise. In the field of computational dentistry, the focus of research groups has been mostly on analyzing a single source data type, such as imaging, molecular, or clinical data. However, the pace of developing integrative models might be increased due to advances in technology and computational platforms and better representation of disease using integrative models. In Bianchi et al. (2020), clinical, biological, and CBCT imaging data were utilized in machine learning models such as Logistic Regression, Random Forest, LightGBM, and XGBoost to diagnose Temporomandibular Joint (TMJ) disorders in its early stage. Twenty imaging, 25 biomolecular, and five clinical features were extracted and the product between each pair of features was calculated and used as additional features. The combination of XGBoost and LightGBM using a five-fold cross-validation approach could outperform other machine learning models when interaction features (i.e., pairs of features) were being used as inputs of the models.

The term biomarker is also widely used to describe patients' data that leads to a precise diagnosis. In the past, there was an effort to represent the best biomarker for specific diseases and conditions; however, with the advances in research and medicine, the diagnosis paradigm has shifted from a single biomarker to integration

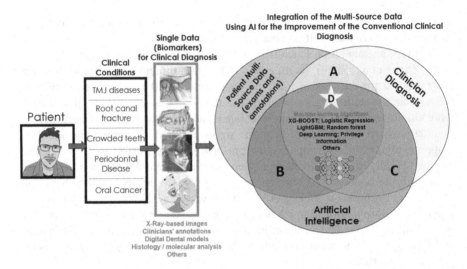

FIGURE 9.4   AI techniques using Integrative multi-source data models. **A.** Representing the diagnosis based only on the clinician's judgment and single exams; this model lacks integrating the data using statistical and computational approaches. **B.** Representing the patient data and artificial intelligence interactions that can give an accurate diagnosis, but it is based on the data only, without clinical validation. **C.** Clinician diagnosis for training AI algorithms; in this scenario, the models lack multi-source data and rely only on clinician diagnosis. **D.** Gold-standard for integrating multi-source data, clinical diagnosis, and AI approaches.

of multiple markers to categorize complex and multifactorial diseases (Strimbu & Tavel, 2010). Within the AI techniques, is it possible to not only integrate those biomarkers, but also, to create computational models that can learn with new data and data interactions. Figure 9.4 is showing a schematic summary of the state-of-the-art for integrating multi-source biomarkers intending to perform a precise diagnosis, toward personalized medicine.

One of the challenges facing many AI applications, and in particular deep learning models, is that they are employed as a black box and cannot illustrate the decision-making process in a medically acknowledgeable format (Shan et al., 2020). This is known as lack of interpretability and transparency not only could change practitioners' trust in the clinical value of AI, but it could also make it difficult to predict failures and generalize specific algorithms for similar contexts (Shan et al., 2020).

9.3   EVALUATION METRICS

Depending on the application, evaluation metrics might vary. For instance, for evaluating the performance of classification learning tasks such as differentiating between disease and control patients, accuracy, sensitivity/recall/true positive rate, specificity/true negative rate, and precision/positive predictive value metrics are commonly used to summarize classification performance over different classes (Figure 9.5). Some examples of classification tasks in dentistry include the

Total Population	True Label		
	Label Positive	Label Negative	
Predicted Positive	True Positive (TP)	False Positive (FP) [Type I Error]	Precision = $\dfrac{\sum TP}{\sum \text{Predicted Positive}}$
Predicted Negative	False Negative (FN) [Type II Error]	True Negative (TN)	Accuracy = $\dfrac{\sum TP + \sum TN}{\sum \text{Total Population}}$
	Sensitivity (Recall) = $\dfrac{\sum TP}{\sum \text{Label Positive}}$	Specificity (Selectivity) = $\dfrac{\sum TN}{\sum \text{Label Negative}}$	F1 Score = $2 * \dfrac{\text{Precision} * \text{Recall}}{\text{Precision} + \text{Recall}}$

(Predicted Label)

FIGURE 9.5 Common performance metrics obtained from the binary class confusion matrix.

classification of teeth in bitewing images to be premolar or molar (Lin et al., 2010) and the classification of dental diseases using Radiovisiography (RVG) x-ray images (Prajapati et al., 2017).

Segmentation can be viewed as a pixel binary classification task, where pixels in the ground truth region are considered as the positive class and other pixels are in the negative class. For both 2D and 3D segmentation algorithms, a popular metric is the Dice similarity score/F1 score which measures the amount of overlap between a ground truth region and a predicted segmentation. Accuracy, sensitivity, specificity, and precision are also commonly reported for segmentation tasks as well.

Other popular segmentation metrics for comparing ground truth manual segmentation with algorithm-predicted segmentation include the false positive error (FPE) and the false negative error (FNE) (Wu et al., 2007). The FPE is the percentage of the area reported positive by the algorithm but is not positive according to the ground truth, corresponding to Type I Error in Figure 9.5. The FNE is the percentage of the area labeled positive by the ground truth but is reported negative by the algorithm, corresponding to Type II Error in Figure 9.5. Similarity and dissimilarity indices, denoted S_{agr} and S_{dis}, are also used to measure the agreement and disagreement between algorithm area A_{alg} and ground truth area A_{man}. The similarity index is equivalent to the Dice similarity score.

$$S_{agr} = 2\frac{A_{mam} \cap A_{alg}}{A_{mam} + A_{alg}},$$
$$S_{dis} = 2\frac{A_{mam} \cup A_{alg} - A_{man} \cap A_{alg}}{A_{mam} + A_{alg}}$$

Segmentation applications in dentistry such as tooth segmentation commonly apply the above metrics to compare between different segmentation algorithms (Gao & Chae, 2010).

9.4   FINAL CONSIDERATIONS

Computational dentistry is a growing field that poses great promise by increasing the knowledge about patient data and hence helping clinicians in their decision-making. AI applications vary from diagnosis, treatment plan, and prognosis. However, a number of factors should be considered before deploying AI models in clinical practice. Even for the same application, different research groups might use different input data and different ways for validating their methods or there might be no validation. Therefore, it is usually impossible to evaluate methods in terms of their applicability in a clinic. Thus, a standard set of outcomes and outcome metrics relevant to stakeholders should be defined while considering some aspects such as applicability, impact on decision-making, safety and utility of AI-based applications. Moreover, replication studies are required to test validity, reproducibility, and generalizability of the AI-based techniques (Jordan & Mitchell, 2015). These techniques should also avoid both alert fatigue and unquestioning acceptance of their prediction and ensure the privacy of patient/practitioner data (Jordan & Mitchell, 2015).

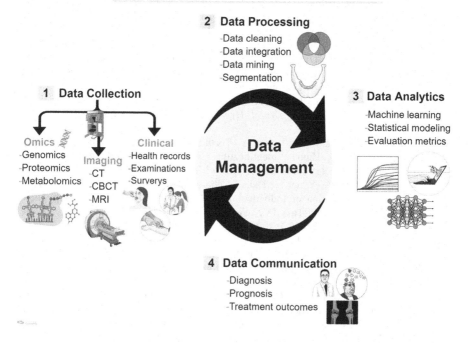

FIGURE 9.6   Data science approaches for clinical decision support systems in dentistry.

One way of improving the performance of AI-based models is to integrate the model with anatomical domain knowledge either in loss functions as a new term (e.g. regularization term in machine learning models) or in designed programming rules in non-learning methods. However, domain knowledge across different disciplines such as imaging, biomolecular and clinical is required to make better use of data and design a computational model. Integrative models are getting more attention due to their better performance (Figure 9.6). However, challenges facing these models such as variations in input data type, dimensionality, noise, and complexity should be considered especially when data come from different centers. Moreover, advanced computational techniques such as incorporating uncertainty labels (e.g., where there is less confidence in diagnosis) and privilege information (e.g. where all patients' data are not available in testing the model) (Sabeti et al., 2019) should be employed in the field of computational dentistry to better represent aspects of the problem that is modeled.

REFERENCES

Arifin, A. Z., Syuhada, F., Ni'mah, A. T., Suryaningrum, D. A., Indraswari, R., & Navastara, D. A. (2019). *Teeth Segmentation Using Gamma Adjustment and Transition Region Filter Based on Wavelet Method.* Proceedings of 2019 4th International Conference on Informatics and Computing, ICIC 2019. Scopus. https://doi.org/10.1109/ICIC47613. 2019.8985725

Balbin, J. R., Banhaw, R. L., Martin, C. R. O., Rivera, J. L. R., & Victorino, J. R. R. (2019). *Caries Lesion Detection Tool Using Near Infrared Image Processing and Decision Tree Learning. 11198.* Scopus. https://doi.org/10.1117/12.2540896

Bianchi, J., de Oliveira Ruellas, A. C., Gonçalves, J. R., Paniagua, B., Prieto, J. C., Styner, M., Li, T., Zhu, H., Sugai, J., Giannobile, W., Benavides, E., Soki, F., Yatabe, M., Ashman, L., Walker, D., Soroushmehr, R., Najarian, K., & Cevidanes, L. (2020). Osteoarthritis of the Temporomandibular Joint Can Be Diagnosed Earlier Using Biomarkers and Machine Learning. *Scientific Reports, 10*(1), 1–14.

Boiko, O., Hyttinen, J., Falt, P., Jasberg, H., Mirhashemi, A., Kullaa, A., & Hauta-Kasari, M. (2019). *Deep Learning for Dental Hyperspectral Image Analysis. 2019-October* (pp. 295–299). Scopus.

Bozkurt, M. H., & Karagol, S. (2020). Jaw and Teeth Segmentation on the Panoramic X-Ray Images for Dental Human Identification. *Journal of Digital Imaging, 33*(6), 1410–1427. Scopus. https://doi.org/10.1007/s10278-020-00380-8

Chen, Y., Du, H., Yun, Z., Yang, S., Dai, Z., Zhong, L., Feng, Q., & Yang, W. (2020). Automatic Segmentation of Individual Tooth in Dental CBCT Images from Tooth Surface Map by a Multi-Task FCN. *IEEE Access, 8,* 97296–97309. Scopus. https://doi.org/10.1109/ACCESS.2020.2991799

Chen, H., Zhang, K., Lyu, P., Li, H., Zhang, L., Wu, J., & Lee, C.-H. (2019). A Deep Learning Approach to Automatic Teeth Detection and Numbering Based on Object Detection in Dental Periapical Films. *Scientific Reports, 9*(1). Scopus. https://doi.org/10.1038/s41598-019-40414-y

Cui, Z., Li, C., Chen, N., Wei, G., Chen, R., Zhou, Y., & Wang, W. (2021). TSegNet: An Efficient and Accurate Tooth Segmentation Network on 3D Dental Model. *Medical Image Analysis, 69.* Scopus. https://doi.org/10.1016/j.media.2020.101949

Datta, S., & Chaki, N. (2020). *Dental X-ray Image Segmentation Using Maker Based Watershed Technique in Neutrosophic Domain.* 2020 International Conference on Computer Science, Engineering and Applications, ICCSEA 2020. Scopus. https://doi.org/10.1109/ICCSEA49143.2020.9132957

Dong, T., Xia, L., Cai, C., Yuan, L., Ye, N., & Fang, B. (2019). Accuracy of In Vitro Mandibular Volumetric Measurements From CBCT of Different Voxel Sizes With Different Segmentation Threshold Settings. *BMC Oral Health, 19*(1). Scopus. https://doi.org/10.1186/s12903-019-0891-5

Gao, H., & Chae, O. (2010). Individual Tooth Segmentation From Ct Images Using Level Set Method With Shape And Intensity Prior. *Pattern Recognition, 43*(7), 2406–2417.

Guijarro-Rodríguez, A. A., Witt-Rodríguez, P. M., Cevallos-Torres, L. J., Contreras-Puco, S. F., Ortiz-Zambrano, M. C., & Torres-Martínez, D. E. (2020). Image Segmentation Techniques Application for the Diagnosis of Dental Caries. *Advances in Intelligent Systems and Computing, 1066,* 312–322. Scopus. https://doi.org/10.1007/978-3-030-32022-5_30

Harrison, J., Chantrel, S., Schmittbuhl, M., & de Guise, J. A. (2019). Segmentation and 3D-Modelling of Single-rooted Teeth from CBCT Data: An Automatic Strategy Based on Dental Pulp Segmentation and Surface Deformation. *IFMBE Proceedings, 68*(1), 201–205. https://doi.org/10.1007/978-981-10-9035-6_36.

Hung, M., Voss, M. W., Rosales, M. N., Li, W., Su, W., Xu, J., Bounsanga, J., Ruiz-Negrón, B., Lauren, E., & Licari, F. W. (2019). Application of Machine Learning for Diagnostic Prediction of Root Caries. *Gerodontology, 36*(4), 395–404. Scopus. https://doi.org/10.1111/ger.12432

Janardanan, R. P., & Logeswaran, R. (2019). Dental Radiograph Segmentation and Classification—A Comparative Study Of Hu's Moments and Histogram of Oriented Gradients. *Journal of Computational and Theoretical Nanoscience, 16*(8), 3612–3616. Scopus. https://doi.org/10.1166/jctn.2019.8334

Jaskari, J., Sahlsten, J., Järnstedt, J., Mehtonen, H., Karhu, K., Sundqvist, O., Hietanen, A., Varjonen, V., Mattila, V., & Kaski, K. (2020). Deep Learning Method for Mandibular Canal Segmentation in Dental Cone Beam Computed Tomography Volumes. *Scientific Reports*, *10*(1). Scopus. https://doi.org/10.1038/s41598-020-62321-3

Jordan, M. I., & Mitchell, T. M. (2015). Machine Learning: Trends, Perspectives, and Prospects. *Science*, *349*(6245), 255–260.

Kwak, G. H., Kwak, E.-J., Song, J. M., Park, H. R., Jung, Y.-H., Cho, B.-H., Hui, P., & Hwang, J. J. (2020). Automatic Mandibular Canal Detection Using a Deep Convolutional Neural Network. *Scientific Reports*, *10*(1). https://doi.org/10.1038/s41598-020-62586-8

Lakshmi, M. M., & Chitra, P. (2020). *Tooth Decay Prediction and Classification from X-Ray Images using Deep CNN.* (pp. 1349–1355). Scopus. https://doi.org/10.1109/ICCSP48568.2020.9182141

Lee, S., Woo, S., Lee, C., Lee, J., & Seo, J. (2019). Fully-Automatic Synthesizing Method of Dental Panoramic Radiograph by Using Internal Curve of Mandible in Dental Volumetric CT. *Electronic Imaging*, *2019*(13). https://doi.org/10.2352/ISSN.2470-1173.2019.13.COIMG-148.

Li, S., Pang, Z., Song, W., Guo, Y., You, W., Hao, A., & Qin, H. (2020). *Low-Shot Learning of Automatic Dental Plaque Segmentation Based on Local-to-Global Feature Fusion.* *2020-April*, 664–668. Scopus. https://doi.org/10.1109/ISBI45749.2020.9098741

Lin, P.-L., Lai, Y.-H., & Huang, P.-W. (2010). An Effective Classification and Numbering System For Dental Bitewing Radiographs Using Teeth Region And Contour Information. *Pattern Recognition*, *43*(4), 1380–1392.

Majanga, V., & Viriri, S. (2020). A Deep Learning Approach for Automatic Segmentation of Dental Images. *Lecture Notes in Computer Science (Including Subseries Lecture Notes in Artificial Intelligence and Lecture Notes in Bioinformatics)*, *11987 LNAI*, 143–152. Scopus. https://doi.org/10.1007/978-3-030-66187-8_14.

Muresan, M. P., Barbura, A. R., & Nedevschi, S. (2020). *Teeth Detection and Dental Problem Classification in Panoramic X-Ray Images using Deep Learning and Image Processing Techniques* (pp. 457–463). Scopus. https://doi.org/10.1109/ICCP51029.2020.9266244

Nishitani, Y., Nakayama, R., Hayashi, D., Hizukuri, A., & Murata, K. (2021). Segmentation of Teeth in Panoramic Dental X-ray Images Using U-net With A Loss Function Weighted on The Tooth Edge. *Radiological Physics and Technology*. Scopus. https://doi.org/10.1007/s12194-020-00603-1

Prajapati, S. A., Nagaraj, R., & Mitra, S. (2017). Classification of Dental Diseases Using CNN and Transfer Learning. *2017 5th International Symposium on Computational and Business Intelligence (ISCBI)*, 70–74.

Prasaanth, S. A., Reddy, T. V. K., Mitthra, S., & Venkatesh, K. V. (2021). Applications of Micro-computed Tomography in Dentistry. *International Journal of Pharmaceutical Research*, *13*(1), 267–272. Scopus. https://doi.org/10.31838/ijpr/2021.13.01.052

Ramesh, A. N., Kambhampati, C., Monson, J. R., & Drew, P. J. (2004). Artificial Intelligence in Medicine. *Annals of the Royal College of Surgeons of England*, *86*(5), 334.

Sabeti, E., Drews, J., Reamaroon, N., Gryak, J., Sjoding, M., & Najarian, K. (2019). Detection of Acute Respiratory Distress Syndrome by Incorporation of Label Uncertainty and Partially Available Privileged Information. *2019 41st Annual International Conference of the IEEE Engineering in Medicine and Biology Society (EMBC*, 1717–1720.

Schwendicke, F., Golla, T., Dreher, M., & Krois, J. (2019). Convolutional Neural Networks for Dental Image Diagnostics: A Scoping Review. *Journal of Dentistry*, *91*. Scopus. https://doi.org/10.1016/j.jdent.2019.103226

Schwendicke, F., Samek, W., & Krois, J. (2020). Artificial Intelligence in Dentistry: Chances and Challenges. *Journal of Dental Research*, *99*(7), 769–774. Scopus. https://doi.org/10.1177/0022034520915714

Shan, T., Tay, F. R., & Gu, L. (2020). Application of Artificial Intelligence in Dentistry. *Journal of Dental Research*. Scopus. https://doi.org/10.1177/0022034520969115

Shomorony, I., Cirulli, E. T., Huang, L., Napier, L. A., Heister, R. R., Hicks, M., Cohen, I. V., Yu, H.-C., Swisher, C. L., Schenker-Ahmed, N. M., Li, W., Nelson, K. E., Brar, P., Kahn, A. M., Spector, T. D., Caskey, C. T., Venter, J. C., Karow, D. S., Kirkness, E. F., & Shah, N. (2020). An Unsupervised Learning Approach to Identify Novel Signatures of Health and Disease From Multimodal Data. *Genome Medicine*, *12*(1). Scopus. https://doi.org/10.1186/s13073-019-0705-z

Silva, G., Oliveira, L., & Pithon, M. (2018). Automatic Segmenting Teeth in X-ray Images: Trends, a Novel Data Set, Benchmarking and Future Perspectives. *Expert Systems with Applications*, *107*, 15–31. Scopus. https://doi.org/10.1016/j.eswa.2018.04.001

Strimbu, K., & Tavel, J. A. (2010). What Are Biomarkers? *Current Opinion in HIV AIDS*, *5*(6), 463–466.

Tian, S., Dai, N., Zhang, B., Yuan, F., Yu, Q., & Cheng, X. (2019). Automatic Classification and Segmentation of Teeth on 3D Dental Model Using Hierarchical Deep Learning Networks. *IEEE Access*, *7*, 84817–84828. Scopus. https://doi.org/10.1109/ACCESS.2019.2924262

Vranckx, M., Van Gerven, A., Willems, H., Vandemeulebroucke, A., Leite, A. F., Politis, C., & Jacobs, R. (2020). Artificial Intelligence (AI)-driven Molar Angulation Measurements to Predict Third Molar Eruption on Panoramic Radiographs. *International Journal of Environmental Research and Public Health*, *17*(10). Scopus. https://doi.org/10.3390/ijerph17103716

Wang, L., Mao, J., Hu, Y., & Sheng, W. (2020). Tooth Identification Based on Teeth Structure Feature. *Systems Science and Control Engineering*, *8*(1), 521–533. Scopus. https://doi.org/10.1080/21642583.2020.1825238

Wu, X., Gao, H., Heo, H., Chae, O., Cho, J., Lee, S., & Lee, Y.-K. (2007). Improved B-Spline Contour Fitting Using Genetic Algorithm for the Segmentation Of Dental Computerized Tomography Image Sequences. *Journal of Imaging Science and Technology*, *51*(4), 328–336.

Xu, X., Liu, C., & Zheng, Y. (2019). 3D Tooth Segmentation and Labeling Using Deep Convolutional Neural Networks. *IEEE Transactions on Visualization and Computer Graphics*, *25*(7), 2336–2348. Scopus. https://doi.org/10.1109/TVCG.2018.2839685

Yang, Y., Xie, R., Jia, W., Chen, Z., Yang, Y., Xie, L., & Jiang, B. (2021). Accurate and Automatic Tooth Image Segmentation Model With Deep Convolutional Neural Networks and Level Set Method. *Neurocomputing*, *419*, 108–125. Scopus. https://doi.org/10.1016/j.neucom.2020.07.110

Zhang, J., Li, C., Song, Q., Gao, L., & Lai, Y.-K. (2020). Automatic 3D Tooth Segmentation Using Convolutional Neural Networks in Harmonic Parameter Space. *Graphical Models*, *109*. Scopus. https://doi.org/10.1016/j.gmod.2020.101071

Zhao, Y., Li, P., Gao, C., Liu, Y., Chen, Q., Yang, F., & Meng, D. (2020). TSASNet: Tooth Segmentation on Dental Panoramic X-ray Images by Two-Stage Attention Segmentation Network. *Knowledge-Based Systems*, *206*. Scopus. https://doi.org/10.1016/j.knosys.2020.106338

Zheng, Z., Yan, H., Setzer, F. C., Shi, K. J., Mupparapu, M., & Li, J. (2020). Anatomically Constrained Deep Learning for Automating Dental CBCT Segmentation and Lesion Detection. *IEEE Transactions on Automation Science and Engineering*. Scopus. https://doi.org/10.1109/TASE.2020.3025871

10 Literature Review of Computer Tools for the Visually Impaired: A Focus on Search Engines

Guy Meyer, Alan Wassyng, Mark Lawford, Kourosh Sabri, and Shahram Shirani

CONTENTS

DOI: 10.1201/9781003120902-10

10.1   INTRODUCTION

This chapter encompasses a collection of attributes of common tools for visually impaired internet users. While specific attributes address the major components of computer interfaces, they are not limited to the visually impaired and can be further extended to visually enabled users as well. The attributes, as shown in later sections (Section 4), are presented as disjoint topics, allowing the reader to identify key concepts of interest and focus on collected insights. Under each attribute, a total of four subsections headings help the reader understand the major issues and progress in the field.

With the growing demand for internet applications, there is a growing need for integrating visually impaired users to the web. In fact, both sighted and non-sighted users would benefit from more accessible and adaptable interfaces.

Many ongoing efforts are actively attempting to bridge the gap between sighted and non-sighted online users, but a lack of standardization for either end-user is still a reminder. Though users have the flexibility to operate on whichever platform they desire (whether a specific OS or Web Browser), no application or device is found to be an obvious choice for the disabled.

In addition to the lack of tooling, relatively few reviews were identified addressing issues for online computer tools. Though most research groups are invested in developing technology, few resources are allocated to gathering information on existing efforts in the field. This chapter does exactly that collect information about an array of existing products, projects, and research to present it in a quick-access format.

While the concepts are presented in a written composition, the reader has the option of accessing the same information in a tabular format, found in the supplementary material. The designed purpose of this chapter is to provide the reader with sufficient background knowledge to confidently initiate their research and developmental efforts.

10.2 RESEARCH METHODS

With a wide range of existing research in the field of visual impairment, the scope was limited to reviewing web and computer-based assistive devices. The initial realignment of scope was due to the abundance of research efforts, along with a lack of existing systematic reviews. A collection of recognized databases were identified that could serve as reliable sources of peer-reviewed papers.

These databases can be categorized into three sectors; Health Science, Engineering, and Other. The Health Science databases are medically focused and contain disability-related studies along with diagnostics. Health Science databases include *OVID Medline*, *PubMed*, *Embase*, and *PsycINFO*. The Engineering related databases, focusing more on products, devices, and applications are *Engineering Village*, composed of *Compendex* and *Inspec*. For additional content, labeled as Other, *Google Scholar* is the desired tool. Google Scholar would often feature papers based on popularity, and has an intuitive interface where the user can often encounter inconsistency with regards to the repeatability of a search.

Once relevant databases are established, appropriate keywords would help further locate studies of interest. Several keywords were identified to be of significant use since they narrow the search substantially; Visually Impaired, Visual Impairment, Blindness, Internet, web-based, Human-Computer Interfaces, Search Engine, optimization, and virtual. Note that the use of the keyword *'Blind'* generated poor search results due to its association with Blind Studies and Double-Blind Experiments.

Other strategies for finding quality publications are by reviewing the reference list of different papers. By browsing the references of each paper, the reader can discover cited works that may advance their own research. Additionally, the reader will often encounter works that are repeatedly cited in separate papers, indicating importance and relevance.

In this chapter, some studies with significant relevance were analyzed throughout, while others were judged on the information presented in the abstracts and conclusions. Papers deemed as irrelevant were rejected based on three criteria metrics; the year of publication, content relevancy, and release of newer studies (where certain research groups would produce newer comprehensive results).

10.3 TOOLS

10.3.1 WHAT ARE TOOLS?

A tool can be described as any computer program or application available to the user. This definition is an extension of the variation used by (Powsner and Roderer, 1994), "nav tools", concerned with web accessibility. For the purposes of this chapter, the term navigational tools is excellent since Search Engines do exactly that. Furthermore, the idea of client-server software is mentioned, relating well to the nature of most websites as an attempt to offload from the client.

Powsner and Roderer also iterate that "The Internet is not 'arranged' in the usual sense of the word" (Powsner and Roderer, 1994), implying that through its development over decades certain demographics are disregarded (an important point-of-interest in HCI).

Some examples of tools are,

- Search Engines
- Web Browsers
- Word Processors
- Social Media Platforms
- Music Applications
- Other Human-Computer Interfaces (via mobile or PC)

The attributes that will be described in the following sections (Section 4) are the fundamental components of most computer tools and apps. By understanding how each attribute influences the user, developers can create more intuitive and robust tools.

10.3.2 SEARCH ENGINES (SEs)

This chapter primarily focuses on the application of web-based Search Engines as a tool for computer users dealing with visual impairment. This particular focus is largely due to the high daily dependence on SEs by computer users. Furthermore, there is a benefit through independence when learning to use SEs efficiently.

The ability to locate desired information online is very useful. But since the internet is so large and complex, the user employs a Search Engine (SE) to sift through potential results and rank them in relevancy. SEs provide a quick and accurate response to most general knowledge questions along with help in the online navigation.

The concept of SEs is to provide the user with a "glimpse" of a web page, along with bits of relevant information. By analyzing this response the user should be knowledgeable enough on the "potential" of the web page (Webpage Potential - describes how likely it is that this website will be useful in answering the user's query) in order to decide if it is worth delving deeper.

The issue for VI users is the inability to quickly and accurately capture a glimpse of the webpage. Additionally, a standard SE relies heavily on the input query in order to retrieve relevant results. Through the understanding of related search terms and proper Boolean Logic (such as AND, OR, and NOT) the SE will provide links that are more accurate to answer the initial query (Yang et al. (2012). These additions to standard search methods allow the user to narrow the search space, and as a result, focus their efforts and reduce the amount of time they spend exploring results (Tsai et al., 2010). SE are extremely useful tools since they help users congregate a collection of relevant sites and data, otherwise difficult to locate.

As of 2019, Google, Baidu, and Bing process 74.80%, 11.32%, and 8.08% of the world's search queries, respectively (NetMarketShare, 2019). Furthermore, Google handles approximately 80.79% of searches on mobile devices, along with 85.43% of searches submitted by tablets (NetMarketShare, 2019). These dominant statistics emphasize the need for a small set of assistive tools to aid the visually impaired.

In order to improve the experience of VI users, various applications have been released to quicken the search process. Yang et al., have created a Specialized Search Engine for the Blind (SSEB) that breaks down the Search Engine Results Page (SERP) (Yang et al., 2012). The paper also references an application by Google called Personalized Search which returns more relevant results to the SERP by basing current searches on past ones performed by the user (Yang et al., 2012). By employing a powerful Application Programming Interface (API) provided by the major SEs (i.e. Custom Search Engine by Google or the Bing Custom Search by Microsoft) the developers do not need to reimplement these algorithms. Google's PageRank, RankBrain, and Hummingbird search engine algorithms are intricate search techniques that require lots of effort to recreate. As a result, when developing new tools it is recommended that the focus remains on elevating the user experience rather than optimizing the search results.

A different application called WhatsOnWeb (WoW) changes the SERP by tailoring it specifically to the user (Mele et al., 2010). VoiceApp is a speech-based web search engine developed by Griol et al. Another useful application is TrailNote that manages the search process for each user to support "complex information seeking" (Sahib et al., 2015). The use of trail-managers is strongly recommended. Once implemented properly, VI users can focus on synthesizing the information at hand rather than memorizing past results. It is also important for the users to have quick accessibility to their trail, regardless of their proficiency level.

It is worth noting that these tools for browsing the web have implications beyond that of accessibility. Count the times you have wanted to recall a website, or search result you encountered. You will quickly run out of fingers. But with the use of an organized trail of breadcrumbs, as proposed by TrailNote, you can quickly recall your past and synthesize deeper.

Several studies have been published, focusing on the effectiveness of SEs, along with helpful concepts. Tsai et al. center on query specification and the minimization of the search space as a method to improve the quality of the SERP (Tsai et al., 2010). The paper also identifies the differences between novice and expert searchers (Tsai et al., 2010). Other papers study the level of brain activity while using an SE (Small et al., 2009), preferred engines amongst users (Hu and Feng, 2015), and principal components (Principal Component Analysis (PCA)) that construct a standard SE search (Ivory et al., 2004; Tsai et al., 2010). Several studies published results on SE metrics (Aliyu and Mabu, 2014), ideal design (Andronico et al., 2006; Baguma and Lubega, 2008), accessibility evaluation (Lewandowski and Kerkmann, 2012), and conformance levels (Andronico et al., 2004). Finally, a study by Sahib et al. has been published documenting how VI users navigate an SE and how they collect information online (Sahib et al., 2012).

SE are applications that will only increase in their commonality due to their ability to reduce the workload of the user. It should be clear that VI users would benefit substantially from highly accessible SEs. Furthermore, sighted users would benefit equally from more efficient SEs. As a result, future developments and research should focus on conformance and adequate design to ensure global accessibility to all user types. Developers should also leverage user-oriented techniques since a user's context, history, or trail can impact future searches.

10.4   PRIMARY ATTRIBUTES OF TOOLS

A collection of attributes is described below that highlight the major elements of computer use for all user types. By accounting for an array of components in the user experience, the reader can focus on the concepts that are relevant to their research.

The order in which the attributes are presented is arbitrary, meaning that the reader can analyze attributes that are relevant to their application. The reader also has the option of accessing the information organized by reference which can be found in the supplementary material. The appendices also include direct quotes and generic summaries of specific papers.

Section Outline. Each attribute is discussed under the following subheadings:

- Scope
- Difficulties faced by VI Users
- Existing Products and ongoing Research
- Necessary Future Research and Consideration

Note that these subheadings aid in separating the concerns of the reader to allow for additional organization. The reader will also find an analysis of important papers embedded in the text to explain the effectiveness of the attribute.

10.4.1   ATTRIBUTE: NAVIGATION

10.4.1.1   Scope

The internet, being a fantastic source of information, is primarily useful for those who know what they are specifically browsing for and feel comfortable with the information. If a user is proficient in their ability to navigate between web pages then it would be natural for that user to skilfully locate important information. The ability to navigate through a computer system or the internet is an invaluable skill that is being taught at increasingly younger ages. Even more so, computer proficiency is a common requirement when applying for most jobs.

10.4.1.2   Difficulties Faced by VI Users

Since most computers, along with their peripherals (mouse, monitors, keyboard, etc.) are designed for sighted users, universal accessibility is not highly prioritized. As a result, the internet is less accessible for non-sighted users that rely on these devices for navigation (more detail regarding information accessibility in Section 4.4). In addition, most web pages are designed to be used as graphical user interface (GUI)s which heavily favor visual elegance over simplicity in navigation. Consequently, it is very difficult for non-proficient visually impaired internet users to interact with the web, resulting in a less stimulating, slower online experience.

A common difficulty found by many visually impaired users is virtual disorientation (Baguma and Lubega, 2008; Ismail and Zaman, 2010). This may result from several situations:

a. Inability to recall current virtual location[1]: *The website currently observed by user or the user's location within a web page.*
b. Inability to recall previously visited web pages (Known as "The Trail" (Sahib et al., 2015): *The recent web pages previously visited by users that are relevant for the current session online.*
c. Indecisiveness regarding future steps: *The websites that the user should visit next.*

Inexperienced users generally cope with this issue by refreshing the web page, restating the search, or closing the browser to restart (more in Section 4.7) (Murphy et al., 2008). This dramatic course of action commonly discourages and frustrates the user, since they are forced to retrace their virtual trail.

10.4.1.3   Existing Products and Ongoing Research

The issues with online navigation, as mentioned above, are important to consider due to their strong impact on the user's experience. Technologies have been developed to resolve some of these problems. Yang et al. developed an SSEB, made to assist with user orientation and access for those who struggle online (Yang et al., 2012). The paper also provides guidelines when adding shortcuts to an application. Hakobyan et al., have developed the AudioBrowser, used to navigate the web on the go (Hakobyan et al., 2013). WoW developed at the University of Rome creates a single sonificated, browsable page that can be more easily accessed by VI users (Mele et al., 2010).

The most commonly used application for navigation is JAWS, developed by Freedom Scientific (Scientific, n.d.a). The application is compatible with most Windows applications, including web browsers. Users can browse the web using their preferred web browser and special JAWS[2] key commands (Scientific, n.d.b). The user receives web information through dictation synthesized by the JAWS application. A comparable solution is a screen reader tool called NVDA that is gaining popularity as a free-to-use reader with great support, and an evolving community.

If using a Mac computer running the macOS by Apple, a helpful tool for web navigation is called the Rotor that comes installed with VoiceOver (VO) (the native Apple screen reader) app (2017). This tool attempts to summarize the links, headings and other page elements into groups. So instead of 'tabbing' around from link to link, the rotor presents all common links in a single menu to increase navigational ease. These different groups are then presented in adjacent menus in the Rotor. This feature can be used on individual sites as well, congregating information into groups to assist the user. This feature is excellent for mac users with visual impairment that want more from VoiceOver.

The VoiceApp Griol et al. (2011) and the Homer Web Browser Dobriˇsek et al. (2002) offer navigation using voice commands alone, the results are also returned via audio (Dobriˇsek et al., 2002; Griol et al., 2011). The Audio Hallway Schmandt (1998) provides navigation using head motions where the user passes 'rooms' as potential selection options. The physical movement allows the user to be immersed in the online experience, resulting in more control and focus.

For users that require multi-session tasks (online tasks that cannot be finished in one sitting), applications such as Search Trail and TrailNote were developed. They are particularly handy to pause their current session, save relevant information locally, and pick up where they left off once they resume activity. There is high importance for managing users between sessions since users may forget the mental map they worked hard to create in their previous session. This idea is also applicable with the use of relevant feedback (Tsai et al., 2010), which allows users to draw information from previous sessions.

As applications become increasingly complex, users are expected to keep up with the versatility of these tools. Even more so, users are expected to work on computers for more than just searching the web. The idea introduced by Sahib et al. that addresses 'the Trail' (Sahib et al., 2014) is a powerful concept that highlights the difficulty of VI users to perform long-term computer tasks. This idea stretches to all types of tools and helps focus on the task at hand.

A portion of active research is dedicated to collecting feedback from the user on desirable features. Common requests are the addition of more feedback from the web application to the user (Murphy et al., 2008), along with an overview and general hints as to where the user is located virtually on the page (Baguma and Lubega, 2008; Murphy et al., 2008). Additional papers study how users with cognitive disabilities navigate the web (Hu and Feng, 2015), how VI users collect information online (Sahib et al., 2012), and what elements are leveraged by VI users to aid in their navigational processes (Ivory et al., 2004).

10.4.1.4   Necessary Future Research and Consideration

A strong need for standardization! With several navigation softwares in circulation, each tends to develop their own set of commands and shortcuts. New developments should aim to minimize the number of commands so that the user is not overwhelmed. Also, the user could simply begin their online tasks efficiently and intuitively. VI users could benefit from an application that would provide a general overview of a web page and allow the user to skim the page similar to sighted users.

The concept of Navigation is closely related to Search Engines (Section 3.2) and Latency (Section 4.5).

10.4.2   ATTRIBUTE: USER INTERFACE

10.4.2.1   Scope

When designing a product that aids in overcoming a disability it is crucial that the technology prioritizes the user. Too often are devices designed and tested by visually capable developers that seem to be counter-intuitive for VI users, in practical settings. The purpose of assistive technology is to allow full accessibility to those in need without compromising the quality of information and the ease of accessibility.

10.4.2.2   Difficulties Faced by VI Users

Technologies that can be categorized with poor user interfaces are most noticeably those that neglect a crucial phase in the user's life-cycle: early learning stages.

If a tech is complex in nature, then the average user is less likely to rely on its recurring usage.

When considering the usage of search engines (more in Section 3.2) by visually capable users, the level of simplicity often goes unnoticed. It is the responsibility of the designer to create a blind interface that conveys the same level of intuition as its graphical duality. This boils down to the ability of the designer to implement their interface in a format that could either be used by all or has the capability of transforming into a non-graphical UI.

The JAWS screen reader is an exceedingly common computer tool that supports nearly all computer tasks in an operating system (OS). A noticeable drawback from its design is the number of keyboard commands available to the user (Scientific, n.d.b), the userspace. This results in a steep learning curve that must be overcome to achieve adequate proficiency (Andronico et al., 2006; Murphy et al., 2008). Additionally, the user must memorize commands which map a keystroke to a visual change on the screen (i.e. buttons for scrolling or jumping between menus or text blocks). This issue forces the user to draw implicit assumptions regarding explicit changes on the screen. Potentially yielding a poor conversion between visual and non-visual interfaces for the same application.

Also remember, the average user does not know the nomenclature of web elements. For example, while using JAWS a user can use the "A" button as a Quick Key to jump to the next radio box. But for a novice user a follow up question would be "well, what is a radio box?" Understanding HTML structure and terminology heightens the learning curve. While for sighted users it can be effectively ignored.

Multiple screen reader options are built custom for a specific OS. NonVisual Desktop Access (NVDA) is a free screen reader (nvd, 2017) that is a part of an initiative to provide access to technology for all. Microsoft narrator is an additional option for those working with the Windows OS (mic, n.d.). The tool is turned on with a keystroke combination available at all times. Users may prefer certain screen readers simply due to their key commands or the intonation of synthesized voice.

Visual authentication interfaces also pose difficulties for VI users. A common automated Turing Test service, CAPTCHA, requires a visually capable user to select or decipher components of images in order to prove the user is not a robot. Although these tests may prove trivial for sighted users, they are nearly impossible if a person struggles with their vision (Murphy et al., 2008). This is addressed by the W3C guidelines for web accessibility where they provide an overview of other approaches [https://www.w3.org/TR/turingtest/].

10.4.2.3 Existing Products and Ongoing Research

Though user interfaces are related to accessibility, there are clear distinctions between them (GUI, AUI, TUI) when evaluating the user's operation within an application. The use of GUIs is extremely common since it is simplest for sighted users. Unfortunately, the GUIs are complex for VI users (Chiang et al., 2005) due to their high visual dependency. As a result several studies have introduced other modalities that could be useful for VI users. The use of Auditory User Interface (AUI)s, Tactile User Interface (TUI)s, and combinations of all three (multi-modal systems) are commonly mentioned in the literature (Dobri˘sek et al., 2002;

Frauenberger et al., 2005; Hakobyan et al., 2013; Macias et al., 2002; Schmandt 1998; Siekierska and McCurdy 2008; Trippas 2016; Yang and Hwang 2007; Yang et al., 2012). Other products may not specifically acknowledge the application of a specific User Interface (UI), although their developments generated a unique non-graphical interface, such as JAWS. Where accessibility is offered to the entire spectrum of Visual Impairment (VI).

Several papers study user interfaces and to develop guidelines (Baguma and Lubega, 2008; Chen et al., 2003; Frauenberger et al., 2005; Murphy et al., 2008; Tsai et al., 2010; Yang et al., 2012), statistics (Chiang et al., 2005; Crossland et al., 2014; Yang et al., 2012), or evaluations (Halimah et al., 2008; Macías et al., 2004; Menzi-Çetin et al., 2017; Murphy et al., 2008; Muwanguzi and Lin, 2012; Trippas, 2016) to improve the usability and intuition behind their respective applications. This extends to proper query formulation for search engines (Tsai et al., 2010) or the acknowledgment of a user's level of experience when developing applications (Net Savvy vs. Net Naive) (Small et al., 2009).

Many product developments have also been well documented in the literature. JAWS is among the most common UIs for VI users. Unfortunately, its complexity is documented resulting in a high learning curve (Murphy et al., 2008). An interesting result noted by Menzi-Çetin et al. is the high preference of JAWS users toward Internet Explorer (IE) web browsers (Menzi- Çetin et al., 2017). The issue with IE is the lack of online community support for the browser. With sighted users the popularity of Internet Explorer is known, creating a gap in development and support between VI and sighted users, where sighted users are tailored to and use platforms with long-term support. Other UIs for web accessibility are Mg Sys Visi (Halimah et al., 2008), Accessibility Kit for the Internet (KAI) (Macias et al., 2002; Macıas et al., 2004), WoW (Mele et al., 2010), and the Homer Web Browser (Dobriˇsek et al., 2002).

Some applications like EasySnap developed by Jayant et al. aid VI users with developing skills in photography as well as sharing their content online (Jayant et al., 2011). Siekierska and McCurdy, developed a product to provide users with an interface for physical world navigation, allowing them to use maps freely (Siekierska and McCurdy, 2008). Sahib et al. have developed a non-visual spelling support system (Sahib et al., 2015). Finally, Audio Hallway is a conceptual AUI product developed for browsing collections using head motions, giving the user an immersive experience (Schmandt, 1998).

10.4.2.4 Necessary Future Research and Consideration
Developers would benefit greatly from referencing and considering these principles when developing applications. The user should not be frustrated with the UI because if developed with all users in mind, these interfaces will become as simple as using a screen.

10.4.3 ATTRIBUTE: INFORMATION ACCESSIBILITY

10.4.3.1 Scope
The process of collecting and synthesizing information from the internet is an important skill to have in order to become efficient in using online applications.

But for synthesis to occur, the information must be quickly and easily accessible to the user. Navigation assistance (Section 4.2) is not enough to interact with information online, the user must also be able to understand and access the media they encounter.

10.4.3.2   Difficulties Faced by VI Users

The internet is designed for sighted users resulting in a highly graphical presentation of information. Furthermore, there is little consideration for VI users that may be equipped with screen readers or assistive aids (Mac´ıas et al., 2004). Consequently, the VI user may read a web page while having to subconsciously guess the contents of information that is inaccessible to them.

As an example, consider a university web page terminal that allows students and staff to check for events and updates around campus. Studies have compared a collection of university sites that are ideally supposed to be accessible to all students and yet include surprising levels of inaccessibility (Harper and DeWaters, 2008; Menzi-Çetin et al., 2017; Muwanguzi and Lin, 2012). After evaluating the compliance levels of each site it becomes clear that most visually impaired students cannot access a substantial percentage of university content. This results in a lack of knowledge and frustration for the students. In addition, these sites did not comply with the Web Accessibility Initiative guidelines (WAI) published by the World Wide Web Consortium (W3C) Consortium (1999).

It is also important to avoid overloading the user when they are browsing for content (Murphy et al., 2008). Since it is quicker to skim through a document visually, it is expected that the online experience is fast. But in the case of VI users, the experience may be slowed down to accommodate for screen readers. If the technological aid reads an excessive amount of information from the web page, then the user will experience a slower consumption rate (the rate at which a user is presented with new information). Conversely, if the aid outputs lots of audio, then the user may feel overwhelmed and is forced to slow down equally.

10.4.3.3   Existing Products and Ongoing Research

The ability to provide accessibility to computer and online apps is increasingly important since virtual media (ie. text, images, videos, etc.) is how the relevant data is commonly represented. As a result, much effort is in converting standard sites to become accessible. As a humanitarian effort to ensure accessibility, the government of Ontario (Canada) has filed the Accessibility for Ontarians with Disabilities Act (AODA) detailing standards and regulations that organizations and individuals should abide by to make their products or services more accessible (aoda, 2019). The AODA also provides good teaching, coding, and design practices that extend past the web to improve the accessibility of public places and schools. Halimah et al. have developed a translator that can convert HTML to multiple mediums, including; voice output, braille, or text. Due to this versatility, the translator can be employed by an array of users including the elderly and other individuals with ranging disabilities (Halimah et al., 2008). Macias et al. developed a product named KAI that is composed of two modules. The first, a markup language designed for the blind, Blind Markup Language (BML) (Macias et al., 2002; Mac´ıas et al.,

2004). Their other is an app called WebTouch, a multi-modal web browser used in conjunction with BML (Macias et al., 2002; Mac´ıas et al., 2004).

Another application, VoiceApp, allows web browsing using voice commands alone (Griol et al., 2011). The VoiceApp generates markup metadata called VoiceXML that indicates relevant voice information to be transmitted (Griol et al., 2011). The Web Access Project, developed by Yang and Hwang, adds captions and audio descriptions to video clips as context for VI users (Yang and Hwang, 2007). SSEB (Yang et al., 2012) also adds to the accessibility of the web, by allowing the user to comfortably search for web pages. Additionally, the paper by Yang et al. indicates a minimum requirement claiming that anyone should be able to understand the contents of any web page (Yang et al., 2012). Though this goal may seem ambitious, it depicts the ideal compliance status of the web. Chen et al. focus on web browsing via, "handheld computers, Personal Digital Assistant (PDA) and smartphones" (Chen et al., 2003). Their application compartmentalizes web content so that it can be accessed using small form-factor devices (Chen et al., 2003).

Wearable technology is used as a method of accessing information. AlterEgo, a smart, non-invasive, wearable computer which sits externally around the human vocal cords. AlterEgo allows the user to communicate with computers without audibly voicing a word (Kapur et al., 2018). This provides human-computer interaction that is totally discreet (more in Section 4.6). Other physical products are used to provide VI users with access to physical graphical information, such as maps (Hakobyan et al., 2013; Roentgen et al., 2009; Siekierska and McCurdy, 2008). These technologies are not only useful when transporting from one location to another but their use declines once the user becomes familiar with the space, indicating the use of successful learning methods (Roentgen et al., 2009).

Additional studies are focused on researching ways to improve accessibility. Several papers study the issue of overloading the user (Baguma and Lubega, 2008; Murphy et al., 2008; Trippas, 2016). Others indicate that the use of multiple modalities (audio, touch or both) are good ways of replacing graphical information (Chiang et al., 2005; Griol et al., 2011; Mac´ıas et al., 2004). Baguma and Lubega, have produced a list of requirements that aid developers to assure accessibility (Baguma and Lubega, 2008).

10.4.3.4 Necessary Future Research and Consideration

In the future, when new web content is generated, it is important to take preemptive measures such as adding alternative text to images and videos, focusing on web page accessibility, and performing proper testing to ensure accessibility with adequate, non-visual computer peripherals (keyboard only). For products and applications that are developed in this field, it is crucial to remember that users have a range of visual impairments along with other disabilities that could also benefit from their product. Current accessibility applications are complex with a large learning curve that is overwhelming for the elderly or naive online users. Developers must consider what is important in terms of accessibility and what can be omitted.

10.4.4 ATTRIBUTE: LATENCY

10.4.4.1 Scope

The efficiency of the web, more specifically search engines, has allowed internet users to spend significantly less time looking for results. Consequently, a standard user is expecting quick retrieval. The latency, also known as "search time" (Yang et al., 2012), or task completion time, of a web search is the time difference between initial formulation of a query and the final intake of information. This could be extended to describe the amount of time a user spends on a website to absorb the information.

Note, this attribute is not limited to SEs since all computer tools are expected to work quickly. Another reason for Latency to be regarded as an important attribute is that it attempts to numerically quantify how useful and accessible a tool is.

10.4.4.2 Difficulties Faced by VI Users

The internet has become an endless pool of knowledge that can be ideally accessed by anyone. The major distinction between VI and sighted users is their ability to consume information quickly. Since a sighted user has higher visual acuity they are comfortable skimming through dense pages with lots of data. Conversely, VI users are forced to examine the same page more carefully, resulting in a slower online experience. Each VI user then spends more time per webpage and therefore experiences more latency between query and result.

Mack and Rock have also addressed the issue of latency but with an attempt to identify its source. They claim that VI users construct explicit perceptions of web pages, rather than visually driven implicit observations (Mack and Rock, 1998). These explicit perceptions are more difficult to comprehend and force VI users to spend more time online.

10.4.4.3 Existing Products and Ongoing Research

Several studies attempt to quantify the difference in the time duration between VI and sighted users. Menzi-Çetin et al., captured the latency of VI users when completing online tasks (Menzi-Çetin et al., 2017). Ivory et al. also focus on speed of information access and collect measurable metrics on user evaluation time (Ivory et al., 2004). Others have highlighted the advantage of using mobile devices (Tsai et al., 2010), and the importance of developing more efficient user interfaces for SE users (Andronico et al., 2006).

Products that have been developed to aid VI users have tackled a variety of issues. AlterEgo, a wearable input device, allows the user to input information to a computer at a faster rate and at any distance since the wearable collects muscle movements directly from the vocal cords (Kapur et al., 2018). EasySnap, an application for sharing pictures and videos, is intuitive and easy to operate (Jayant et al., 2011). This allows users to become faster with simple activities, reducing latency in processes such as sharing media. Search Trail, a multi-session assistant for VI users, reduces the resumption time between sessions by allowing the user to revisit their virtual trail and pick up where they left off (Sahib et al., 2014). Wikipedia, the online encyclopedia, is a good source for quick descriptions (Griol et al., 2011).

With immediate accessibility to the web, many applications are capable of answering questions, providing guidance, and helping manage personal devices. Examples of these applications include; "Siri" by Apple, "Hey Google" by Google, "Alexa" by Amazon, and "Cortana" by Microsoft (Yoffie et al., n.d.). Other examples specifically for SEs include the Featured Results at the top of the Google SERP that attempts to determine the most confident result.

10.4.4.4 Necessary Future Research and Consideration

Through the interaction with online apps, users generally favor those that reach solutions quicker. Whether it is the start-up time, resumption time, or time spent completing tasks, developers and engineers must focus on minimizing the latency of the overall experience.

10.4.5 ATTRIBUTE: DISCREETNESS

10.4.5.1 Scope

In an age where privacy is a growing concern user may favor devices that operate in a discreet manner. The user is then free to explore the web as they please, without the fear of stigmatization or the negative social implications of using obtrusive devices. Note that the concept of discreetness extends past the context of computer usage, and into all fields of assistive devices.

10.4.5.2 Difficulties Faced by VI Users

During standard computer usage, in public settings, it often goes unnoticed that a sighted user is capable of using their devices discreetly. The user enjoys privacy via speechless text entries, auditory feedback through headphones, and compact touch screens that can be hidden from others. The user can then reduce their noticeability and blend with the local surroundings (ie. library, coffee shop, waiting room, etc.). This concept is most prevalent in youth, where phones and computers are plentiful. Modern culture relies on discreet phone use in all settings.

A lack of discretion occurs when the user's access modality cannot be kept private. An example of this may be the use of Apple's Siri, where a user must audibly voice their query. In order to aid users to employ the technology more comfortably, it would be beneficial if the operation is physically hidden (or at the least discreet).

As an example, imagine designing a new controller for VI users which improves control of their electronic devices. If the new modality requires the user to swing their arm violently, then it may be uncomfortable for public use and subject to stigmatization. Although the device may operate with increased efficacy, it will likely be rejected by the end-user. Therefore, it is the designer's responsibility to forecast where it will be used, along with the effects on its users.

Finally, users should not be required to disclose their handicap. As a result, VI users may appear as an ordinary user to others. This form of confidentiality has positive emotional implications (more in Section 4.7).

10.4.5.3   Existing Products and Ongoing Research

Discreetness can be expanded into components; user-voicing, and audio feedback. For user-voicing application discreetness is more difficult to achieve since the user is forced to audibly operate the application. For audio feedback applications, the user may use computer peripherals such as a keyboard, mouse or touchpad to silently interact with the computer or device. An example of user voicing applications can be found in (Griol et al., 2011; Halimah et al., 2008; Ismail and Zaman, 2010; Trippas, 2016). Examples of audio feedback applications can be found in (Dobri˘sek et al., 2002; Frauenberger et al., 2005; Griol et al., 2011; Hakobyan et al., 2013; Halimah et al., 2008; Ismail and Zaman, 2010; Jayant et al., 2011; Kapur et al., 2018; Mac´ıas et al., 2004; Murphy et al., 2008; Schmandt, 1998; Siekierska and McCurdy, 2008; Trippas, 2016; Wu and Adamic, 2014).

A wearable, non-invasive device named AlterEgo, is designed to allow the user to communicate with a computer without audible pronunciation (Kapur et al., 2018). When utilizing audio feedback systems, VI users must be focused by listening for long periods of time without distraction. As a result, the development of tactile or multi-modal interfaces is beneficial (Frauenberger et al., 2005). Siekierska and McCurdy have developed map interfaces for physical world navigation that focus on tactile modalities, allowing the user to have their ears listening for dangers or physical threats in their commute (i.e. cars, other pedestrians, traffic light signals, and more) (Siekierska and McCurdy, 2008).

The AlterEgo device developed in MIT motivated a fully discreet system. By allowing the user to quickly send text phrases as input, the system may be operated silently in public spaces. Similar to how SMS messages are a discreet form of communication, with the help of devices like AlterEgo, VI users will now be able to send and accept computer data without being noticed. The current methods for discreet text entry include a standard 'QWERTY' keyboard, mobile keypads with predictive text, and other forms of non-verbal entries.

10.4.5.4   Necessary Future Research and Consideration

With a growing dependence on internet accessibility, there is a growing need for discreetness in computer and web applications. Along with a need for quick and quiet access (what is now only present for the average or 'ideal' user, and Net Savvy users (Small et al., 2009). Once VI users can comfortably operate their devices and participate in online activities regularly, they will be able to develop independence and social awareness.

10.4.6   ATTRIBUTE: EMOTIONAL IMPLICATIONS

10.4.6.1   Scope

The emotions that a user feels when facing challenges in an unfamiliar environment are important for the success of a product. Whether it is happiness, confusion, frustration, or despair, the response of a user to the functionality of a product is a definite indication of its usability and accuracy.

10.4.6.2   Difficulties Faced by VI Users

Negative reactions by individuals employing web browsers are frustrating since the user is unable to determine their virtual location or recall previously acquired knowledge. Depending on the experience of the user, a potential solution would be to close the program and retrace their steps (discussed in Section 4.2) (Murphy et al., 2008).

When content is found to be inaccessible, additional frustrations and confusions can set in. Furthermore, partially accessible web pages are equally problematic since text may be easily understood via screen readers, but images and videos are unobservable. This form of inaccessibility is critical since many sites rely on visual content to convey critical information.

The use of standard web pages and search engines constitutes a large part of computer usage. Furthermore, participating in social media platforms has become a standard for many. Although there are conflicting views on the effects of social media, Jayant et al. have noted that social media is beneficial for VI users (Jayant et al., 2011). Since social media allows for anonymity in the public sphere, the visually impaired can freely express themselves without stigmatization or additional confrontation (WHO, 2011). VI users can share their experiences with others almost instantaneously and receive positive feedback from their peers online.

10.4.6.3   Existing Products and Ongoing Research

Studies in the literature have focused their research to understand the human factors of accessible products. Menzi-Cetin et al. have noted the importance of usable products since they make its users happy (Menzi-Çetin et al., 2017), these factors are often overlooked. Tsai et al. attempt to quantify the amount of time it takes for a VI user to get frustrated while using the web (Tsai et al., 2010). Murphy et al. also indicate the frustration behind online application (Murphy et al., 2008), Andronico et al. describe the need for less frustrating user interfaces that are more user-oriented (Andronico et al., 2006). Other papers will describe the dependence of VI users (Hersen et al., 1995) and the increased brain activity used by Net Savvy (experienced) users (Small et al., 2009), emphasizing the non-intuitive nature of online platforms.

The World Health Organization (WHO) highlights the importance of online communities as a method to overcome barriers experienced by face-to-face interaction (WHO, 2011). Wu et al. studied a text-based online social network platform, Twitter, displaying the levels of influence by specific VI groups (Wu et al., 2011). Wu and Adamic analyzed the social network density, size, and usage of VI and sighted Facebook users (Wu and Adamic, 2014). VI users on Facebook were identified by their use of Apple's iOS voiced accessibility feature, VoiceOver, these users statistically received more feedback from peers (Wu and Adamic, 2014). Jayant et al. found that sharing pictures and videos online had a positive effect on the individual (Jayant et al., 2011). Other research indicated that VI users feel as though they are missing out on a perceptual experience online

(Murphy et al., 2008), referencing the stimulus of visual content. In the educational setting, Muwanguzi and Lin have studied the reactions of VI students when accessing web-based educational content, as well as, their ability to communicate with professors and colleagues virtually (Muwanguzi and Lin, 2012).

Ismail and Zaman present, in 2010, the disappointment of VI users when using voice-activated browsers. Hakobyan et al. studied the motivation behind the development of a Mobile Assistive Technology (MAT), discovering that, "individuals feel less stigmatized or labelled", when using these products. This topic relates well to the concept of discreetness discussed in Section 4.6. Search Trail, an application that aids VI users with multi-session tasks, provides the users with confidence knowing that the program tracks their virtual trail (Sahib et al., 2014). The most frequent reasons for not using applications or devices is the lack of interest, cost, or simply being unaware of its availability (Crossland et al., 2014).

10.4.6.4   Necessary Future Research and Consideration

There is a need for humane considerations of emotional implications while developing future products, devices, and applications. It is not enough to make a product that it is accessible, but also one that allows the user to enjoy the online experience. It should not be a burden for the user to interface with online applications but rather an integral part of a person's life. If an application causes users to become frustrated, the likelihood of repetitive use declines dramatically, resulting in abandoned devices and products. Developers and engineers must consider the end user in their entirety.

10.5   UNDERSTANDING THE USERSPACE

When designing a UI there's a constant battle between user freedom and providing an overwhelming amount of information. By binding user commands to keystroke combinations, the user spends less time while achieving more functionality. But first they must spend time memorizing commands (Murphy et al., 2008). This concept of user preference for 'recognition over recall' is exemplified by Scott MacKenzie (MacKenzie, 2012), in his analysis of menus. So where is the balance? The easy answer is, it depends. Primarily on the users of the application along with their preferences and capabilities.

First let's define what the user wants to reach, perhaps everything. Since this chapter is centralized around search engines, let's focus on the Internet. An environment that encompasses all which the web has to offer is known as the 'cyberspace', coined by William Gibson in 1982 (Gibson, 2014). This idea encapsulates the struggle of UI design. Users want to traverse cyberspace as quickly as possible without memorizing steps.

Let's hone it down by concentrating on a single webpage. By forgetting (for a moment) all the places this page could take you, it becomes easier to see how a single webpage is a defined cyberspace. More specifically, a webpage has a better-defined set of possibilities, being a subset of the total cyberspace. It's like counting the leaves of a tree versus the entire forest.

As an example consider the SERP. Being a results page it holds a finite number of search results. By manipulating how the user interacts, they might find a more efficient method of accessing the same information. In the case of VI users, this is true, primarily due to a large majority of web pages being designed for sighted users.

To conclude, it is important to consider the users' space of possibilities when designing tools. In the context of SEs the user's end result is unpredictable since the space is infinite. But for other tools like word processors or music applications that allow the user to interact with different content the space may be very well defined. As a result, by looking at the whole picture designers may find interfaces that benefit users of all types regardless of their visual acuity.

10.6 EVALUATIONS AND SURVEYS

10.6.1 EVALUATIONS

As a tool developer, it is important to consider the usability of your implementation. Simply satisfying the requirements does not necessarily prove success. Structuring evaluations that provide statistical backing to your application can be achieved by focusing on one of the attributes analyzed above (Section 4).

For example, examiners can focus on latency as their main quantitative factor. By doing so they are able to select an independent variable more easily. For latency this could be a timed variable. While for information accessibility the metric could be a percentage of inaccessible items in a site. Scott MacKenzie (MacKenzie, 2012) describes the field of human-computer interaction accurately and provides techniques for structuring, defining, and carrying out evaluations and experiments.

10.6.2 VISUAL QUESTION ANSWERING SURVEYS

A VQA survey provides its participants with an image. They are then asked specific questions regarding that image so it can be alternatively described. This data is recorded and can later be used to develop vision algorithms that can annotate web content automatically. This technology can provide VI users with a "personal assistant" that answers questions like a human.

When considering key web page components, images and videos are often the entire focus (think of Youtube, Instagram or Facebook). This can be seen by the UI features like Youtube's Autoplay and Facebook's endless scroll. Successful VQA methods are extremely useful since they allow VI users to generate annotations upon request. Several VQA papers are described in this text that attempt to resolve this issue.

Several VQA surveys have been conducted historically, our attempt is to congregate a sample of relevant surveys (Agrawal et al., 2017; Gurari et al., 2018; Wang et al., 2015). The success of these algorithms is still unclear due to the complexity and variability of the initial problem. Regardless this issue is quite important since the annotation of every picture online is an impossibly exhaustive task. If this aligns with your field of interest it would be well advised to delve deeper into Machine Learning (ML) algorithms concerned with classification.

The results of VQA surveys are useful in understanding what users are interested in knowing. By analyzing commonalities between different results, surveys and content types, researchers can synthesize data to determine common concepts of intrigue. Additionally, predictive software, such as machine learning techniques, can be employed by VI users to answer simple questions discreetly. This provides the user with independence, confidence, and excitement.

10.7 UNDERSTANDING THE BIG PICTURE

Highly accessible and user-friendly computer tools are difficult to design. Furthermore, due to the wide spectrum of visual disabilities, it is impossible to tailor an ideal program for each user. As a result, designers and engineers must employ the characteristics and attributes that are most concerned with the needs of the target audience.

Common results uncovered in this chapter include; the lack of user-oriented accessible design, incompatibility with existing technologies, and the difficulty of becoming proficient as a regular user (the learning curve).

When faced with navigation, VI users are forced to construct abstract layouts of the page since most developments are targeted to the visually enabled. Additionally, information is less accessible than originally perceived, most images and videos include little, if any, alternative descriptive text, while standard menus and options are difficult to find, mostly, to enhance visual attractiveness.

The online community lacks standardization causing each individual to build their own distinct portfolio of preferred programs. Furthermore, the interaction of VI users with these interfaces, particularly in public settings, is rather obtrusive and loud. The lack of overall discretion attracts unwanted attention to a demographic that simply wants to use common applications.

This chapter attempts to merge the attributes most commonly addressed in the literature, while focusing on Search Engines (SEs) as the assistive tool. SEs have become a natural starting point for any web search due to their simplicity and usefulness. Unfortunately, these entry-level platforms include ingrained barriers that restrict VI users from browsing freely.

Sections explaining tools, along with common examples of their use can be found in Tools, Section 3. More analysis of issues and difficulties attributed to the online experience are described in depth starting in Section 4, Primary Attributes of Tools.

NOTES

1 VI users, both in the physical and cyber worlds, encounter the added requirement of storing the following in their working memory; where they have been (ie. their Trail) (Sahib et al., 2014) their current location, and where they intend to go next. As a result, VI users would benefit from assistive devices that manage their Trail and aid in the conceptualization of the internet.

2 **Analogy:** Imagine the internet to be a physical interface that a person can traverse (similar to how a sighted user analyzes a single webpage on a screen). This is analogous to a person walking across their home. Since the space (their home) has been thoroughly navigated by

the person it would be trivial for them to complete their journey. But if the space (or interface) is foreign to the person (ex. stranger's home) then the lack of sight would result in a significant disadvantage.

REFERENCES

(2019). The Act (AODA). https://www.aoda.ca/the-act/

Agrawal, A., Lu, J., Antol, S., Mitchell, M., Lawrence Zitnick, C., Parikh, D., & Batra, D. (2017). VQA: Visual Question Answering. *International Journal of Computer Vision*, *123*, 1 (May 2017), 4–31. https://doi.org/10.1007/s11263-016-0966-6

Aliyu, F. M. & Mabu, A. M. (2014). Google Query Optimization Tool. 2014 IEEE 6th International Conference on Adaptive Science Technology (ICAST). 1–5. https://doi.org/10.1109/ICASTECH.2014.7068150

Andronico, P., Buzzi, M., Castillo, C., & Leporini, B. (2006). Improving Search Engine Interfaces for Blind Users: A Case Study. *Universal Access in the Information Society*, *5*, 1 (June 2006), 23–40. https://doi.org/10.1007/s10209-006-0022-3

Andronico, P., Buzzi, M., & Leporini, B. (2004). Can I Find What I'm Looking for?. Proceedings of the 13th International World Wide Web Conference on Alternate Track Papers & Posters (WWW Alt. '04). ACM, New York, NY, USA, 430–431. https://doi.org/10.1145/1013367.1013510

Baguma, R. & Lubega, J. T. (2008). Web Design Requirements for Improved Web Accessibility for the Blind. In J. Fong, R. Kwan, & F. L. Wang (Eds.), *Hybrid Learning and Education (Lecture Notes in Computer Science)* (pp. 392–403). Springer Berlin Heidelberg.

Chen, Y., Ma, W.-Y., & Zhang, H.-J. (2003). Detecting Web Page Structure for Adaptive Viewing on Small Form Factor Devices. Proceedings of the 12th International Conference on World Wide Web (WWW '03). ACM, New York, NY, USA, 225–233. https://doi.org/10.1145/775152.775184

Chiang, M. F., Cole, R. G., Gupta, S., Kaiser, G. E., & Starren, J. B. (2005). Computer and World Wide Web Accessibility by Visually Disabled Patients: Problems and Solutions. *Survey of Ophthalmology*, *50*, 4 (July 2005), 394–405. https://doi.org/10.1016/j.survophthal.2005.04.004

Crossland, M. D., Silva, R. S., & Macedo, A. F. (2014). Smartphone, Tablet Computer and E-reader Use by People with Vision Impairment. *Ophthalmic and Physiological Optics*, *34*, 5 (Sept. 2014), 552–557. https://doi.org/10.1111/opo.12136

Dobrišek, S., Gros, J., Vesnicer, B., Mihelič, F., & Pavešić, N. (2002). A Voice-Driven Web Browser for Blind People. In P. Sojka, I. Kopeček, & K. Pala (Eds.), *Text, Speech and Dialogue (Lecture Notes in Computer Science)* (pp. 453–459). Springer Berlin Heidelberg.

Frauenberger, C., Stockman, T., Putz, V., & Holdrich, R. (2005). Mode Independent Interaction Pattern Design. Ninth International Conference on Information Visualisation (IV'05). 24–30. https://doi.org/10.1109/IV.2005.80

Freedom Scientific. (n.d.a). JAWS Information Webpage. https://www.freedomscientific.com/Products/Blindness/JAWS

Freedom Scientific. (n.d.b). JAWS Web Browsing Keystrokes. https://www.freedomscientific.com/Content/Documents/Manuals/JAWS/Keystrokes.pdf

Gibson, W. (2014). *Burning Chrome*. Harper Collins.

Griol, D., Molina, J. M., & Corrales, V. (2011). The VoiceApp System: Speech Technologies to Access the Semantic Web. In J. A. Lozano, J. A. G´amez, & J. A. Moreno (Eds.), *Advances in Artificial Intelligence (Lecture Notes in Computer Science)*(pp. 393–402). Springer Berlin Heidelberg.

Gurari, D., Li, Q., Stangl, A. J., Guo, A., Lin, C., Grauman, K., Luo, J., & Bigham, J. P. (2018). VizWiz Grand Challenge: Answering Visual Questions from Blind People. (Feb. 2018). https://arxiv.org/abs/1802.08218

Hakobyan, L., Lumsden, J., O'Sullivan, D., & Bartlett, H. (2013). Mobile Assistive Technologies for the Visually Impaired. *Survey of Ophthalmology, 58*, 6 (Nov. 2013), 513–528. https://doi.org/10.1016/j.survophthal.2012.10.004

Halimah, B. Z., Azlina, A., Behrang, P., & Choo, W. O. (2008). Voice Recognition System for the Visually Impaired: Virtual Cognitive Approach. 2008 International Symposium on Information Technology, Vol. 2. 1–6. https://doi.org/10.1109/ITSIM.2008.4631738

Harper, K. A. & DeWaters, J. (2008). A Quest for Website Accessibility in Higher Education Institutions. *The Internet and Higher Education, 11*, 3 (Jan. 2008), 160–164. https://doi.org/10.1016/j.iheduc.2008.06.007

Hersen, M., Kabacoff, R. I., Hasselt, V. B. V., Null, J. A., and et al. (1995). Assertiveness, Depression, and Social Support in Older Visually Impaired Adults. *Journal of Visual Impairment & Blindness, 89*, 6 (1995), 524–530.

Hu, R. & Feng, J. H. (2015). Investigating Information Search by People with Cognitive Disabilities. *ACM Transactions on Computer Systems, 7*, 1 (June 2015), 1:1–1:30. https://doi.org/10.1145/2729981

Ismail, N. & Zaman, H. B. (2010). Search Engine Module in Voice Recognition Browser to Facilitate the Visually Impaired in Virtual Learning (MGSYS VISI-VL). 4, 11 (2010), 5.

Ivory, M. Y., Yu, S., & Gronemyer, K. (2004). Search Result Exploration: A Preliminary Study of Blind and Sighted Users' Decision Making and Performance. In *CHI '04 Extended Abstracts on Human Factors in Computing Systems (CHI EA '04)* (pp. 1453–1456). ACM, New York, NY, USA. https://doi.org/10.1145/985921.986088

Jayant, C., Ji, H., White, S., & Bigham, J. P. (2011). Supporting Blind Photography. The Proceedings of the 13th International ACM SIGACCESS Conference on Computers and Accessibility (ASSETS '11). ACM, New York, NY, USA, 203–210. https://doi.org/10.1145/2049536.2049573

Kapur, A., Kapur, S., & Maes, P. (2018). AlterEgo: A Personalized Wearable Silent Speech Interface. 23rd International Conference on Intelligent User Interfaces (IUI '18). ACM, New York, NY, USA, 43–53. https://doi.org/10.1145/3172944.3172977

Lewandowski, D. & Kerkmann, F. (2012). Accessibility of Web Search Engines: Towards a Deeper Understanding of Barriers for People With Disabilities. *Library Review, 61*, 8/9 (Aug. 2012), 608–621. https://doi.org/10.1108/00242531211292105

Macías, M., Reinoso, A., González, J., García, J. L., Díaz, J. C., & Sánchez, F. (2004). WebTouch: An Audio-tactile Browser for Visually Handicapped People. In E. O'Neill, P. Palanque, & P. Johnson (Eds.), *People and Computers XVII — Designing for Society* (pp. 339–347). Springer London.

Macias, M., Gonzalez, J., & Sanchez, F. (2002). On Adaptability of Web Sites for Visually Handicapped People. 2nd International Conference on Adaptive Hypermedia and Adaptive Web-Based Systems, AH 2002, May 29, 2002 - May 31, 2002 (Lecture Notes in Computer Science (including subseries Lecture Notes in Artificial Intelligence and Lecture Notes in Bioinformatics), Vol. 2347). Springer Verlag, 264–273.

Mack, A. & Rock, I. (1998). *Inattentional Blindness*. A Bradford Book, http://libaccess.mcmaster.ca/login?url=http://search.ebscohost.com/login.aspx?direct=true&db=nlebk&AN=1439&site=ehost-live&scope=site

MacKenzie, I. S. (2012). *Human-Computer Interaction: An Empirical Research Perspective*. Morgan Kaufmann.

Mele, M. L., Federici, S., Borsci, S., & Liotta, G. (2010). Beyond a Visuocentric Way of a Visual Web Search Clustering Engine: The Sonification of WhatsOnWeb. In K. Miesenberger, J. Klaus, W. Zagler, & A. Karshmer (Eds.), *Computers Helping People with Special Needs (Lecture Notes in Computer Science)* (pp. 351–357). Springer Berlin Heidelberg.

Menzi-Çetin, N., Alemdağ, E., Tüzün, H., & Yıldız, M. (2017). Evaluation of a University Website's Usability for Visually Impaired Students. *Universal Access in the Information Society*, *16*, 1 (March 2017), 151–160. https://doi.org/10.1007/s10209-015-0430-3

Murphy, E., R. Kuber, G. McAllister, P. Strain, & W. Yu. (2008). An Empirical Investigation into the Difficulties Experienced by Visually Impaired Internet Users. *Universal Access in the Information Society*, *7*, 1 (April 2008), 79–91. https://doi.org/10.1007/s10209-007-0098-4

Muwanguzi, S. & Lin, L. (2012). Coping with Accessibility and Usability Challenges of Online Technologies by Blind Students in Higher Education. *Intelligent Learning Systems and Advancements in Computer-Aided Instruction: Emerging Studies* (2012), 269–286. https://doi.org/10.4018/978-1-61350-483-3.ch016.

NetMarketShare. (2019). Search Engine Market Share. https://netmarketshare.com/search-engine-market-share.aspx

Powsner, S. M. & Roderer, N. K. (1994). Navigating the Internet. *Bulletin of the Medical Library Association*, *82*, 4 (Oct. 1994), 419–425. https://www.ncbi.nlm.nih.gov/pmc/articles/PMC225968/

Roentgen, U. R., Gelderblom, G. J., Soede, M., & de Witte, L. P. (2009). The Impact of Electronic Mobility Devices for Persons Who Are Visually Impaired: A Systematic Review of Effects and Effectiveness. *Journal of Visual Impairment & Blindness; New York*, *103*(11) (Dec. 2009), 743–753. https://search.proquest.com/docview/222041535/abstract/C34D798C66C14B11PQ/1

Sahib, N. G., Thani, D. A., Tombros, A., & Stockman, T. (2012). Accessible Information Seeking. (2012), 3.

Sahib, N. G., Thani, D. A., Tombros, A., & Stockman, T. (2014). Investigating the Behavior of Visually Impaired Users for Multi-session Search Tasks. *Journal of the Association for Information Science and Technology*, *65*, 1 (Jan. 2014), 69–83. https://doi.org/10.1002/asi.22955

Sahib, N. G., Thani, D. A., Tombros, A., & Stockman, T. (2015). Evaluating a Search Interface for Visually Impaired Searchers. *Journal of the Association for Information Science and Technology*, *66*, 11 (Nov. 2015), 2235–2248. https://doi.org/10.1002/asi.23325

Schmandt, C. (1998). Audio Hallway: A Virtual Acoustic Environment for Browsing. Proceedings of the 11th Annual ACM Symposium on User Interface Software and Technology (UIST '98). ACM, New York, NY, USA, 163–170. https://doi.org/10.1145/288392.288597

Siekierska, E. & McCurdy, W. (2008). Internet-Based Mapping for the Blind and People With Visual Impairment. *Lecture Notes in Geoinformation and Cartography*, 9783540720287. 283–300. https://doi.org/10.1007/978-3-540-72029-4_19.

Small, G. W., Moody, T. D., Siddarth, P., & Bookheimer, S. Y. (2009). Your Brain on Google: Patterns of Cerebral Activation during Internet Searching. *The American Journal of Geriatric Psychiatry*, *17*, 2 (Feb. 2009), 116–126. https://doi.org/10.1097/JGP.0b013e3181953a02

Trippas, J. R. (2016). Spoken Conversational Search: Speech-only Interactive Information Retrieval. Proceedings of the 2016 ACM on Conference on Human Information Interaction and Retrieval (CHIIR '16). ACM, New York, NY, USA, 373–375. https://doi.org/10.1145/2854946.2854952

Tsai, C., Lin, W., & Hung, C. (2010). Mobile Web Search by Query Specification: An example of Google Mobile. The 40th International Conference on Computers Industrial Engineering. 1–4. https://doi.org/10.1109/ICCIE.2010.5668288

Wang, P., Wu, Q., Shen, C., van den Hengel, A., & Dick, A. (2015). Explicit Knowledge-Based Reasoning for Visual Question Answering. *arXiv:1511.02570 [cs]* (Nov. 2015). http://arxiv.org/abs/1511.02570 arXiv:1511.02570.

World Health Organization WHO (2011). *World Report on Disability.* http://whqlibdoc.who.int/publications/2011/9789240685215_eng.pdf

World Wide Web Consortium. (1999). Web Content Accessibility Guidelines 1.0. (May 1999). http://travesia.mcu.es/portalnb/jspui/handle/10421/2544

Wu, S. & Adamic, L. A. (2014). Visually Impaired Users on an Online Social Network. *Proceedings of the SIGCHI Conference on Human Factors in Computing Systems (CHI'14).* ACM, New York, NY, USA, 3133–3142. https://doi.org/10.1145/2556288.2557415

Wu, S., Hofman, J. M., Mason, W. A., & Watts, D. J. (2011). Who Says What to Whom on Twitter. Proceedings of the 20th International Conference on World Wide Web (WWW'11). ACM, New York, NY, USA, 705–714. https://doi.org/10.1145/1963405.1963504

Yang, Y.-F. & Hwang, S.-L. (2007). Specialized Design of Web Search Engine for the Blind People., 997–1005 pages.

Yang, Y.-F., Hwang, S.-L., & Schenkman, B. (2012). An improved Web Search Engine for Visually Impaired Users. *Universal Access in the Information Society, 11,* 2 (June 2012), 113–124. https://doi.org/10.1007/s10209-011-0250-z

Yoffie, D. B., Wu, L., Sweitzer, J., & Eden, D. (n.d.). Voice War: Hey Google vs. Alexa vs. Siri. https://hbr.org/product/voice-war-hey-google-vs-alexa-vs-siri/718519-PDF-ENG

11 Tensor Methods for Clinical Informatics

Cristian Minoccheri, Reza Soroushmehr,
Jonathan Gryak, and Kayvan Najarian

CONTENTS

11.1 INTRODUCTION

Tensors and tensor methods are a generalization of classic linear algebra techniques for vectors and matrices to higher-order, multiway data. While the foundational ideas have been known for more than a century, it was only in the Sixties that tensor techniques became known as powerful tools due to newly found applications in psychometrics and chemometrics (Carroll & Chang, 1970; Harshman, 1927; Kruskal, 1977; Tucker, 1966). In the past two decades, partly because of advances in computational power, these ideas have attracted even more interest and seen a wide variety of applications.

There are already several excellent surveys that discuss applications of tensor methods to many machine learning problems in different areas. Kolda & Bader (2009) is a thorough introduction to tensor decompositions and an overview of many papers published until 2009. More recently, to name a few,

DOI: 10.1201/9781003120902-11

Cichocki et al. (2015) is an overview of tensor methods with emphasis on signal processing; Papalexakis et al. (2016) is a comprehensive discussion of tensor methods, algorithms, computational aspects, and applications; Sidiropoulos et al. (2017) is a more technical overview which includes more advanced algorithms an optimization methods; Song et al. (2019) is a recent survey focused on tensor completion and many applications in big data analytics. With respect to biomedical applications, Luo et al. (2017) is a recent brief survey of how tensors have been used in precision medicine. Finally, the book Taguchi (2020) covers both the mathematical theory of tensor decompositions and many applications to genomic science. In this survey, we focus on two areas: tensor decomposition and tensor completion. Our goal is to provide a short introduction to the mathematics behind tensor techniques, as well as resources for the reader interested in more technical details, and a comprehensive overview of how these methods have been used in clinical informatics. With respect to applications, we focus on papers published approximately during the past 10 years.

11.2 DEFINITIONS AND NOTATION

A tensor is a multi-way extension of a matrix, and it can be thought of as a multi-dimensional array. We will denote (column) vectors as $v, w, \ldots \in \mathbb{R}^I$, matrices as $A, B, \ldots \in \mathbb{R}^{I \times J}$, tensors as $\mathscr{X}, \mathscr{Y}, \ldots \in \mathbb{R}^{I_1 \times \cdots \times I_N}$ for $N \geq 3$. A tensor $\mathscr{X} \in \mathbb{R}^{I_1 \times \cdots \times I_N}$ is called an N−dimensional tensor, or a tensor with N modes, or an N−way tensor, or an Nth order tensor. Matrices are simply 2−way tensors. Each entry of an N−way tensor is determined by N indices. We will denote the entry in position $(i_1, \ldots, i_N)$ by $\mathscr{X}(i_1, \ldots, i_N)$ or $x_{i_1, \ldots, i_N}$.

Consider for example a third-order tensor. Entries of a third-order tensor $\mathscr{X} \in \mathbb{R}^{I \times J \times K}$ depend on three indices x_{ijk}, with $i = 1, \ldots, I, j = 1, \ldots, J, k = 1, \ldots, K$. Fixing two of the indices, say j and k, we get mode−1 fibers $x_{:jk}$ (which are vectors) of the tensor. Fixing one of the indices, say k, we get mode−3 slices $x_{:jk}$ (which are matrices) of the tensor.

It is often useful to reshape tensors, for example in the form of vectors or matrices, to be able to apply known linear algebraic methods. The vectorization $vec(\mathscr{X})$ of a tensor is obtained by stacking all mode−1 fibers into a column vector. The flattening (or matricization) of an N−way tensor along the n−th mode is a matrix of size $I_n \times \prod_{j \neq n} I_j$ obtained by juxtaposing the mode−n slices of the tensor.

To introduce the key notion of rank of a tensor we need to define the *outer product of vectors*: given u, v, w vectors of lengths I, J, K, their outer product $u \circ v \circ w$ is a 3-way tensor $\mathscr{X}$ with

$$x_{ijk} = u_i v_j w_k.$$

A tensor of the form $u \circ v \circ w$ is said to have rank 1. For matrices, $u \circ v = uv^t$. Outer products are related to the notion of rank. A factorization of a matrix $X \in \mathbb{R}^{I \times J}$ is a sum of rank 1 matrices $X = AB^t = \sum_{i=1}^{R} a_i \circ b_i$ for some $A \in \mathbb{R}^{I \times R}$ and $B \in \mathbb{R}^{J \times R}$ called factor matrices; the rank of a matrix can be described as the minimum number of rank 1 summands for which we can write such a factorization. We say that a factorization is essentially unique if the factor matrices are unique up to scaling and permuting the columns. Unless the matrix X has rank 1, matrix factorization is never essentially unique, even if R is minimal. For this reason, one often requires additional constraints on the factor matrices (such as orthogonality, triangularity, noon-negativity, etc.) to obtain essential uniqueness. For example, requiring orthogonality we obtain the Singular Value Decomposition (SVD), which is essentially unique provided that the singular values are distinct.

We extend this definition to that of the rank of a tensor: the rank of a tensor $\mathscr{X}$ is the minimal number of rank 1 summands for which we can write $\mathscr{X}$ as a sum of rank 1 terms. This notion of rank turns out to be difficult to work within the practice, so several other definitions of rank are more manageable. For example, the mode$-n$ rank of a tensor $\mathscr{X}$ is the (matrix) rank of the mode$-n$ flattening $X_{(n)}$, and the multilinear rank of $\mathscr{X}$ is the $n-$tuple of mode$-n$ ranks.

11.3 MATRIX AND TENSOR PRODUCTS

We will be interested in several ways of multiplying matrices and tensors.

Kronecker product: given two matrices $A \in \mathbb{R}^{I \times H}$ and $B \in \mathbb{R}^{J \times K}$, their Kronecker product $A \otimes B$ is the $(IJ) \times (HK)$ matrix

$$\begin{bmatrix} a_{11}B & a_{12}B & \dots & a_{1H}B \\ a_{21}B & a_{22}B & \dots & a_{2H}B \\ \vdots & \vdots & \ddots & \vdots \\ a_{I1}B & a_{I2}B & \dots & a_{IH}B \end{bmatrix}$$

Khatri-Rao product: given two matrices $A \in \mathbb{R}^{I \times K}$ and $B \in \mathbb{R}^{J \times K}$ with columns $a_1, \dots, a_K$ and $b_1, \dots, b_K$, $A \odot B$ is a $(IJ) \times K$ matrix with columns

$$[a_1 \otimes b_1 a_2 \otimes b_2 \dots a_K \otimes b_K].$$

Hadamard product: given two tensors of the same size $\mathscr{X}$ and $\mathscr{Y}$, their Hadamard product is a tensor $\mathscr{X} * \mathscr{Y}$ with entry in position $(i_1, \dots, i_N)$ given by $\mathscr{X}(i_1, \dots, i_N) \mathscr{Y}(i_1, \dots, i_N)$.

Inner product and Frobenius norm: given two tensors of the same size $\mathscr{X}$ and $\mathscr{Y}$, their inner product is

$$< \mathscr{X}, \mathscr{Y} > = \sum_{i_1, \dots, i_N} \mathscr{X}(i_1, \dots, i_N) \cdot \mathscr{Y}(i_1, \dots, i_N).$$

The Frobenius norm of a tensor is defined as $\|\mathscr{X}\|_F = \sqrt{<\mathscr{X}, \mathscr{X}>}$.

It is possible to multiply all entries of a tensor by a scalar, or to add two tensors (of the same size) entry-wise.

n-mode matrix-tensor product: $A \times_n \mathscr{X}$. Multiply each mode$-n$ fiber by A (equivalently, multiply the mode$-n$ flattening $X_{(n)}$ by A).

Often, a tensor $\mathscr{X}$ is multiplied in each mode by matrices $A^{(1)}, \ldots, A^{(N)}$, in which case one obtains a full multilinear product which is denoted as

$$[[\mathscr{X}; A^{(1)}, \ldots, A^{(N)}]] = \mathscr{X} \underset{1}{\times} A^{(1)} \underset{2}{\times} \cdots \underset{N}{\times} A^{(N)}.$$

If $\mathscr{X}$ is superdiagonal, with entries $\mathscr{X}(i_1, \ldots, i_N) = 1$ if $i_1 = \ldots = i_N$, and zero otherwise, we usually omit it from the full multilinear product, which we then write $[[A^{(1)}, \ldots, A^{(N)}]]$.

A useful property of the Kronecker and Khatri-Rao products is that they allow us to interpret the full multilinear product in terms of matrices. One can show that $\mathscr{Y} = [[\mathscr{X}; A^{(1)}, \ldots, A^{(N)}]]$ is equivalent to

$$Y_{(n)} = A^{(n)} X_{(n)} (A^{(N)} \otimes \ldots \otimes A^{(n+1)} \otimes A^{(n-1)} \otimes \ldots \otimes A^{(1)})^t;$$

similarly, $\mathscr{Y} = [[A^{(1)}, \ldots, A^{(N)}]]$ is equivalent to $Y_{(n)} = A^{(n)} (A^{(N)} \odot \ldots \odot A^{(n+1)} \odot A^{(n-1)} \odot \ldots \odot A^{(1)})^t$.

11.4 TENSOR DECOMPOSITIONS

There are several forms of tensor decomposition, each with different properties and scope.

11.4.1 CP DECOMPOSITION

A polyadic decomposition of a tensor $\mathscr{X} \in \mathbb{R}^{I_1 \times \cdots \times I_N}$ is a decomposition of $\mathscr{X}$ as a sum of rank 1 tensors:

$$\mathscr{X} = \sum_{i=1}^{R} \sigma_i a_i^{(1)} \circ \ldots \circ a_i^{(N)} = [[A^{(1)}, \ldots, A^{(N)}]].$$

One can think of this as a generalization to tensors of matrix factorization. If R is minimal, we call this a Canonical Polyadic Decomposition (CPD) and we call such R the rank of the tensor (Carroll & Chang, 1970; Harshman, 1927). Note that the definition of tensor rank extends that of matrix rank, but it can behave quite differently. For example, the rank of a tensor can be larger than the dimension of each of its modes, and the problem of computing the rank of a given tensor is NP-hard (Håstad, 1990). Furthermore, unlike matrices, a best rank–k approximation of a tensor might not exist (De Silva & Lim, 2008): there are degenerate tensors that can be approximated arbitrarily well in the Frobenius norm by tensors of lower rank.

On the other hand, CP decompositions of tensors are often essentially unique (i.e., up to scaling and permutation). This is a substantial difference from the matrix case, where to achieve uniqueness one requires additional constraints. One sufficient condition, due to Kruskal (Kruskal, 1977) for third-order tensors and Sidiropoulos and Bro (Sidiropoulos & Bro, 2000) for higher-order ones, is that

$$\sum_{i=1}^{N} k_{X_{(i)}} \geq 2r + N - 1,$$

where k_A is the Kruskal rank of a matrix A (the largest integer k such that any k columns of A are linearly independent), r is the rank of the given tensor $\mathscr{X}$, and $X_{(i)}$ are the factor matrices in its CPD. This is especially relevant in applications, since uniqueness implies that the decomposition reveals latent factors, which can be used to explain and interpret the data. We should point out that the above criterion is known to be only sufficient: finding unique conditions is an ongoing area of research, with the most general condition to date given by (Chiantini & Ottaviani, 2012).

In practice, the CP decomposition is difficult to compute exactly, due to the NP-hardness of determining tensor rank and the presence of noise in the data. However, in applications, it is often enough to estimate the rank and compute a CPD approximation, for which there are several algorithms. It is also worth noting that many tensors arising in applications turn out to have low rank, which makes CPD a very powerful tool (the recent article by Udell & Townsend (2019) gives a theoretical explanation for this phenomenon in the case of matrices).

A simple and efficient algorithm for computing a CPD is Alternating Least Squares (ALS): for a third-order tensor $\mathscr{X}$ and a fixed rank r, one computes factor matrices A, B, C such that $\|\mathscr{X} - [[A, B, C]]\|_F$ is minimal by alternatingly minimizing with respect to each matrix until a convergence criterion is satisfied. For example, keeping B and C fixed,

$$\min_{A} \|\mathscr{X} - [[A, B, C]]\|_F = \min_{A} \|X_{(1)} - A(C \odot B)^t\|_F,$$

which can be solved in closed form.

In Tomasi & Bro (2006), several algorithms for CP decomposition are discussed and compared. Shortcomings of ALS are that if the number of estimated components is larger than the true number of components errors can be significant, and in the case of missing data convergence may be poor. In these cases, optimizing all factors at once (for example with a gradient approach) usually yields better results.

11.4.2 TUCKER DECOMPOSITION

A Tucker decomposition (Tucker, 1966) of a tensor $\mathscr{X}$ amounts to writing $\mathscr{X}$ as a core tensor $\mathscr{G} \in \mathbb{R}^{R_1 \times \dots \times R_N}$, with $J_n \leq I_n$ for $n = 1, \dots, N$, multiplied in each mode by a factor matrix $A^{(n)} \in \mathbb{R}^{I_n \times R_n}$:

$$\mathscr{X} = \sum_{i_1=1}^{R_1} \dots \sum_{i_N=1}^{R_N} g_{i_1, \dots, i_N} a_{i_1}^{(1)} \circ \dots \circ a_{i_N}^{(N)} = [[\mathscr{G}; A^{(1)}, \dots, A^{(N)}]].$$

The entries of the core represent the level of interaction among different components. Note that Tucker is a sum of rank-one tensors like CPD, however, it captures the interaction of a whole factor matrix with every other factor matrix. Typically, one requires extra conditions, like the factor matrices being orthogonal. Note that a Tucker decomposition is not unique as the factor matrices are not rotation invariant. However, the column space defined by each factor matrix is unique. A CP decomposition can be thought of as a special case of a Tucker decomposition (if we allow the factor matrices to be not orthogonal), with a superdiagonal core tensor. However, due to the lack of uniqueness, Tucker decompositions have very different applications. For example, Tucker decompositions can be thought of as higher-order Principal Component Analysis (PCA): by choosing a core of a small dimension we can think of it as a compressed version of the original tensor. While CP decompositions are often unique, easier to interpret, and can yield better compression for low-rank tensors, Tucker decompositions generally yield better compression for tensors of higher rank.

Using the Tucker decomposition, we can introduce a different notion of rank, the multilinear rank (sometimes referred to as n−rank), which has the advantage of being easy to compute (unlike CP rank). The multilinear rank $(r_1, ..., r_N)$ of a tensor $\mathscr{X}$ is the smallest tuple $(R_1, ..., R_N)$ for which $\mathscr{X}$ admits a Tucker decomposition $[[\mathscr{G}; A^{(1)}, ..., A^{(N)}]]$ as above, and one can compute that $r_n = rank(X_{(n)})$.

Typically, like with matrix factorizations, one requires additional constraints on the factor matrices to help determine unique bases for the column spaces of the factor matrices. One very relevant example is that of Higher-Order Singular Value Decomposition, or HOSVD (De Lathauwer et al., 2000a) where the factor matrices are orthogonal matrices consisting of the leading left singular vectors of the matricizations of the tensor, yielding a core of size the multilinear rank of the tensor. HOSVD is not the optimal approximation in terms of the Frobenius norm but is a good generalization of Singular Value Decomposition for matrices (in fact, HOSVD for a second-order tensor reduces to SVD). For example, it can be used to obtain a low multilinear rank approximation of a tensor, by truncating singular vectors with small singular values from the factor matrices, which leads to a smaller core.

The decomposition $[[\mathscr{G}; U^{(1)}, ..., U^{(N)}]]$ with orthogonal factor matrices that gives the best approximation in Frobenius norm and has multilinear rank no larger than a bound $(R_1, ..., R_N)$, can be obtained with an ALS algorithm. Starting from the HOSVD, one iteratively multiplies the core by all factor matrices except $U^{(n)}$, and then updates $U^{(n)}$ as the matrix with the leading R_n left singular vectors of the mode−n matricization of the tensor. This method is known as Higher-Order Orthogonal Decomposition (HOOI) (De Lathauwer et al., 2000b).

11.4.3   OTHER DECOMPOSITIONS

There exist a variety of other decomposition that can be more suitable for a specific goal. Block Term Decompositions (De Lathauwer, 2008a, 2008b) aim at being more versatile than CP decompositions while retaining some of their interpretability. In the case of a third-order tensor, for example, one can replace rank 1 matrices $a_i \circ b_i$ with low-rank matrices $A_i B_i^t$, obtaining a decomposition

$$\mathcal{X} = \sum_{i=1}^{R} \sigma_i A_i B_i^t \circ c_i.$$

In the case of high order tensors, it is often desirable to approximate them with lower-order ones, to avoid the curse of dimensionality. This can be achieved for example with the Hierarchical Tucker Decomposition, or H-Tucker (Grasedyck, 2010; Grasedyck et al., 2013; Hackbusch & Kühn, 2009): one creates a binary tree by splitting the modes hierarchically, where the hierarchy usually comes from the concrete application, and computes factor matrices from the SVD of a generalized matricization whose rows correspond to the modes in the given node. The leaves of the tree provide factor matrices for the H-Tucker decomposition.

Another way to approximate a tensor of high order is the Tensor Train Decomposition, or TT-decomposition (Oseledets, 2011), which approximates it with respect to terms of matrices and third-order tensors. For an N−way tensor $\mathcal{X}$, a tensor train decomposition consists in writing an entry $\mathcal{X}(i_1, \ldots, i_N)$ as

$$\sum_{j_1=1}^{R_1} \ldots \sum_{j_{N-1}=1}^{R_{N-1}} G_1(i_1, j_1) \cdot \mathcal{G}_2(j_1, i_2, j_2) \cdot \ldots \cdot \mathcal{G}_{N-1}(j_{N-2}, i_{N-1}, j_{N-1}) \cdot G_N(j_{N-1}, i_N),$$

for some $R_1, \ldots, R_{N-1}$. This is equivalent to writing the tensor $\mathcal{X}$ as the product of a matrix G_1, N–2 third-order tensors $\mathcal{G}_2, \ldots, \mathcal{G}_{N-2}$, and a final matrix G_N (where the product is the mode−n multiplication and occurs along hidden modes of the decomposition). The smallest values $(R_1, \ldots, R_{N-1})$ for which a TT-decomposition exists defines the TT-rank of the tensor. The TT-decomposition allows to store a tensor very efficiently, and it is possible to perform operations on tensors using their TT-decomposition. Tensor train decompositions are also known as linear tensor networks, and more general tensor networks built on graphs are possible, but few of them enjoy good properties. Finally, unlike H-Tucker, the TT-decomposition doesn't require the presence or the knowledge of a hierarchy among the modes.

In applications, one often has extra auxiliary information that comes in the form of coupled matrices or tensors. For example, if we consider a matrix whose entries represent the level of interaction between drugs (rows) and some targets (columns), we typically have extra information in the form of similarity between two drugs and similarity between two targets. This information provides two matrices in the form of drug-drug interaction and target-target interaction, which we say are coupled with the initial drug-target interaction matrix.

Using extra information not only improves the results of the task at hand, but it also provides additional constraints which make our model better behaved mathematically, for example by enforcing essential uniqueness of a decomposition (Acar et al., 2011; Lahat et al., 2015; Sorber et al., 2015). For example, if two third-order tensors $\mathcal{X}$ and $\mathcal{Y}$ are coupled in the first mode, a coupled CP decomposition will be of the form $\mathcal{X} = [[A, B, C]]$ and $\mathcal{Y} = [[A, D, E]]$, with the same factor matrix A for the coupled-mode; a coupled Tucker decomposition will be of the form $\mathcal{X} = [[\mathcal{G}; U^{(1)}, U^{(2)}, U^{(3)}]]$ and $\mathcal{Y} = [[\mathcal{G}; U^{(1)}, U^{(4)}, U^{(5)}]]$. Algorithms for coupled factorizations are similar to the uncoupled case, such as ALS and gradient-based methods.

11.5   TENSOR COMPLETION

It is often the case in applications that the matrix or tensor of interest is not fully known: several or even most entries may be missing. The problem of finding these missing values is known as matrix or tensor completion. Generally speaking, this problem is unsolvable – even for matrices, since there is no restriction on the missing values of a matrix. However, in practice, the datasets enjoy additional properties and constraints that lower the number of degrees of freedom and can be leveraged to impute the missing values.

A common theoretical assumption that is often consistent with the data is to assume that the matrix or tensor has a low rank. The low-rank matrix completion problem is still non-convex and NP-hard (Chistov & Grigoriev, 1984). In the case of matrices, the notion of rank is well understood, and one efficient method which is also well understood theoretically is that of convex relaxation: one uses the nuclear norm as a convex relaxation of the rank of a matrix and finds a matrix that minimizes the nuclear norm by semidefinite programming (Candés & Recht, 2008). Other approaches use for example matrix factorization and non-convex optimization by means of gradient descent (Rennie & Srebro, 2005).

The case of tensor completion is generally more complicated than the matrix case and is in general NP-hard (Peeters, 1996) as are the calculations of the tensor nuclear norm (Friedland & Lim, 2018), and its dual, the spectral norm (Hillar & Lim, 2013). However, there is ongoing research on how to estimate them, such as Tokcan et al. (2021) that provides computationally efficient approximations of the spectral norm of a tensor, which in turn can be used to approximate the nuclear norm. As with matrices, approximate tensor completion algorithms often start with the assumption that the tensor is low rank in a suitable sense. However, there are several notions of rank for tensors, and no choice seems to be always better than the other ones. Each choice leads to a different optimization problem. One possible approach to the problem is to choose a notion of tensor rank (e.g., CP rank, multilinear rank, n−rank, or TT-rank) and impute the missing values so that the chosen notion of rank is minimized. This idea is however often computationally unfeasible; in practice, one often fixes values of the chosen notion of rank and imputes the missing entries so that the distance from the known tensor is minimized. Unfortunately, even estimating the rank in advance is often a difficult task; see (Song et al., 2019) and the references therein for a more detailed discussion of possible formulations of the general completion problem.

Some of the first algorithms for tensor completion were based on tensor decompositions, such as CP (Acar et al., 2011; Tomasi & Bro, 2005) and Tucker (Walczak & Massart, 2001). The main idea is to iteratively and alternatingly fill in the missing values, decompose the tensor, and replace the known values.

In (Tomasi & Bro, 2005), the authors consider a single imputation CP-ALS algorithm. Say that, for simplicity, $\mathscr{X}$ is a third-order tensor to complete, $\mathscr{W}$ is a tensor (of the same size as $\mathscr{X}$) whose entries are 1 or 0 according to whether they are known or unknown entries of $\mathscr{X}$, and $\mathscr{U}$ is a tensor of the same size of $\mathscr{X}$ whose entries are all 1s. Then one can construct a tensor

$$\mathscr{X}^{z(s)} = \mathscr{X} * \mathscr{W} + \mathscr{Y}^{(s)} * (\mathscr{U} - \mathscr{W}),$$

where s is the number of the iteration, and $\mathscr{Y}^{(s)}$ is the interim model computed at the sth iteration. Since $\mathscr{X}^{z(s)}$ contains no missing values, factor matrices A^s, B^s, and C^s for the sth step can be computed with the usual ALS method for CP decomposition. One then computes $\mathscr{Y}^{(s)}$ repeats the process. $\mathscr{Y}^{(0)}$ is often assigned in terms of a suitable average of the known entries.

In Acar et al. (2011), the authors develop a CP weighted optimization (CP-WOPT) algorithm that uses first-order optimization to solve a weighted least squares problem on the known entries of the tensor. With the same notation as above, the objective function to minimize is

$$f(A, B, C) = \frac{1}{2}\|\mathscr{X} - [[A, B, C]]\|^2 = \frac{1}{2}\|\mathscr{W} * \mathscr{X} - \mathscr{W} * [[A, B, C]]\|^2,$$

with the minimization performed by computing the gradients with respect to A, B, and C.

These core methods for both performing a CP decomposition in the case of missing data and estimating missing entries can be similarly applied to other decompositions, including Tucker, H-Tucker, and HOOI.

A different approach is to apply convex relaxation by replacing the rank of the tensor with a convex surrogate, in a similar fashion to the matrix case. Unlike with matrices, computing the tensor nuclear norm is NP-hard, and for this reason, it is replaced by computationally tractable variants. In Gandy et al. (2011), the authors suggest minimizing the multilinear rank of the tensor, leading to the optimization problem

$$argmin_{\mathscr{X}} \sum_{i=1}^{N} rank(X_{(i)}), \ \mathscr{X} * \mathscr{W} = \mathscr{T},$$

where $\mathscr{T}$ is the tensor with the known entries. This low n-rank tensor completion problem is still non-convex and difficult to solve, and is therefore relaxed to a convex problem by replacing the ranks of the flattenings with their nuclear norms:

$$argmin_{\mathscr{X}} \sum_{i=1}^{N} \|X_{(i)}\|_*, \ \mathscr{X} * \mathscr{W} = \mathscr{T}.$$

In Liu et al. (2012), a tensor trace norm is defined as a weighted sum of the nuclear norms of the flattenings and used to formulate a convex optimization problem. The authors then propose three algorithms to solve the problem: one using block

TABLE 11.1

Papers Using Brain, Heart, and EHR Sata by Core Tensor Method Adopted

Data	Core method	Papers
EEG	CP	(Acar et al., 2007; Mørup et al., 2011)
EEG + fMRI	CP	(Mørup et al., 2008)
EEG	CP and Tucker	(Latchoumane et al., 2012; Cong et al., 2013)
EEG	Tucker	(Thanh et al., 2020)
EEG	MDA	(Frølich et al., 2018)
EEG + MEG	Coupled CP	(Naskovska et al., 2020)
EEG + fMRI	Coupled CP	(Hunyadi et al., 2016)
ECG	HOSVD	(Padhy & Dandapat, 2015; 2017)
ECG	CP and HOSVD	(Boussé et al., 2017; Goovaerts et al., 2015; 2017; Goovaerts et al., 2017; Luo et al., 2018; Hernandez et al. 2021)
EHR	CP	(Dauwels 2011; Ho et al., 2014; Ho, Ghosh, & Sun, 2014; Wang et al., 2015; Zhao et al., 2019; Ma et al., 2019)
EHR	Tucker	(Luo et al., 2015)
EHR	PARAFAC2	(Perros et al., 2019)

coordinate descent, one using a smoothing scheme, and one using the alternating direction method of multipliers.

As with tensor decompositions, it is often the case that the matrices and tensors to be completed are coupled with other matrices or tensors (which could themselves be incomplete). One approach to this problem (Li & Yeung, 2009; Narita et al., 2012) takes into account object similarity by introducing similarity matrices and factorizes the tensors with two regularization methods: within-mode regularization (using graph Laplacians to force similar objects to have similar factors) and cross-mode regularization (for very sparse tensors).

Another approach is to use coupled tensor decompositions (similarly to the uncoupled case), like coupled CP or coupled Tucker (Acar et al., 2011; Ermiş et al., 2015) (Table 11.1).

More recently, Wimalawarne et al. (2018) propose a convex coupled tensor completion method by introducing norms for coupled tensors that are mixtures of trace norms and coupled trace norms for matrices.

11.6 TENSOR DECOMPOSITION APPLICATIONS – A FEW EXAMPLES

11.6.1 BRAIN DATA

EEG data are nowadays often collected using several electrodes: since each electrode records a time series, several ones provide an additional mode corresponding to space. In practice, additional modes besides time and channel are often natural to consider, such as frequency, or the condition of the trial (in case the experiment is related to one or more stimuli), or – in case of multiple subjects – the persons from which the readings are collected. This type of data is therefore an obvious candidate for tensor methods, and both CP and Tucker have been successfully applied in a variety of tasks. We refer the reader to the survey (Cong et al., 2015) for a review of a variety of applications of EEG tensor analysis. A different type of brain data consists of magnetic resonance imaging (MRI) and functional magnetic resonance imaging (fMRI) scans, which can be made into tensorial data for example by stacking these images. Tensor methods on MRI and fMRI data often outperform vector-based ones since they take advantage of this natural spatial structure and can extract features by looking at multiple scans simultaneously. The survey (Hunyadi et al., 2015) reviews many successful applications of the tensorial analysis of EEG and fMRI data in the study of epilepsy, as well as methods that combine the two types of data.

One of the first applications of CP decomposition to EEG data from a single subject is Acar et al. (2007). The authors construct a three-way Epilepsy Tensor with modes time, scales, and electrodes; they use CP decomposition to localize seizures, by analyzing the coefficients of factors corresponding to the third mode and identifying which electrodes are involved. They also show how to use CP to define and identify artifacts in the data. Subsequent works take into account the natural time shifts arising from measurements of EEG data. Shift-CP is a shift-invariant CP model that introduces a time shift in the factor matrix for the second mode, leading to more accurate results in latent components identification both for EEG and fMRI data (Mørup et al., 2008). Conv-CP is a further generalization that allows multiple time shifts per trial via a convolutive filter (Mørup et al., 2011).

Latchoumane et al. (2012) create third-order tensors from multiple subjects – with modes corresponding to patients, frequency, and electrodes – and use CP and Tucker to extract features used to diagnose Alzheimer's disease. Tensor decompositions have been used to analyze event-related potentials (ERPs): for example, Cong et al. (2013) construct a fourth-order tensor with modes frequency, time, channel, and subject, and use non-negative Tucker and CP to study visual mismatch negativity elicited by pictures of facial expressions in healthy adults.

Tucker decomposition can also be used as a possible way of extending Linear Discriminant Analysis to the multilinear case, and several forms of Multilinear Discriminant Analysis (MDA) for classification of tensorial data have been proposed. Recently, in Frølich et al. (2018), the authors compare several MDA methods with a CP-decomposition-based MDA method and manifold optimization techniques to solve MDA optimization problems; these methods are tested on the classification of letters across two subjects based on EEG data.

Thanh et al. (2020) devise a simultaneous multilinear low-rank approximation algorithm for automatic detection of EEG epileptic spikes, by classifying third-order (scale, channel, time) tensors according to whether they contain epileptic spikes.

Naskovska et al. (2020) used both Magnetoencephalography (MEG) and EEG recordings to extract and differentiate physiologically meaningful signal sources during intermittent photic stimulation. The authors construct third-order tensors (with modes frequency, time, channels) from both MEG and EEG recordings, and – under the assumption that they have a common frequency mode – they use a coupled CP decomposition via simultaneous matrix diagonalization to extract features.

Data coming from fMRI images has been used in a variety of ways. Batmanghelich et al. (2011) construct a third-order tensor with modes voxel, subject, fMRI, and define an optimization problem which provides a factorization used to classify cognitive brain function as normal or declining.

Wang et al. (2014) develop a multilinear sparse logistic regression model: by assuming that the output depends multilinearly on the input, they define a loss function and impose a sparsity regularization term and propose a block proximal gradient descent method to solve the resulting optimization problem. The method is tested on fMRI data for early prediction of Alzheimer's disease.

Developing kernel methods that would make full use of the tensor structure of the data has also attracted interest. In He et al. (2014) the authors introduce a kernelized version of SVM that uses CP to embed the data in a way that preserves their multilinear structure and test the proposed method on a brain fMRI classification problem; data consist of third-order tensors obtained by concatenating fMRI images from different angles.

In Ma et al. (2016), the problem of classifying brain disorders using fMRI data is also considered. Fourth-order tensors are constructed by concatenating 3D images along a fourth dimension given by time, then the authors use a shifted CP decomposition to extract features and a tensorial kernel to embed the tensors in a higher-dimensional space before classification.

Hunyadi et al. (2016) combine EEG and fMRI data to gain new insights on epilepsy, by formulating a coupled matrix-tensor factorization problem. EEG data are organized as a three-way tensor with modes participants, time, and channel, coupled with a matrix with modes participants and voxels representing the fMRI data. The resulting coupled factorization can be used to analyze the dataset both spatially and temporally, revealing new connections between an interictal epileptic discharge and fMRI activation clusters.

Finally, several articles tested their proposed methods on MRI data for reconstruction and image recovery: Acar et al. (2011) and Liu et al. (2015) use CP-completion and a regularized CP-completion method; Bazerque et al. (2013) develop a CP-based probabilistic method with a Bayesian framework, which is tested on corrupted three-dimensional MRI scans as well as on RNA-sequencing data. In Yokota et al. (2016), the authors develop a smooth CP tensor completion model especially suited for the recovery of visual data, such as three-dimensional images and color images. The idea is to add smoothness constraints in terms of total variation and quadratic variations. The method is tested, among other datasets, on 3D MRI image recovery.

11.6.2 HEART DATA

Tensor methods have proven fruitful in studying various aspects of ECG signal processing. Many methods we now describe are discussed in the recent survey article (Padhy et al., 2020). Given a multi-lead ECG signal, one can segment individual heartbeats and construct a third-order tensor with modes channels, time, and heartbeats. The advantage of a tensorial approach is that one can analyze all available data – both from different channels and different times – simultaneously, which often leads to capturing patterns and structures that would otherwise remain hidden.

In this setting, HOSVD has been successfully used to compress the data and thus reduce the number of samples needed to store the signal, by truncating the core tensor and only retaining the most relevant terms (Padhy & Dandapat, 2015). HOSVD has also been used to detect and localize Myocardial Infarction (Padhy & Dandapat, 2017): after compression, the multilinear singular values can be used as features for an SVM classifier. Singular values in modes 1 and 2 can be used for detection, and singular values in mode 3 for localization.

CP decomposition can be used for irregular heartbeat detection (Boussé et al., 2017). The rank 1 CP approximation results in a rank 1 tensor $a_1 \circ a_2 \circ a_3$ with interpretable loading vectors: a spatial factor vector a_1 capturing differences in heartbeat morphology over different channels, a temporal factor vector a_2 representing the average heartbeat, and a heartbeat factor vector a_3 capturing differences over different heartbeats. Irregular heartbeats can then be identified from the heartbeat factor vector.

Further applications include detection of T-wave alternans (Goovaerts et al., 2015; 2017) and analysis of changes in heartbeat morphology (Goovaerts et al., 2017).

Luo et al. (2018) overviews recent results in using tensor techniques to study Heart Failure with Preserved Ejection Fraction. The authors stress how, compared to other machine learning methods, tensorial methods allow to better deal with heterogeneous datasets such as those coming from precision medicine, and how they are often more easily interpretable.

11.6.3   EHR Data

Electronic health records (EHRs) are often a rich source of valuable information about patients, but they can be difficult to use due to their own nature: variance among patients, heterogeneity of the data, etc. Tensor methodologies are natural candidates for this type of data: by dealing efficiently with heterogeneous data, they can capture multilinear dependences among sources in the data and provide interpretable results.

Applications to EHR data include imputing missing information and extracting phenotypes. Dauwels et al. (2011) consider the problem of handling missing data in medical questionnaires with a CP-based tensor completion method and demonstrate how it outperforms several non-tensor-based ones. In (Ho et al., 2014), the authors use a nonnegative tensor factorization method to extract phenotypes from electronic health records. A third-order tensor with modes patient, diagnosis, and medication is constructed from EHR data, and a nonnegative CP decomposition is computed. Each rank 1 term of the decomposition can be used to identify a phenotype. New patients' phenotype memberships can be also computed, by projecting observed features to the space of known phenotypes. Ho et al. (2014) propose a decomposition into an interaction tensor (the sum of rank 1 terms corresponding to phenotypes) and a bias tensor (which captures baseline characteristics common to the entire population), while also imposing sparsity to extract only relevant and non-redundant phenotypes.

In Luo et al. (2015), the authors consider the problem of extracting information from narrative text within health records. Narrative sentences are converted to graphs, from which subgraphs are extracted via frequent subgraph mining. The authors then construct a tensor with modes patients, atomic features (words), and higher-order features (subgraphs). By applying a non-negative Tucker factorization, one can simultaneously cluster patients and identify higher-order features to be associated with each cluster. The method is used to cluster lymphoma subtypes based on pathology reports.

Wang et al. (2015) also consider the problem of identifying phenotypes from EHR data. To deal with noise and missing data, the algorithm combines tensor completion and tensor factorization, using an alternating direction method of multipliers to solve the optimization problem. The method can also discover sub-phenotypes by adding knowledge guidance constraints, which consist in imposing that a subset of the columns of a factor matrix is close to the columns represented by prior knowledge.

Zhao et al. (2019) propose an unsupervised method to extract phenotypes via constrained non-negative factorization on a third-order tensor with modes pheno-types, times, and individuals, by taking into account temporal patterns. The method is used to extract 14 subphenotypes among patients with the complexity of the cardiovascular disease, as well as their evolution during the 10 years prior to the diagnosis.

Temporal phenotyping with a tensorial approach is studied in Perros et al. (2019). For each patient, available data are structured as a matrix X_k of size $I_k \times J$ where each row represents one of the I_k encounters and each column one of the J clinical features considered; entry (i, j) of the matrix X_k is 1 if the kth patient has been diagnosed with feature j on the ith visit, and 0 otherwise. The number of encounters for different patients can vary, therefore the matrices X_k's cannot be stacked together as a tensor, but using a PARAFAC2 decomposition (a variant of CP) one can decompose X_k as a product of matrices $U_k S_k V^t$, with S_k square $R \times R$ diagonal matrix, R being the number of extracted phenotypes. Rows of V^t correspond to phenotypes (extracted from the aggregated data and common to all patients), and each column of a matrix U_k provides a temporal signature of the corresponding phenotype across encounters.

Ma et al., (2019) introduce a privacy-preserving collaborative tensor factor-ization method that allows hospitals to keep their EHR databases private while collaboratively using them to extract phenotypes. Each hospital is assumed to create a third-order tensor with modes patient, procedure, and diagnosis, and that the factor matrices corresponding to procedure and diagnosis are the same among hospitals (so that the phenotypes are also the same). Non-patient factor matrices are then shared in a privacy-preserving way with the server back and forth for the collaborative learning algorithm.

11.6.4 DRUG REPURPOSING

A natural application of completion methods is drug repurposing, and more generally imputing unknown values of interactions between proteins and chosen targets. The survey article (Bagherian et al., 2021) gives an overview of machine learning algorithms to predict drug-target interactions, and the article by Gachloo et al. (2019) provides an overview of the research on drug-related knowledge discovery, including tensor-based methods. Some advantages of matrix and tensor-based methods are that they can efficiently deal with a lot of missing data and can integrate diverse datasets by coupling or adding extra modes.

In Arany et al. (2015), the authors combine different measurement types into a single tensor together with proteins and compounds, develop a Bayesian factorization approach based on Gibbs sampling, and incorporate high dimensional side information with a noise injection sampler. The article (Wang et al., 2019) also studies drug repurposing, by constructing a three-way tensor with modes drug, target, and disease, and decomposing it to find latent factors, which are then clustered using topological data analysis. Then the authors examine the ability to predict new associations of the proposed method. The latent factors derived from the association tensor are analyzed to uncover functional patterns of drugs, targets, and diseases. (Iwata et al., 2019) use a tensor-train weighted optimization algorithm (in the spirit of the CP-WOPT algorithm we discussed earlier) to predict unknown values in tensor-structured gene expression data. The method is applied to a third-order tensor with modes 16 cell lines, 261 drugs, 978 genes. One advantage of using a tensor train decomposition is that it would also work well on higher-order tensor, as results on a fourth-order tensor with an additional mode consisting of for time points show.

A recent, different approach is an application of invariant theory to coupled matrix and tensor completion. The invariant theory is an area of pure mathematics that has found successful applications in many areas of computer science such as coding theory and, more recently, scaling algorithms. In Bagherian et al. (2020), the authors consider several drug-drug and target-target similarity matrices and stack them together as two tensors with modes drug-drug-similarity and target-target-similarity, to integrate the various ways in which similarity can be measured. Then they set up a coupled tensor completion problem in the form of a matrix with modes drug and target, coupled with the similarity tensors described before. The invariant theory is used to find optimal coordinates in each mode and impute the missing values by solving an optimization problem with respect to these coordinates. The method is used to effectively predict drug-target interaction by coupled tensor completion, taking into account side information consisting of multiple types of similarity scores.

11.6.5 OTHER APPLICATIONS

Semerci et al. (2014) consider the problem of reconstructing X-ray attenuation distribution in multi-energy computed tomography. The authors construct a 3-way tensor with two spatial dimensions and a third energy dimension and set up a convex optimization problem via nuclear norm regularizers.

An application of H-Tucker can be found in Perros et al. (2015). The authors consider a large healthcare database containing the disease history of almost 30,000 patients and more than 300,000 diagnostic events and construct a high order tensor with modes given by the diagnoses and entries the number of co-occurrences of diagnoses in a given multi-index across all patients. Given the large size of the dataset and the high order – 18 – of the resulting tensor, the authors introduce Sparse Hierarchical Tucker, a variant of H-Tucker which preserves sparsity of the data and scales almost linearly in the number of non-zero tensor elements. The study identified connections within diagnostic families and disease co-occurrences compatible with known clinical results.

In Hernandez et al. (2021), the authors study the problem of predicting life-threatening complications in patients recovering from cardiovascular surgeries. Several physiological signals, as well as EHR data, are processed and combined in tensor form. After dimensionality reduction, relevant features are extracted with tensor decomposition, and five different machine learning methods were trained on these features, with the goal of predicting hemodynamic decompensation from half an hour to 12 hours before the event. The results provide further evidence of the effectiveness of multimodal approaches in clinical informatics, which can be implemented by means of tensor techniques.

11.7 CONCLUSION

Tensor decomposition and tensor completion methods are well-established techniques to be used either by themselves or as a preprocessing tool in combination with classical machine learning methods. Many of them are more powerful extensions of matrix methods and formulating problems in tensorial form often leads to better-behaved solutions. Compared to other machine learning techniques, tensor methods tend to be more easily interpretable. Furthermore, they often allow to efficiently combine heterogeneous datasets, which is especially relevant in the current age of big data. Within clinical informatics, there are types of data (EEG data, ECG data, EHR data) these techniques have been successfully applied to, resulting in a rich body of research, with ongoing attention as improvements are achieved.

There are currently still problems that have not been addressed in an entirely satisfactory manner, for example in relation to temporal data, as already discussed in Papalexakis et al. (2016) and (Luo et al. (2017): it remains an issue how to properly encode the time dimension, as well as how to allow efficient updating of a tensorial method as new data is acquired. In certain situations, another limitation is the level of preprocessing required to reshape the data in tensor form, as well as determining a priori the most effective way of tensorizing data when multiple options are available.

However, as new applications are found and novel mathematical and algorithmic methodologies are developed, this is an active and ongoing area of research with the potential to achieve yet better results.

REFERENCES

Acar, E., Bingol, C. A., Bingol, H., Bro, R., & Yener, B. (2007). Multiway analysis of epilepsy tensors. *Bioinformatics, 23*(13), 10–18.

Acar, E., Dunlavy, D., Kolda, T., & Mørup, M. (2011) Scalable tensor factorizations for incomplete data. *Chemometrics and Intelligent Laboratory Systems, 106*(11), 41–56.

Acar, E., Kolda, T., & Dunlavy, D. (2011). All-at-once optimization for coupled matrix and tensor factorizations. Proceedings of Mining and Learning with Graphs. arXiv:1105.3422.

Arany, A., Simm, J., Zakeri, P., Haber, T., Wegner, J. K., Chupakhin, V., Ceulemans, H., & Moreau, Y. (2015). Highly scalable tensor factorization for prediction of drug-protein interaction type. MLCB/MLSB 2015. arXiv:1512.00315.

Bagherian, M., Kim, R. B., Jiang, C., Sartor, M. A., Derksen, H., & Najarian, K. (2020). Coupled matrix-matrix and coupled tensor-matrix completion methods for predicting drug-target interactions. *Briefings in Bioinformatics, 22*(2), 2161–2171.

Bagherian, M., Sabeti, E., Wang, K., Sartor, M. A., Nikolovska-Coleska, Z., & Najarian, K. (2021). Machine learning approaches and databases for prediction of drug-target interaction: a survey paper. *Briefings in Bioinformatics, 22*(1), 247–269.

Batmanghelich, N., Dong, A., Taskar, B. & Davatsikos, C. (2011). Regularized tensor factorization for multi-modality medical image classification. *Medical Image Computing and Computer-Assisted Intervention - MICCAI 2011* (6893). 17–24.

Bazerque, J. A., Mateos, G., & Giannakis, G. B. (2013). Rank regularization and Bayesian inference for tensor completion and extrapolation. *IEEE transactions on signal processing, 61*, 5689–5703.

Boussé, M., Goovaerts, G., Vervliet, N., Debals, O., Van Huffel, S., & De Lathauwer, L. (2017). *Irregular heartbeat classification using Kronecker product equations*. 39th Annual International Conference of the IEEE Engineering in Medicine and Biology Society (EMBC 2017), 438–441.

Candés, E. J. & Recht, B. (2008). Exact matrix completion via convex optimization. *Foundations of Computational Mathematics, 9*(6), 717–772.

Carroll, J. & Chang, J.-J. (1970). Analysis of individual differences in multidimensional scaling via an N-way generalization of Eckart-Young decomposition. *Psychometrika, 35*(3), 283–319.

Chiantini, L. & Ottaviani, G. (2012). On generic identifiability of 3-tensors of small rank. *SIAM Journal of Matrix Analysis & Applications, 33*(3), 1018–1037.

Chistov, A.L. & Grigoriev, D.Y. (1984). *Complexity of quantifier elimination in the theory of algebraically closed fields*. International Symposium on Mathematical Foundations of Computer Science (pp. 17–31).

Cichocki, A., Mandic, D., De Lathauwer, L., Zhou, G., Zhao, Q., Caiafa, C., & Anh Phan, H. (2015). Tensor decompositions for signal processing applications: From two- way to multiway component analysis. *IEEE Signal Processing Magazine, 32*(2), 145–163.

Cong, F., Lin, Q.-H., Kuang, L.-D., Gong, X.-F., Astikainen, P., & Ristaniemi, T. (2015). Tensor decomposition of EEG signals: A brief review. *Journal of Neuroscience Methods, 248*, 59–69.

Cong, F., Phan, A.-H., Astikainen, P., Zhao, Q., Wu, Q., Hietanen, J., Ristaniemi, T., & Cichocki, A. (2013). Multi-domain feature extraction for small event-related potentials through nonnegative multi-way array decomposition from low dense array EEG. *International Journal of Neural Systems, 23*(2).

Dauwels, J., Garg, L., Earnest, A., & Pang., L. K. (2011). Handling missing data in medical questionnaires using tensor decompositions. Proceedings of ICICS. https://doi.org/10.1109/ICICS.2011.6174300

De Lathauwer, L., De Moor, B., & Vandewalle, J. (2000a). A multilinear singular value decomposition. *SIAM J. Matrix Anal. Appl.*, *21*(4), 1253–1278.

De Lathauwer, L., De Moor, B., & Vandewalle, J. (2000b). On the best rank-1 and rank-$(r_1,r_2,...,r_n)$ approximation of higher-order tensors. *SIAM Journal on Matrix Analysis and Applications*, *21*(4), 1324–1342.

De Lathauwer, L. (2008a). Decompositions of a higher-order tensor in block terms—Part I: Lemmas for partitioned matrices. *SIAM Journal on Matrix Analysis and Applications*, *30*(3), 1022–1032.

De Lathauwer, L. (2008b). Decompositions of a higher-order tensor in block terms—Part II: Definitions and uniqueness. *SIAM Journal on Matrix Analysis and Applications*, *30*(3), 1033–1066.

De Silva, V. & Lim, L.-H. (2008). Tensor rank and the Ill-posedness of the best low-rank approximation problem. *SIAM Journal on Matrix Analysis and Applications*, *30*(3), 1084–1127.

Ermiş, B., Acar, E., & Taylan, C. A. (2015). Link prediction in heterogeneous data via generalized coupled tensor factorization. *Data Mining and Knowledge Discovery*, *29*(1), 203–236.

Friedland, S., & Lim, L. H. (2018). Nuclear norm of higher-order tensors. *Mathematics of Computation*, *87*(311), 1255–1281.

Frølich, L., Andersen, T. S., & Mørup, M. (2018). Rigorous optimisation of multilinear discriminant analysis with Tucker and PARAFAC structures. *BMC Bioinformatics*, *19*(1).

Gachloo, M., Wang, Y., & Xia, J. (2019). A review of drug knowledge discovery using BioNLP and tensor or matrix decomposition. *Genomics & Informatics*, *17*(2).

Gandy, S., Recht, B., & Yamada, I. (2011). Tensor completion and low-n-rank tensor recovery via convex optimization. *Inverse Problems*, *27*(2).

Goovaerts, G., Vandenberk, B., Willems, R., Van Huffel, S. (2015). Tensor-based detection of T wave alternans using ECG. Proc. of the 37th Annual International Conference of the IEEE Engineering in Medicine and Biology Society of the IEEE (EMBC), Milan, Italy, Aug. 2015, 6991–6994.

Goovaerts, G., Vandenberk, B., Willems, R., & Van Huffel, S. (2017). Automatic detection of T wave alternans using tensor decompositions in multilead ECG signals. *Physiological Measurement*, *38*(8), 1513–1528.

Goovaerts, G., Van Huffel, S., & Hu, X. (2017). Tensor-based Analysis of ECG changes prior to in-hospital cardiac arrest. Proc. of the 44rd Annual Computing in Cardiology (CinC 2017), CinC, Rennes, France, Sep. 2017.

Grasedyck, L. (2010). Hierarchical singular value decomposition of tensors. *SIAM Journal on Matrix Analysis and Applications*, *31*(4), 2029–2054.

Grasedyck, L., Kressner, D., & Tobler, C. (2013). *A literature survey of low-rank tensor approximation techniques*. *GAMM-Mitteilungen*, *36*(1), 53–78.

Hackbusch, W. & Kühn, S. (2009). A new scheme for the tensor representation. *Journal of Fourier Analysis and Applications*, *15*(5), 706–722.

Harshman, F. (1927). Foundations of the PARAFAC procedure: Models and conditions for an explanatory multimodal factor analysis. *UCLA Working Papers in Phonetics*, *16*, 1–84.

Håstad, J. (1990). Tensor rank is NP-complete. *Journal of Algorithms*, *11*(4), 644–654.

He, L., Kong, X., Philip, S. Y., Ragin, A. B., Hao, Z. & Yang, X. (2014). Dusk: A dual structure-preserving kernel for supervised tensor learning with applications to neuroimages. Proceeding in the Thirteenth SIAM International Conference on Data Mining, 127–135.

Hernandez, L., Kim, R., Tokcan, N., Derksen, H., Biesterveld, B. E., Croteau, A., Williams, A. M., Mathis, M., Najarian, K. & Gryak, J. (2021). Multimodal tensor-based method for integrative and continuous patient monitoring during postoperative cardiac care. *Artificial Intelligence In Medicine, 113.* https://doi.org/10.1016/j.artmed.2021.102032

Hillar, C. J., & Lim, L. H. (2013). Most tensor problems are NP-hard. *Journal of the ACM (JACM), 60*(6), 1–39.

Ho, J. C., Ghosh, J., Steinhubl, S. R., Stewart, W. F., Denny, J. C., Malin B. A., & Sun, J. (2014). Limestone: High-throughput candidate phenotype generation via tensor factorization. *Journal of Biomedical Informatics, 52,* 199–211.

Ho, J. C., Ghosh, & Sun, J. (2014). Marble: high-throughput phenotyping from electronic health records via sparse nonnegative tensor factorization. Proceedings of the 20th ACM SIGKDD International Conference on Knowledge Discovery and Data Mining. https://doi.org/10.1145/2623330.2623658

Hunyadi, B., Dupont, P., Van Paesschen, W. V., & Van Huffel, S. (2015). Tensor decompositions and data fusion in epileptic electroencephalography and functional magnetic resonance imaging data. *WIREs Data Mining and Knowledge Discovery, 7*(1). https://doi.org/10.1002/widm.1197.

Hunyadi, B., Van Paesschen, W. V., De Vos, M., & Van Huffel, S. (2016). Fusion of electroencephalography and functional magnetic resonance imaging to explore epileptic network activity. 2016 24th European Signal Processing Conference (EUSIPCO), 240–244.

Iwata, M., Yuan, L., Zhao, Q., Tabei, Y., Berenger, F., Sawada, R., Akiyoshi, S., Hamano, M., & Yamanishi, Y. (2019). Predicting drug-induced transcriptome responses of a wide range of human cell lines by a novel tensor-train decomposition algorithm. *Bioinformatics, 35*(14), i191–i199.

Kolda, T. & Bader, B. (2009). Tensor decompositions and applications. *SIAM Review, 51*(3), 455–500.

Kruskal, J. (1977). Three-way arrays: Rank and uniqueness of trilinear decompositions, with application to arithmetic complexity and statistics. *Linear Algebra and its Applications, 18*(2), 95–138.

Lahat, D., Adalî, T., & Jutten, C. (2015). Multimodal data fusion: An overview of methods, challenges, and prospects. *Proceedings of the IEEE, 103*(9), 1449–1477.

Latchoumane, C.-F., Vialatte, F.-B., Solé-Casals, J., Maurice, M., Wimalaratna, S. R., Hudson, N., Jeong, J., Cichocki, A. (2012). Multiway array decomposition analysis of EEGs in Alzheimer's disease. *Journal of Neuroscience Methods, 207,* 41–50.

Li, W.-J. & Yeung, D.-Y. (2009). Relation regularized matrix factorization. Proceedings of the 21st International Joint Conference on Artificial Intelligence, 1126–1131.

Liu, J., Musialski, P., Wonka, P., & Ye, J. (2012). Tensor completion for estimating missing values in visual data. *IEEE Transactions on Pattern Analysis and Machine Intelligence, 35*(1), 208–220.

Liu, Y., Shang, F., Jiao, L., Cheng, J., & Cheng, H. (2015). Trace norm regularized CANDECOMP/PARAFAC decomposition with missing data. *IEEE transactions on cybernetics, 45,* 2437–2448.

Luo, Y., Ahmad, F. S., & Shah, S. J. (2018). Tensor Factorization for precision medicine in heart failure with preserved ejection fraction. *Journal of Cardiovascular Translational Research, 10*(3), 305–312.

Luo, Y., Wang, F., & Szolovits, P. (2017). Tensor factorization toward precision medicine. *Briefings in Bioinformatics, 18*(3), 511–514.

Luo, Y., Xin, Y., Hochberg, E., Joshi, R., Uzuner, O., & Szolovits, P. (2015). Subgraph augmented non-negative tensor factorization (SANTF) for modeling clinical narrative text. *Journal of the American Medical Informatics Association, 22,* 1009–1019.

Ma, G., He, L., Lu, C.-T., Yu, P. S., Shen, L., & Ragin, A. B. (2016). Spatio-Temporal Tensor Analysis for Whole-Brain fMRI Classification. SIAM International Conference on Data Mining (SDM). https://doi.org/10.1137/1.9781611974348.92

Ma, J., Zhang, Q., Lou, J., Ho, J. C., Xiong, L, & Jiang, X. (2019). Smooth PARAFAC Decomposition for Tensor Completion. *Proceedings of the 28th ACM International Conference on Information and Knowledge Management*, 1291–1300.

Mørup M., Hansen L. K., Arnfred S. M., Lim L., & Madsen K. H. (2008). Shift-invariant multilinear decomposition of neuroimaging data. *NeuroImage*, *42*, 1439–1450.

Mørup M., Hansen L. K., & Madsen K. H. (2011). Modeling latency and shape changes in trial based neuroimaging data. *IEEE Conference Record of the 45th Asilomar Conference on Signals, Systems and Computers (ASILOMAR 11)*, 439–443.

Narita, A., Hayashi, K., Tomioka, R., & Kashima, H. (2012). Tensor factorization using auxiliary information. *Data Mining and Knowledge Discovery*, *25*(2), 298–324.

Naskovska, K., Lau, S., Korobkov, A. A., Haueisen, J., & Haardt, M. (2020). Coupled CP Decomposition of Simultaneous MEG-EEG Signals for Differentiating Oscillators During Photic Driving. *Frontiers in Neuroscience*, *14*(261), 298–324.

Oseledets, I. (2011). Tensor-train decomposition. *SIAM Journal on Scientific Computing*, *33*(5), 2295–2317.

Padhy, S. & Dandapat, S. (2015). Exploiting multi-lead electrocardiogram correlations using robust third-order tensor decomposition. *Healthcare Technology Letters*, *2*(5), 112–117.

Padhy, S. & Dandapat, S. (2017). Third-order tensor based analysis of multilead ECG for classification of myocardial infarction. *Biomedical Signal Processing and Control*, *31*, 71–78.

Padhy, S., Goovaerts, G., Boussé, M., de Lathauwer, L., & van Huffel, S. (2020). The power of tensor-based approaches in cardiac applications. In Naik G. (Eds.), *Biomedical Signal Processing. Series in BioEngineering* (pp. 291–323). Springer.

Papalexakis, E., Faloutsos, C., & Sidiropoulos, N. (2016). Tensors for data mining and data fusion: Models, applications, and scalable algorithms. *ACM Transactions on Intelligent Systems and Technology*, *8*, 2(16).

Peeters, R. (1996). Orthogonal representations over finite fields and the chromatic number of graphs. *Combinatorica*, *16*(3), 417–431.

Perros, I., Chen, R., Vuduc, R., & Sun, J. (2015). Sparse hierarchical tucker factorization and its application to healthcare. *2015 IEEE 15th International Conference on Data Mining*, 943–948.

Perros, I., Papalexakis, E. E., Vuduc, R., Searles, E., & Sun, J. (2019). Temporal phenotyping of medically complex children via PARAFAC2 tensor factorization. *Journal of Biomedical Informatics*, *93*. https://doi.org/10.1016/j.jbi.2019.103125

Rennie, J. & Srebro, N. (2005). Fast maximum margin matrix factorization for collaborative prediction. *Proceedings of the 22nd International Conference on Machine Learning*, 713–719.

Semerci, O., Hao, N., Kilmer, M. E., & Miller E. L. (2014). Tensor-based formulation and nuclear norm regularization for multienergy computed tomography. *IEEE Transactions on Image Processing*, *23*(4), 1678–1693.

Sidiropoulos, N. & Bro, R. (2000). On the uniqueness of multilinear decomposition of N-way arrays. *Journal of Chemometrics*, *14*(3), 229–239.

Sidiropoulos, N., De Lathauwer, L., Fu, X., Huang, K., Papalexakis, E., & Faloutsos, C. (2017). Tensor decomposition for signal processing and machine learning. *IEEE Transactions on Signal Processing*, *65*(13), 3551–3582.

Song, Q., Ge, H., Caverlee, J., & Hu, X. (2019). Tensor completion algorithms in big data analytics. *ACM Transactions on Knowledge Discovery from Data*, *13*(1)(6).

Sorber, L., Van Barel, M., & De Lathauwer, L. (2015). Structured data fusion. *IEEE Journal of Selected Topics in Signal Processing*, *9*(4), 586–600.

Taguchi, Y.-h. (2020). *Unsupervised Feature Extraction Applied to Bioinformatics (A PCA Based and TD Based Approach)*. Springer International Publishing: Cham, Switzerland, 2020.

Thanh, L. T., Dao, N. T. A., Dung, N. V.,Trung, N. L., & Karim, A. M. (2020). Multi-channel EEG epileptic spike detection by a new method of tensor decomposition. *Journal of Neural Engineering, 17*(1).

Tokcan, N., Gryak, J., Najarian, K., & Derksen, H. (2021). Algebraic methods for tensor data. *SIAM Journal on Applied Algebra and Geometry, 5*(1), 1–27.

Tomasi, G., & Bro, R. (2005). Parafac and missing values. *Chemometrics and Intelligent Laboratory Systems, 75*, 163–180.

Tomasi, G. & Bro, R. (2006). A comparison of algorithms for fitting the PARAFAC model. *Computational Statistics & Data Analysis, 50*(7), 1700–1734.

Tucker, L. (1966). Some mathematical notes on three-mode factor analysis. *Psychometrika, 31*(3), 279–311.

Udell, M. & Townsend, A. (2019). Why are big data matrices approximately low rank? *SIAM Journal on Mathematics of Data Science, 1*(1), 144–160.

Walczak, B. & Massart, D. (2001) Dealing with missing data: Part i. *Chemometrics and Intelligent Laboratory Systems, 59*, 15–27.

Wang, Y., Chen, R., Ghosh, J., Denny, J. C., Kho, A., Chen, Y., Malin, B. A., & Sun, J. (2015). Rubik: Knowledge guided tensor factorization and completion for health data analytics. Proceedings of the 21th ACM SIGKDD International Conference on Knowledge Discovery and Data Mining, 1265–1274.

Wang, F., Li, S., Cheng, L., Wong, M. H., & Leung, K. S. (2019). Predicting associations among drugs, targets, and diseases by tensor decomposition for drug repositioning. *BMC Bioinformatics, 20*(628).

Wang, Y., Zhang, P., Qian, B., Wang, X., & Davidson, I. (2014). Clinical risk prediction with multilinear sparse logistic regression. Proceedings of the 20th ACM SIGKDD International Conference on Knowledge Discovery and Data Mining, 145–154.

Wimalawarne, K., Yamada, M., & Mamitsuka, H. (2018). Convex coupled matrix and tensor completion. *Neural Computation, 30*, 3095–3127.

Yokota, T., Zhao, Q., & Cichocki, A. (2016). Smooth PARAFAC decomposition for tensor completion. *IEEE transactions on signal processing, 64*(20), 5423–5436.

Zhao, J., Zhang, Y., Schlueter, D. J., Wu, P., Kercheberger, V. E., ..., Wei, W.-Q. (2019). Detecting time-evolving phenotypic topics via tensor factorization on electronic health records: Cardiovascular disease case study. *Journal of Biomedical Informatics, 98*(103270).

Index

Printed in the United States
by Baker & Taylor Publisher Services